SCHAUM'S OUTLINE OF

THEORY AND PROBLEMS

of

BASIC
ELECTRICAL
ENGINEERING

•

JIMMIE J. CATHEY, Ph.D.
SYED A. NASAR, Ph.D.

Professors of Electrical Engineering
University of Kentucky

SCHAUM'S OUTLINE SERIES

McGRAW-HILL, INC.

New York   St. Louis   San Francisco   Auckland   Bogotá   Caracas
Hamburg   Lisbon   London   Madrid   Mexico   Milan   Montreal
New Delhi   Paris   San Juan   São Paulo   Singapore
Sydney   Tokyo   Toronto

JIMMIE J. CATHEY earned the Ph.D. degree from Texas A&M University and has 13 years of industrial experience in design and development of electric drive systems. Since 1980, he has been at the University of Kentucky, with research and teaching interest in the areas of power electronics, electric machines, and robotics. He is a Registered Professional Engineer.

SYED A. NASAR is Professor of Electrical Engineering at the University of Kentucky. Having earned the Ph.D. degree from the University of California at Berkeley, he has been involved in teaching, research, and consulting in electric machines for over 25 years. He is the author, or co-author, of five other books and over 100 technical papers.

 This book is printed on recycled paper containing a minimum of 50% total recycled fiber with 10% postconsumer de-inked fiber.

Schaum's Outline of Theory and Problems of
BASIC ELECTRICAL ENGINEERING

9 10 11 12 13 14 15 SH SH 9 8 7 6 5 4 3 2

ISBN 0-07-010234-1

Sponsoring Editor, David Beckwith
Editing Supervisor, Marthe Grice
Production Manager, Nick Monti

Cover design by Amy E. Becker.

Library of Congress Cataloging in Publication Data

Cathey, J.J.
  Schaum's outline of theory and problems of basic
electrical engineering.
  (Schaum's outline series)
  Includes index.
  1. Electric engineering. 2. Electronics.
I. Nasar, S.A. II. Title.
TK146.C226 1983   621.3   83-1155
ISBN 0-07-010234-1

Preface

In most schools, a survey course in electrical engineering is required for engineering students not majoring in electrical engineering. Such a course will treat passive electrical networks, electronic devices and circuits, magnetic circuits and electric machines, linear systems, control systems, and instrumentation—or at least most of these topics—in a two-semester sequence. The present Outline is designed to supplement the standard textbooks for this survey course; it could also function as a textbook in its own right. It should find further use as a review guide for those (electrical engineers and others) who wish to take the Professional Engineer's Examination.

All the above-mentioned areas of electrical engineering, with the exception of instrumentation (not problem-oriented), are covered in this book. As in other Schaum's Outlines, only a brief review of the subject at hand is given in each chapter. The chief emphasis is on solving pertinent problems, of which over 620 are presented. SI units, as specified in the IEEE Standards, are employed throughout.

The authors would like to thank the staff of the Schaum's Outline Series for their editorial assistance.

J.J. CATHEY
S.A. NASAR

Contents

CONTENTS

Chapter 1

Definitions and Basic Concepts

1.1 INTRODUCTION

In this chapter, we introduce the basic field and circuit concepts pertinent to the study of electrical engineering. Although field concepts and circuit concepts complement each other in an exact description of a process (such as current flow in a conductor) or a phenomenon (for example, propagation of electromagnetic waves), the field-theory approach is basic. Using Coulomb's law as a starting point, we define the various field quantities in scalar form (that is, in magnitude only but with direction implied) and develop the ideas in the Solved Problems.

1.2 ELECTRIC CHARGE AND COULOMB'S LAW

We know that an atom of an element is electrically neutral, with a positively charged nucleus surrounded by orbiting, negatively charged electrons. The charge of an electron is -1.6×10^{-19} coulomb (C). The unit *coulomb* may be derived from *Coulomb's law*, according to which the force F between two point charges Q_1 and Q_2 in free space, separated by a distance r, is given by

$$F = k \frac{Q_1 Q_2}{r^2} \quad \text{(N)} \qquad (1.1)$$

The notation in (1.1) is meant to indicate that F will be in *newtons* (N), provided Q_1 and Q_2 are in coulombs, r is in meters (m), and k has the assigned value

$$k \equiv \frac{1}{4\pi\epsilon_0} \approx 9 \times 10^9 \text{ N} \cdot \text{m}^2/\text{C}^2$$

Thus, if a charge of 1 C is placed at a distance of 1 m from an equal charge of the same sign in free space, they produce a repulsive force of 9×10^9 N. The constant ϵ_0 is known as the *permittivity* of free space.

1.3 ELECTRIC CURRENT

Unidirectional transfer of electric charge (specifically, of electrons) through a conductor constitutes a *direct electric current*, or *dc*. Electric current I is measured in *amperes* (A), where the ampere is defined as follows: a 1-A current maintained in two straight, infinitely long conductors of negligible cross sections, placed 1 m apart in free space, will produce between them a force of exactly 2×10^{-7} N/m. In the SI, electric current is taken as a primary quantity, and the unit of charge is derived as

$$1 \text{ C} = 1 \text{ A} \cdot \text{s}$$

That is, a 1-A current transports charge through a fixed area at the rate 1 C/s.

1.4 VOLTAGE OR POTENTIAL DIFFERENCE

Voltage (*potential difference*) between two points is defined as the work required to move a unit charge from one point to the other. The unit of potential difference is the *volt* (V), where

$$1 \text{ V} = 1 \text{ N} \cdot \text{m/C} = 1 \text{ J/C}$$

The *joule* (J) is the SI unit of work or energy.

1.5 POWER AND ENERGY

The *power* of a source of supply is defined as the energy supplied per unit time. Since electric energy U is given by

$$U = QV \quad \text{(J)} \tag{1.2}$$

electric power P is given by

$$P = \frac{dU}{dt} = V\frac{dQ}{dt} = VI \quad \text{(W)} \tag{1.3}$$

where the *watt* (W) is 1 J/s. From (*1.3*), 1 V = 1 W/A; this is, in fact, the official definition of the volt in the SI.

Table 1-1 summarizes the units and dimensions (M ≡ mass, L ≡ length, T ≡ time, A ≡ current) of the various electrical quantities appearing in this chapter. In practical work, decimal multiples or submultiples of the SI units are often employed; Table 1-2 shows how these are indicated by letters prefixed to the unit symbol.

Table 1-1

Quantity	Symbol	Dimensions	SI Unit
Electric charge	Q	$[A][T]$	coulomb, C
Electric potential	V	$[M][A]^{-1}[L]^2[T]^{-3}$	volt, V
Current	I	$[A]$	ampere, A
Electric field intensity	E	$[M][A]^{-1}[L][T]^{-3}$	volt/meter, V/m
Electric flux density	D	$[A][L]^{-2}[T]$	coulomb/meter², C/m²
Permittivity	ϵ	$[M]^{-1}[A]^2[L]^{-3}[T]^4$	farad/meter, F/m
Resistance	R	$[M][A]^{-2}[L]^2[T]^{-3}$	ohm, Ω
Capacitance	C	$[M]^{-1}[A]^2[L]^{-2}[T]^4$	farad, F
Conductivity	σ	$[M]^{-1}[A]^2[L]^{-3}[T]^3$	siemens/meter, S/m

Table 1-2

Multiplication Factor	Letter Prefix	Multiplication Factor	Letter Prefix
10^{18}	E (exa)	10^{-1}	d (deci)
10^{15}	P (peta)	10^{-2}	c (centi)
10^{12}	T (tera)	10^{-3}	m (milli)
10^{9}	G (giga)	10^{-6}	μ (micro)
10^{6}	M (mega)	10^{-9}	n (nano)
10^{3}	k (kilo)	10^{-12}	p (pico)
		10^{-15}	f (femto)
		10^{-18}	a (atto)

Solved Problems

1.1   Two point charges, each $50 \, \mu$C, are held 1.5 m apart in free space. Find the force of repulsion between the charges.

By Coulomb's law, (*1.1*),

$$F = \frac{Q_1 Q_2}{4\pi\epsilon_0 r^2} = \frac{(50 \times 10^{-6} \times 50 \times 10^{-6})}{4\pi(10^{-9}/36\pi)(1.5)^2} = 10 \text{ N}$$

Notice that $\epsilon_0 \approx (10^{-9}/36\pi)$ F/m (more exactly, $\epsilon_0 = 8.8542 \times 10^{-12}$ F/m). Also, we must realize that the Coulomb force is a vector quantity, having in this case a 10-N magnitude.

1.2   From Coulomb's law it is clear that a charge Q_2 will experience a force if brought in a region containing the charge Q_1. We may then say that Q_1 produces a *force field*, measurable by the force on Q_2. If Q_2 is supposed infinitesimally small, then the force per unit Q_2 is defined as the *electric field intensity* (or, simply, *electric field*), *E*, at the location of Q_2. Using this definition, find the electric field intensity at a distance 1.5 m away from a 50-μC charge in free space.

$$E = \lim_{Q_2 \to 0} \frac{F}{Q_2} \tag{1.4}$$

together with (*1.1*) yields

$$E = \frac{Q_1}{4\pi\epsilon_0 r^2} = \frac{50 \times 10^{-6}}{4\pi(10^{-9}/36\pi)(1.5)^2} = 200 \text{ kN/C} = 200 \text{ kV/m}$$

Since *F* is actually a vector quantity, it follows that *E* is also a vector.

1.3   Two point charges, $Q_1 = 50 \, \mu$C and $Q_2 = 25 \, \mu$C, are separated by a distance of 1 m in air. At what distance(s) from Q_1 will the electric field intensity be zero?

Let the required distance be *r*. Because of the vectorial nature of the field, it can vanish only on the line joining the two charges. Thus,

$$E_1 = \frac{Q_1}{4\pi\epsilon_0 r^2} = \frac{50 \times 10^{-6}}{4\pi(10^{-9}/36\pi)r^2} = \frac{450 \times 10^3}{r^2} \quad \text{(V/m)}$$

$$E_2 = \frac{Q_2}{4\pi\epsilon_0(1-r)^2} = \frac{25 \times 10^{-6}}{4\pi(10^{-9}/36\pi)(1-r)^2} = \frac{225 \times 10^3}{(1-r)^2} \quad \text{(V/m)}$$

For a zero resultant field,

$$\frac{450 \times 10^3}{r^2} = \frac{225 \times 10^3}{(1-r)^2}$$

from which $r = 0.585$ m (or else $r = \infty$).

1.4   Determine the work done in moving a 50-μC charge through a distance of 50 cm in the direction of a uniform electric field of 50 kV/m.

From (*1.1*) and (*1.4*),

$$F = QE \tag{1.5}$$

that is,

$$F = (50 \times 10^{-6} \times 50 \times 10^3) = 2.5 \text{ N}$$

Then the work done on the charge by the field is

$$W = \text{force} \times \text{distance} = (2.5 \times 0.50) = 1.25 \text{ J}$$

1.5   What is the potential difference between two points 50 cm apart in the electric field of Problem 1.4?

$$\text{potential difference} = \frac{\text{work}}{\text{charge}} = \frac{1.25}{50 \times 10^{-6}} = 25 \text{ kV}$$

1.6   From Problems 1.4 and 1.5, relate the electric field intensity to potential difference if the field is uniform.

Consider two points, A and B, a distance ℓ apart, such that the uniform electric field is directed along the straight line from A to B. Let V_{AB} be the potential difference between A and B, or, as we shall say, the *potential of A with respect to B*. The work done on a charge Q in moving along the line from A to B is

$$W = \text{force} \times \text{distance} = QE\ell$$

But work done *on* a body represents a *decrease* in its potential energy (think of a falling stone), and so we also have $W = QV_{AB}$. Thus

$$V_{AB} = E\ell \qquad \text{or} \qquad E = \frac{V_{AB}}{\ell} \qquad\qquad (1.6)$$

and the electric field is seen to be a measure of the potential gradient.

Equation (*1.6*) may be rewritten in a form that readily extends to nonuniform fields:

$$V_{AB} = \int_A^B E \, d\ell \qquad \text{or} \qquad E = -\frac{dV_{AB}}{d\ell} \qquad\qquad (1.7)$$

1.7   The quantity ϵ (or, for free space, ϵ_0) is known as the *permittivity* of a material medium and relates the electric field intensity E to the *electric flux density D*, through $D = \epsilon E$. Find the relationship between the electric flux density and the electric charge producing the flux.

From Problem 1.2, for a point charge Q in a medium of permittivity ϵ, the electric field E at a distance r from Q is given by

$$E = \frac{Q}{4\pi\epsilon r^2}$$

Hence,

$$D = \frac{Q}{4\pi r^2} \qquad \text{or} \qquad D(4\pi r^2) \equiv \psi = Q \qquad\qquad (1.8)$$

where ψ is the total outward flux (since $4\pi r^2$ is the surface area of the sphere at each point of which the flux density is D). The statement that the total outward flux equals the charge enclosed is known as *Gauss's law*; it holds for any distribution of charges within any closed surface.

1.8   Defining *capacitance* by $C \equiv Q/V$, find the capacitance between a pair of parallel planes, each of surface area A and separated by a distance ℓ. The material between the parallel planes has a permittivity ϵ.

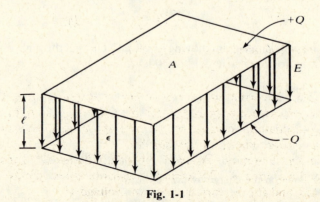

Fig. 1-1

The configuration is shown in Fig. 1-1. The electric flux leaving the upper surface (assumed uniformly charged) is

$$\psi = DA$$

According to Gauss's law, $\psi = Q$, the charge on the surface. Thus, $Q = DA$ and $D = \epsilon E$ yield

$$Q = \epsilon EA$$

But $E = V/\ell$, so that

$$Q = \frac{\epsilon A V}{\ell} \qquad \text{and} \qquad C = \frac{Q}{V} = \frac{\epsilon A}{\ell} \qquad\qquad (1.9)$$

From (1.9) it is seen why permittivity is quoted in F/m.

1.9 In Problem 1.8, when a potential difference is established between the two planes, electric charges accumulate on these two surfaces, and an electric field is produced in between. Work is done in setting up the charge distributions, and we may say that this work is stored as energy in the electric field. Determine the amount of stored energy for Problem 1.8. (This energy is also called *the energy stored in the capacitance*.)

First, we will obtain a general result, and then apply it to Problem 1.8. Consider two point charges, Q_1 and Q_2, located at infinity. First, Q_1 is brought to a point *1*, resulting in an electric field. No work was done in establishing this field, since Q_1 was moved in a field-free region. Next, Q_2 is moved to a point *2*, through the potential difference $V_{2\infty} \equiv V_2$ arising from the field of Q_1. The work done on Q_2 in the process is $W_2 = Q_2 V_2$, and so the energy stored in the resultant field is

$$W_E = W_1 + W_2 = 0 + Q_2 V_2 \qquad\qquad (1.10)$$

Now, if the process is reversed, so that Q_2 is brought to the point *2* first and then Q_1 is brought to the point *1*, we find in the same way

$$W_E = W_1 + W_2 = Q_1 V_1 + 0 \qquad\qquad (1.11)$$

Adding (1.10) and (1.11) yields

$$W_E = \frac{1}{2}(Q_1 V_1 + Q_2 V_2)$$

In general, going over to a continuous distribution of charge,

$$W_E = \frac{1}{2}\sum Q_n V_n \rightarrow \frac{1}{2}\int V(Q)\, dQ \qquad\qquad (1.12)$$

Now, for the configuration of Fig. 1-1, the integration in (1.12) extends over the upper surface, at each point of which the potential is V. Hence,

$$W_E = \frac{1}{2} QV = \frac{1}{2} CV^2 = \frac{\epsilon A V^2}{2\ell} \quad \text{(J)} \qquad\qquad (1.13)$$

Observe that the *energy density* in the region between the two planes,

$$\frac{W_E}{A\ell} = \frac{\epsilon V^2}{2\ell^2} = \frac{\epsilon}{2} E^2 \quad \text{(J/m}^3) \qquad\qquad (1.14)$$

involves only field quantities; (1.14) holds generally.

1.10 We defined electric current as the rate of flow of electric charge. We also have shown that a charge Q will experience a force $F = QE$ in an electric field E. In a constant E, the electrons move through a material medium with a constant average velocity, the *drift velocity* $u = kE$, where k is a constant known as the *mobility*.

With the above in mind, assume a potential difference V across a conductor of length ℓ and cross-sectional area A. Obtain relationships between (*a*) the current density J and the electric field E, and (*b*) the current I and the voltage V.

(a) If ρ (C/m^3) is the volume charge density, then $J = \rho u$ (A/m^2), which, with $u = kE$, gives

$$J = \sigma E \tag{1.15}$$

where $\sigma \equiv k\rho$ (S/m) is the *conductivity* of the conductor.

(b) Substituting $I = JA$ and $V = E\ell$ in (1.14), we find

$$V = RI \tag{1.16}$$

where $R = \ell/\sigma A$ (Ω) is the *resistance* of the conductor.

Equations (1.15) and (1.16) are two forms of *Ohm's law*—(1.15) in terms of field quantities, and (1.16) in terms of circuit quantities.

1.11 A potential difference V exists across the terminals of a conductor of resistance R. Determine the power dissipated in the conductor.

From (1.3) and (1.16),

$$P = \frac{V^2}{R}   \text{(W)} \tag{1.17}$$

1.12 A copper conductor 5 m long, of circular cross section 2 mm in diameter, has a potential difference of 6 V across its terminals. Calculate (a) the resistance of the conductor, (b) the current, and (c) the power dissipated. The *resistivity* (defined as the reciprocal conductivity) for copper at 20 °C is 1.72×10^{-8} $\Omega \cdot$ m.

(a)
$$R = \frac{\ell}{\sigma A} = \frac{5(1.72 \times 10^{-8})}{\pi (2 \times 10^{-3})^2/4} = 27.4 \text{ m}\Omega$$

(b)
$$I = \frac{V}{R} = \frac{6}{27.4 \times 10^{-3}} = 219.2 \text{ A}$$

(c)
$$P = VI = 6(219.2) = 1315 \text{ W}$$

1.13 Find the electric field and the current density in the conductor of Problem 1.12. Also, determine the conductivity using (1.15) and verify that the result is consistent with that of Problem 1.12.

$$E = \frac{V}{\ell} = \frac{6}{5} = 1.2 \text{ V/m}$$

$$J = \frac{I}{A} = \frac{219.2}{\pi \times 10^{-6}} = 69.77 \text{ MA/m}^2$$

$$\sigma = \frac{J}{E} = \frac{69.77 \times 10^6}{1.2} = 58.14 \text{ MS/m}$$

$$\text{resistivity} = \frac{1}{\sigma} = \frac{1}{58.14 \times 10^6} = 1.72 \times 10^{-8} \ \Omega \cdot \text{m}$$

1.14 A certain amount of water is boiled by immersing a current-carrying resistor in water. The heat energy required to boil the water is 99 kJ. The current taken by the resistance is 5 A at 110 V. Find (a) the value of the resistance, and (b) the time required to boil.

(a)
$$R = \frac{V}{I} = \frac{110}{5} = 22 \ \Omega$$

(b) Using $P = IV$ and $U = Pt$,

$$t = \frac{U}{P} = \frac{99 \times 10^3}{5(110)} = 180 \text{ s} = 3 \text{ min}$$

1.15 (*a*) Calculate the current taken by a 100-W/110-V light bulb. (*b*) At the rate of 5 cents per kWh, determine the cost to operate the bulb 10 hours a day for 30 days.

(*a*)
$$I = \frac{P}{V} = \frac{100}{110} = 0.909 \text{ A}$$

(*b*)
$$U = Pt = (100 \times 10^{-3})(10)(30) = 30 \text{ kWh}$$
$$\text{cost of operation} = (30)(\$0.05) = \$1.50$$

1.16 A cube of a material medium, 8 cm on a side, has $\epsilon = 9000\epsilon_0$ and $\sigma = 11$ kS/m. Determine the resistance, R, and the capacitance, C, between any two opposite faces of the cube. Evaluate the product RC and find its units.

$$R = \frac{\ell}{\sigma A} = \frac{8 \times 10^{-2}}{(11 \times 10^3)(8 \times 10^{-2})^2} = 1.136 \text{ m}\Omega$$

$$C = \frac{\epsilon A}{\ell} = 9000(10^{-9}/36\pi)(8 \times 10^{-2}) = 6.36 \text{ nF}$$

$$RC = (1.136 \times 10^{-3})(6.36 \times 10^{-9}) = 7.22 \times 10^{-12} \; \Omega \cdot \text{F} = 7.22 \times 10^{-12} \text{ s}$$

where we have used the conversion

$$(1 \; \Omega)(1 \text{ F}) = \left(1 \frac{\text{V}}{\text{C/s}}\right)\left(1 \frac{\text{C}}{\text{V}}\right) = 1 \text{ s}$$

Supplementary Problems

1.17 Two point charges, Q_1 and Q_2, each 75 μC, and a third point charge, Q_3, of 60 μC, are located at the vertices of an equilateral triangle in free space. If a side of the triangle is 30 cm, calculate the electric force on Q_3.   *Ans.*   779.4 N

1.18 What is the magnitude of the electric field at the center of the triangle of Problem 1.17, due to (*a*) Q_1 and Q_2, and (*b*) all three charges?   *Ans.*   (*a*) 22.5 MV/m; (*b*) 4.5 MV/m

1.19 Determine the work done by a uniform electric field of 3 kV/m in moving a 40-μC point charge through a distance of 20 mm in the direction of the field.   *Ans.*   2.4 mJ

1.20 Calculate the potential difference between two points 3 cm apart in a uniform electric field of 3 kV/m.   *Ans.*   90 V

1.21 Using Gauss's law, repeat Problem 1.2.

1.22 Show that a concentration of electric flux implies a concentration of electric charges.

1.23 Determine the capacitance, per unit length, of a coaxial cable filled with a dielectric of permittivity ϵ. The outer conductor is of radius r_o and the radius of the inner conductor is r_i.
Ans. $2\pi\epsilon/\ln(r_o/r_i)$ (F/m)

1.24 A 2-km cable is made of two concentric cylindrical conductors, of radii 5 mm and 20 mm. The region between the conductors is filled with a dielectric having $\epsilon = 10^3\epsilon_0$. If the cable operates at 1.1 kV, calculate the energy stored in the cable.   *Ans.*   48.5 J

1.25 Find the capacitance between two concentric spherical shells, of radii r_i and $r_o > r_i$, which are separated by a dielectric of permittivity ϵ.   *Ans.* $4\pi\epsilon \left/ \left(\frac{1}{r_i} - \frac{1}{r_o}\right)\right.$ (F)

1.26   (*a*) The plates of a parallel-plate capacitor each have a surface area 500 mm^2 and are 2 mm apart. The space between the plates is occupied by a dielectric having $\epsilon = 10^3 \epsilon_0$. If a voltage of 110 V is established across the plates, how much energy is stored in the capacitor? (*b*) Recalculate the stored energy if the voltage is halved and ϵ is doubled.   *Ans.*   (*a*) 13.37 μJ; (*b*) 6.68 μJ

1.27   Determine the magnitude of the electric flux density at the inner surface of one of the plates of the capacitor of Problem 1.26, for the two cases.   *Ans.*   (*a*) 0.486 μC/m^2; (*b*) 0.486 μC/m^2

1.28   A capacitor is made from two parallel plates separated from each other by a 2-mm-thick dielectric material of permittivity $\epsilon = 600\epsilon_0$. The surface area of each plate is 25 cm^2. A potential difference of 600 V is established across the plates. Calculate (*a*) the charge on a plate, (*b*) the electric field between the plates, (*c*) the electric flux density at a point within the dielectric, and (*d*) the energy stored in the capacitance.   *Ans.*   (*a*) $\pm$3.98 μC; (*b*) 300 kV/m; (*c*) 0.00159 C/m^2; (*d*) 1.19 mJ

1.29   A 20-m metallic conductor, of cross-sectional area 1 mm^2, has a resistance of 6 Ω. Calculate the conductivity of the metal.   *Ans.*   3.33 MS/m

1.30   A cube of an alloy, of resistivity 1.12 $\mu\Omega \cdot$m, is 20 mm on a side. Calculate the resistance between any two opposite faces of the cube.   *Ans.*   56 $\mu\Omega$

1.31   Verify that the dimensions given in Table 1-1 are correct.

1.32   A 110-V light bulb takes 1.1 A in current. Calculate the energy consumed by the bulb in 1 day.   *Ans.*   10.454 MJ = 2.904 kWh

Chapter 2

Circuit Elements and Laws

2.1 CIRCUIT NOTIONS

An electric circuit, which provides for the flow of electric current, is formed by an interconnection of various circuit elements. The basic circuit elements are: resistance, R [see Problem 1.10(b)]; inductance, L; capacitance, C [see Problem 1.8]; voltage source, v; and current source, i. These are symbolized in Fig. 2-1.

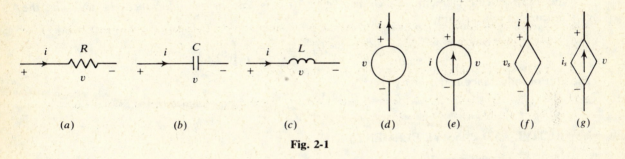

(a)	(b)	(c)	(d)	(e)	(f)	(g)

Fig. 2-1

An interconnection of circuit elements to form an electric circuit, or *network*, is shown in Fig. 2-2 in a generalized form. Notice that each element has two *terminals*; a junction of two or more elements is called a *node*. A circuit element with its leads is said to constitute a *branch* of a circuit. A circuit *loop* is a simple closed path in which nodes and branches alternate; a loop may contain *meshes*, as does that carrying current i_3 in Fig. 2-2.

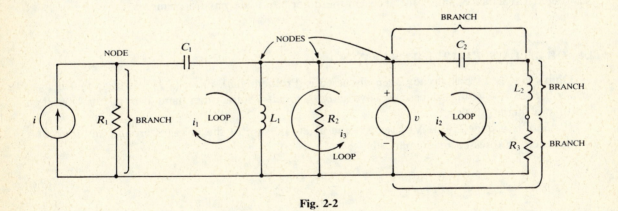

Fig. 2-2

2.2. ELEMENT VOLTAGE-CURRENT RELATIONSHIPS

For the resistance R, Fig. 2-1(a), we have [from (1.16)] the voltage-current relationship

$$v = Ri \qquad \text{or} \qquad i = Gv \qquad\qquad (2.1)$$

where $G \equiv 1/R$ is the *conductance*, measured in siemens (S).

We have defined capacitance C in Problem 1.8 by the relationship

$$C = \frac{Q}{V} = \frac{q}{v} \qquad (2.2)$$

See Fig. 2-1(b). In (2.2), q and v denote time-varying charge and voltage respectively, whereas Q and V are time-stationary (dc) quantities. From Section 1.3, using the definition of current i, we may write

$$i = \frac{dq}{dt} \qquad (2.3)$$

Equations (2.2) and (2.3) yield the voltage-current relationship for a capacitor:

$$v = \frac{1}{C} \int i \, dt \qquad \text{or} \qquad i = C \frac{dv}{dt} \qquad (2.4)$$

The third circuit element, shown in Fig. 2-1(c), is known as *inductance, L*, and has the voltage-current relationship

$$v = L \frac{di}{dt} \qquad \text{or} \qquad i = \frac{1}{L} \int v \, dt \qquad (2.5)$$

The unit of inductance is the *henry* (H).

2.3   ACTIVE AND PASSIVE ELEMENTS

An *ideal voltage source*, Fig. 2-1(d), has a terminal voltage v which is independent of the current i through the source. Similarly, an *ideal current source*, Fig. 2-1(e), is such that the current i through it is independent of the voltage v across it. By contrast, in a *dependent* (or *controlled*) *voltage source*, Fig. 2-1(f), the voltage across the source depends upon the voltage across some other element in the network. Likewise, the current in a *dependent* (or *controlled*) *current source*, Fig. 2-1(g), depends upon the current through another element in the network.

Voltage and current sources are called *active* elements; they can supply energy to the network. Resistance, capacitance, and inductance, on the other hand, are *passive* elements.

2.4   OHM'S LAW; KIRCHHOFF'S LAWS

Ohm's law, (2.1), has already been discussed in Problem 1.10. From (2.1) it is clear that if $i = 0$ but $v \neq 0$, then $R = \infty$; in such case we have an *open circuit*. On the other hand, if $v = 0$ but $i \neq 0$, then $R = 0$, implying a *short circuit*.

Kirchhoff's voltage law (KVL) reads: the algebraic sum of the voltages around any loop of a circuit is zero. Expressed mathematically,

$$\sum_{k=1}^{n} v_k = 0 \qquad (2.6)$$

where n is the number of voltages. In applying (2.6) we must be consistent in accounting for the signs of the various voltages (see Problems 2.15 and 2.16).

Kirchhoff's current law (KCL) states that the algebraic sum of all the currents entering, or of all the currents leaving, any node of a circuit is zero; that is,

$$\sum_{k=1}^{m} i_k = 0 \qquad (2.7)$$

where m is the number of currents at the node under consideration. Again, in applying (2.7), we must be consistent in assigning directions of current flow (see Problem 2.17).

2.5 SERIES AND PARALLEL CIRCUITS

Two or more circuit elements are said to be connected *in series* if they carry the same current (and not merely equal currents). On the other hand, the elements are connected *in parallel* if they have the same voltage across them.

If n resistances are connected in series, it may be shown (see Problem 2.18) that the equivalent resistance, R_{es}, is given by

$$R_{es} = \sum_{k=1}^{n} R_k \tag{2.8}$$

For n resistances in parallel,

$$\frac{1}{R_{ep}} = \sum_{k=1}^{n} \frac{1}{R_k} \tag{2.9}$$

where R_{ep} is the equivalent resistance for the parallel combination (see Problem 2.19). For $n = 2$, (2.9) specializes to $R_{ep} = R_1 R_2/(R_1 + R_2)$.

If the voltages, currents, and resistances in a series circuit can be respectively transformed to currents, voltages, and conductances of a parallel circuit, such that the KVL equations of the series circuit go over into the KCL equations of the parallel circuit, then the circuits are said to be *duals* of each other (see Problems 2.20 and 2.21).

The concept of *voltage division* in a series circuit is often very useful in circuit analysis. If there is a voltage v across the series combination of resistances in (2.8), then the voltage across R_k is given by

$$v_k = \frac{R_k}{R_{es}} v \qquad (k = 1, 2, \ldots, n) \tag{2.10}$$

Analogously, if current i enters the parallel combination of resistances in (2.9), the *current-division rule* for the current in R_k is

$$i_k = \frac{R_{ep}}{R_k} i = \frac{\Pi_k}{\Pi_1 + \Pi_2 + \cdots + \Pi_n} i \qquad (k = 1, 2, \ldots, n) \tag{2.11}$$

where Π_k is the product of all resistances except R_k.

Solved Problems

2.1   Electrical conductivity, and hence resistance, is a slowly-varying function of temperature. To a good approximation, the resistance R_T at temperature T (°C) is related to the resistance R_0 at 0 °C by

$$R_T = R_0(1 + \alpha T)$$

where α is a constant known as the *temperature coefficient of resistance* of the material. If a coil has a resistance of 4.0 Ω at 20 °C and 4.52 Ω at 80 °C, calculate (*a*) the temperature coefficient of resistance of the material; (*b*) the resistance of the coil at 100 °C.

(*a*)
$$R_{20} = R_0(1 + 20\alpha) = 4.0 \ \Omega$$
$$R_{80} = R_0(1 + 80\alpha) = 4.52 \ \Omega$$

Solving simultaneously, $\alpha = 2.27 \times 10^{-3} \ °C^{-1}$ and $R_0 = 3.83 \ \Omega$.

(*b*)
$$R_{100} = (3.83)[1 + 100(2.27 \times 10^{-3})] = 4.69 \ \Omega$$

2.2   Determine the current through, and the voltages across, three resistances, of ohmic values
5 Ω, 7 Ω, and 8 Ω, connected in series and across a 100-V source.

$$\text{total resistance} \equiv R_{es} = 5 + 7 + 8 = 20 \text{ Ω}$$

$$\text{circuit current} \equiv I = \frac{V}{R_{es}} = \frac{100}{20} = 5 \text{ A}$$

$$\text{voltage across the 5 Ω} = 5I = 25 \text{ V}$$

$$\text{voltage across the 7 Ω} = 7I = 35 \text{ V}$$

$$\text{voltage across the 8 Ω} = 8I = 40 \text{ V}$$

2.3   (*a*) Determine the voltage across, and the currents through, three resistances, of 5 Ω, 10 Ω,
and 20 Ω, all connected in parallel and across a 100-V source. (*b*) Find the current and power
drawn from the source.

(*a*)   There is 100 V across each resistor.

$$\text{current through the 5 Ω} = \frac{100}{5} = 20 \text{ A}$$

$$\text{current through the 10 Ω} = \frac{100}{10} = 10 \text{ A}$$

$$\text{current through the 20 Ω} = \frac{100}{20} = 5 \text{ A}$$

(*b*)

$$\text{total current from source} = 20 + 10 + 5 = 35 \text{ A}$$

$$\text{power supplied by source} = VI = 100 \times 35 = 3500 \text{ W}$$

2.4   Reduce the circuit between the terminals *a* and *b*, Fig. 2-3, to a single resistance.

From the law of parallel resistances,

$$\frac{1}{R_{cd}} = \frac{1}{2} + \frac{1}{3} + \frac{1}{6} \qquad \text{or} \qquad R_{cd} = 1 \text{ Ω}$$

The series resistance between *a* and *e* is then $1 + 1 + 6 = 8$ Ω, giving a net resistance

$$R_{ae} = \frac{(8)(8)}{8+8} = 4 \text{ Ω}$$

Finally, $R_{ab} = 4 + 16 = 20$ Ω.

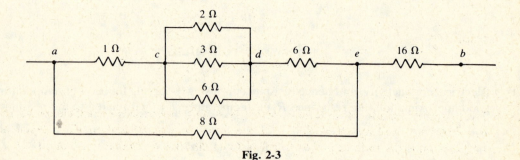

Fig. 2-3

2.5   Calculate the resistances of 110-V light bulbs rated at 25 W, 60 W, 75 W, and 100 W.

From $P = V^2/R$:

$$R_{25w} = \frac{(110)^2}{25} = 484 \text{ Ω} \qquad R_{75w} = \frac{(110)^2}{75} = 161.3 \text{ Ω}$$

$$R_{60w} = \frac{(110)^2}{60} = 201.67 \text{ Ω} \qquad R_{100w} = \frac{(110)^2}{100} = 121 \text{ Ω}$$

2.6 An electric heating pad rated at 110 V and 55 W is to be used at a 220-V source. It is proposed to connect the heating pad in series with a series-parallel combination of light bulbs, each rated at 110 V; bulbs are available having ratings of 25 W, 60 W, 75 W, and 100 W. Obtain a possible scheme of the pad-bulbs combination. At what rate will heat be produced by the pad with this modification?

From Problem 2.5 we know the resistances of the various light bulbs. The resistance of the heating pad is

$$R_p = \frac{(110)^2}{55} = 220 \ \Omega$$

We must combine the bulbs to obtain a total resistance of 220 Ω; then, by voltage division, the pad voltage will be the required 110 V. One possibility is a 100-W bulb in series with a parallel combination of two 60-W bulbs:

$$R_b = R_{100} + \frac{1}{2} R_{60} = 121 + \frac{1}{2} (201.67) = 221.83 \ \Omega$$

which is on the safe side. Then

$$R_p + R_b = 220 + 221.83 = 441.83 \ \Omega \qquad\qquad I_p = \frac{220}{441.83} = 0.498 \ \text{A}$$

and so the heat output of the pad is $I_p^2 R_p = (0.498)^2 (220) = 54.54$ W.

2.7 Four resistors, of ohmic values 5 Ω, 10 Ω, 15 Ω, and 20 Ω, are connected in series and a 100-V source is applied across the combination. How is this voltage divided among the various resistors?

Using the voltage-division rule, we have

$$V_5 = \left(\frac{5}{5 + 10 + 15 + 20} \right)(100 \ \text{V}) = 10 \ \text{V}$$

Similarly, $V_{10} = 20$ V, $V_{15} = 30$ V, $V_{20} = 40$ V.

2.8 Formulate the law of current division among three resistances R_1, R_2, and R_3 connected in parallel. The total input current is i.

The common voltage across the resistances is $V = i R_{ep}$, where

$$\frac{1}{R_{ep}} = \frac{1}{R_1} + \frac{1}{R_2} + \frac{1}{R_3}$$

Hence,

$$i_1 = \frac{V}{R_1} = \frac{R_{ep}}{R_1} i \qquad\qquad i_2 = \frac{R_{ep}}{R_2} i \qquad\qquad i_3 = \frac{R_{ep}}{R_3} i$$

2.9 Two resistors, made of different materials having temperature coefficients of resistance $\alpha_1 = 0.004 \ °\text{C}^{-1}$ and $\alpha_2 = 0.005 \ °\text{C}^{-1}$, are connected in parallel and consume equal power at 10 °C. What is the ratio of power consumed in resistance R_2 to that in R_1 at 60 °C?

At 10 °C, $R_1 = R_2$, which implies

$$R_{01}(1 + 10\alpha_1) = R_{02}(1 + 10\alpha_2) \qquad \text{or} \qquad \frac{R_{01}}{R_{02}} = \frac{1 + 10\alpha_2}{1 + 10\alpha_1} \qquad\qquad (1)$$

Consequently, the power ratio at 60 °C is

$$\frac{V^2/R_2}{V^2/R_1} = \frac{R_1}{R_2} = \frac{R_{01}(1 + 60\alpha_1)}{R_{02}(1 + 60\alpha_2)} = \frac{(1 + 10\alpha_2)(1 + 60\alpha_1)}{(1 + 10\alpha_1)(1 + 60\alpha_2)}$$

Substituting the numerical values of α_1 and α_2 yields the value 0.963.

2.10 A 50-mH inductance carries a current of 5 A which reverses in 25 ms. What is the average voltage induced in the inductance because of this current reversal?

The induced voltage v is zero except at the instant of reversal, when it is infinite. However, over an interval $T = 25$ ms, we have (see figure below)

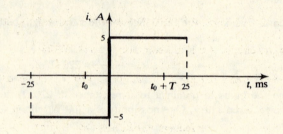

$$i(t_0 + T) - i(t_0) = \frac{1}{L}\int_{t_0}^{t_0+T} v\, dt \equiv \frac{T}{L} v_{avg}$$

or

$$v_{avg} = L\frac{i(t_0 + T) - i(t_0)}{T} = (50 \times 10^{-3})\frac{5 - (-5)}{25 \times 10^{-3}} = 20 \text{ V}$$

2.11 Plot the voltage across an inductive coil if the coil current is as shown in Fig. 2-4(a).

$v = L(di/dt)$; see Fig. 2-4(b).

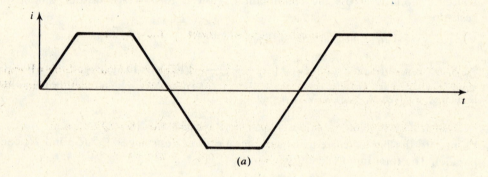

(a)

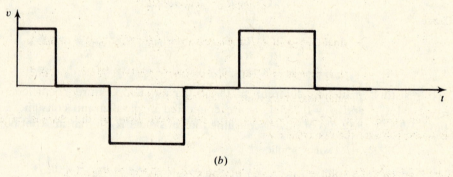

(b)

Fig. 2-4

2.12 (a) Inductances L_1 and L_2 are connected in series. Determine the equivalent inductance. (b) If these inductances are connected in parallel, what is the equivalent inductance? Generalize (a) and (b) to the case of n inductances.

(a) For the series circuit, with common current i, we have

$$v = L_{es}\frac{di}{dt} = v_1 + v_2 = L_1\frac{di}{dt} + L_2\frac{di}{dt} \qquad \text{whence} \qquad L_{es} = L_1 + L_2$$

In general,

$$\textbf{\textit{n} inductances in series} \qquad L_{es} = \sum_{k=1}^{n} L_k$$

(b) For the parallel circuit, if v is the common voltage, we have

$$i = \frac{1}{L_{ep}}\int v\, dt = i_1 + i_2 = \frac{1}{L_1}\int v\, dt + \frac{1}{L_2}\int v\, dt \qquad \text{whence} \qquad \frac{1}{L_{ep}} = \frac{1}{L_1} + \frac{1}{L_2}$$

In general,

$$\textbf{\textit{n} inductances in parallel} \qquad \frac{1}{L_{ep}} = \sum_{k=1}^{n}\frac{1}{L_k}$$

2.13 Obtain the equivalent capacitance for two capacitances C_1 and C_2 connected (a) in series and (b) in parallel. Generalize (a) and (b) to the case of n capacitances.

(a)
$$v = \frac{1}{C_{es}}\int i\, dt = v_1 + v_2 = \frac{1}{C_1}\int i\, dt + \frac{1}{C_2}\int i\, dt \qquad \text{whence} \qquad \frac{1}{C_{es}} = \frac{1}{C_1} + \frac{1}{C_2}$$

$$\textbf{\textit{n} capacitances in series} \qquad \frac{1}{C_{es}} = \sum_{k=1}^{n}\frac{1}{C_k}$$

(b)
$$i = C_{ep}\frac{dv}{dt} = i_1 + i_2 = C_1\frac{dv}{dt} + C_2\frac{dv}{dt} \qquad \text{whence} \qquad C_{ep} = C_1 + C_2$$

$$\textbf{\textit{n} capacitances in parallel} \qquad C_{ep} = \sum_{k=1}^{n} C_k$$

2.14 A 40-μF capacitance is charged to store 0.2 J of energy. An uncharged, 60-μF capacitance is then connected in parallel with the first one through perfectly conducting leads. What is the final energy of the system?

The initial charge on the 40-μF capacitance is obtained from $U = Q^2/2C$; thus,

$$0.2 = \frac{Q^2}{2(40 \times 10^{-6})} \qquad \text{or} \qquad Q = 4 \times 10^{-3}\text{ C}$$

When the capacitances are connected in parallel, the common voltage V is given by

$$V = \frac{\text{total } Q}{\text{total } C} = \frac{4 \times 10^{-3}}{(40 + 60)10^{-6}} = 40\text{ V}$$

Then:

$$\text{final energy in 40-}\mu\text{F capacitance} = \frac{1}{2}(40 \times 10^{-6})(40)^2 = 0.032\text{ J}$$

$$\text{final energy in 60-}\mu\text{F capacitance} = \frac{1}{2}(60 \times 10^{-6})(40)^2 = 0.048\text{ J}$$

$$\text{final total energy} = 0.032 + 0.048 = 0.08\text{ J}$$

The energy lost, $0.2 - 0.08 = 0.12$ J, represents work done by the charges on one another in spreading out over the two capacitors.

2.15 Apply Kirchhoff's voltage law, (2.6), to the circuit shown in Fig. 2-5.

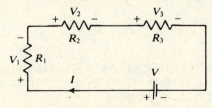

Fig. 2-5

The direction of current I is arbitrarily chosen as indicated in Fig. 2-5. The voltage across each resistor is assigned a polarity; it is understood that Ohm's law yields $V = IR$ if I enters the positive terminal of a resistor, and $V = -IR$ otherwise. Application of KVL to Fig. 2-5 leads to

$$V_1 + V_2 + V_3 - V = 0$$

where $V_1 = IR_1$, $V_2 = IR_2$, and $V_3 = IR_3$.

2.16 Write the Kirchhoff's voltage equations for the two indicated loops of the network of Fig. 2-6. Assume polarities as marked. Also, express V_2 in terms of I_1, I_2, and R_2.

$$\text{loop 1:} \qquad -V_\alpha + V_1 + V_2 + V_\beta = 0$$
$$\text{loop 2:} \qquad -V_\beta - V_2 + V_3 + V_4 + V_\gamma = 0$$
$$\text{voltage across } R_2: \qquad V_2 = (I_1 - I_2)R_2$$

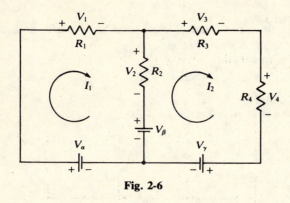

Fig. 2-6

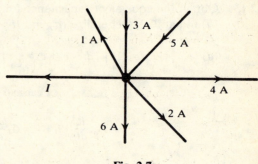

Fig. 2-7

2.17 A node of a network is shown in Fig. 2-7, with currents flowing in the directions shown. Determine the magnitude and direction of I.

Assume the indicated direction for I. Then from (2.7) we have

$$-I - 1 + 3 + 5 - 4 - 2 - 6 = 0$$

Hence $I = -5$ A; i.e. 5 A into the node.

2.18 Derive an expression for the equivalent resistance formed by n resistances $R_1, R_2, \ldots, R_n$ connected in series.

Let V be the voltage across the series circuit combination, I the current through it, and R_{es} the equivalent resistance. Then, from Ohm's law,

$$V = I(R_1 + R_2 + \cdots + R_n) = IR_{es} \qquad \text{whence} \qquad R_{es} = \sum_{k=1}^{n} R_k$$

2.19 Resistances $R_1, R_2, \ldots, R_n$ are connected in parallel. Obtain an expression for the equivalent resistance.

Let I be the total current into the parallel combination, V the voltage across it, and R_{ep} the equivalent resistance. Then, from KCL,

$$I = V\left(\frac{1}{R_1} + \frac{1}{R_2} + \cdots + \frac{1}{R_n}\right) = \frac{V}{R_{ep}} \qquad \text{or} \qquad \frac{1}{R_{ep}} = \sum_{k=1}^{n} \frac{1}{R_k}$$

2.20 Three resistances, R_1, R_2, and R_3, are connected in series with a voltage source V. Construct the dual network.

The network is shown in Fig. 2-8(a). By Section 2.5, we draw the dual network, Fig. 2-8(b), by replacing the series elements by parallel elements. The equations for the two networks are:

(a)
$$V = IR_1 + IR_2 + IR_3$$

(b)
$$I = \frac{V}{R_1} + \frac{V}{R_2} + \frac{V}{R_3} = VG_1 + VG_2 + VG_3$$

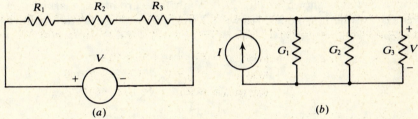

Fig. 2-8

2.21   A two-mesh network is shown in Fig. 2-9(a). Draw its dual network.

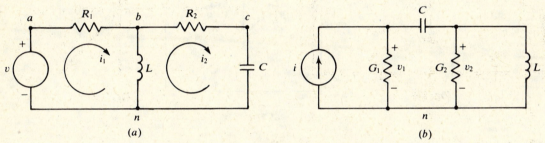

Fig. 2-9

The network equations are:

$$R_1 i_1 + L\frac{di_1}{dt} - L\frac{di_2}{dt} = v$$

$$R_2 i_2 + \frac{1}{C}\int i_2\, dt + L\frac{di_2}{dt} - L\frac{di_1}{dt} = 0$$

To construct the dual, replace i_1 and i_2 in the above equations by two voltages v_1 and v_2, and replace v by i, giving

$$R_1 v_1 + L\frac{dv_1}{dt} - L\frac{dv_2}{dt} = i$$

$$R_2 v_2 + \frac{1}{C}\int v_2\, dt + L\frac{dv_2}{dt} - L\frac{dv_1}{dt} = 0$$

Now, if we replace resistances by conductances and interchange inductances and capacitances, we obtain

$$G_1 v_1 + C\frac{dv_1}{dt} - C\frac{dv_2}{dt} = i$$

$$G_2 v_2 + \frac{1}{L}\int v_2\, dt + C\frac{dv_2}{dt} - C\frac{dv_1}{dt} = 0$$

These last equations are represented by the circuit of Fig. 2-9(b), which is the dual network.

2.22   Two resistances, $R_1 = 15\ \Omega$ and $R_2 = 25\ \Omega$, are connected in parallel and are supplied by a current source, $I = 5$ A. Determine the power consumed by each resistance.

From (2.11), with $n = 2$,

$$P_1 = I_1^2 R_1 = \frac{R_1 R_2^2}{(R_1 + R_2)^2} I^2 = \frac{(15)(25)^2}{(40)^2}(5)^2 = 146.5 \text{ W}$$

$$P_2 = I_2^2 R_2 = \frac{R_1^2 R_2}{(R_1 + R_2)^2} I^2 = \frac{(15)^2(25)}{(40)^2}(5)^2 = 87.9 \text{ W}$$

2.23 A bridge circuit is shown in Fig. 2-10. With the currents as marked, (a) write Kirchhoff's current law at the four nodes; (b) write Kirchhoff's voltage law around the loops $abda$, $bcdb$, and $adca$.

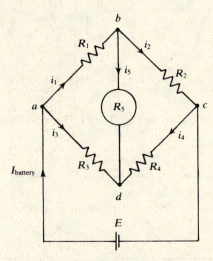

Fig. 2-10

(a)

node a:	$I = i_1 + i_3$
node b:	$i_1 = i_2 + i_5$
node c:	$i_2 = I + i_4$
node d:	$0 = i_3 + i_4 + i_5$

(b)

loop $abda$:	$i_1 R_1 + i_5 R_5 = i_3 R_3$
loop $bcdb$:	$i_5 R_5 = i_2 R_2 + i_4 R_4$
loop $adca$:	$i_3 R_3 - i_4 R_4 = E$

2.24 For the circuit of Problem 2.23, consider the special case of the balanced bridge ($i_5 = 0$). (a) If $R_1 = 10 \ \Omega$, $R_2 = 20 \ \Omega$, and $R_3 = 30 \ \Omega$, determine R_4. (b) If $E = 45$ V, calculate the current supplied by the battery.

(a) Since $i_5 = 0$, we have $i_1 = i_2$ and $i_3 = i_4$. Also, nodes b and d are at the same potential. Thus,

$$i_1 R_1 = i_3 R_3 \tag{1}$$
$$i_2 R_2 = i_1 R_2 = i_4 R_4 = i_3 R_4 \tag{2}$$

From (1) and (2) we obtain

$$\frac{R_1}{R_2} = \frac{R_3}{R_4} \qquad \text{or} \qquad R_4 = \frac{R_2 R_3}{R_1} = \frac{(20)(30)}{10} = 60 \ \Omega$$

(b) The effective resistance, R_e, across the battery becomes

$$R_e = \frac{(10 + 20)(30 + 60)}{10 + 20 + 30 + 60} = 22.5 \ \Omega \qquad \text{and} \qquad I_{\text{battery}} = \frac{E}{R_e} = \frac{45}{22.5} = 2.0 \text{ A}$$

2.25 Determine the currents i_x and i_y in the network shown in Fig. 2-11.

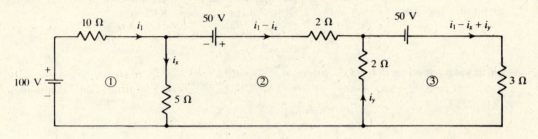

Fig. 2-11

On the basis of KCL, the currents in the remaining branches are also marked in Fig. 2-11. By KVL for meshes 1, 2, and 3:

$$100 = 10i_1 + 5i_x$$
$$50 = 2(i_1 - i_x) - 2i_y - 5i_x$$
$$50 = 3(i_1 - i_x + i_y) + 2i_y$$

Solving these simultaneous equations yields $i_x = -6.13$ A, $i_y = 0.51$ A. The negative sign on i_x implies that the actual current flows in the direction opposite to that given in Fig. 2-11.

2.26 This problem relates to the concept of *source transformation*. Replace the voltage source v and its internal (series) resistance R_v, in Fig. 2-12(a), by a current source i with internal (shunt) resistance R_i, in Fig. 2-12(b), such that the current through R remains unchanged.

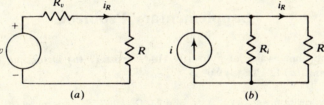

Fig. 2-12

From Fig. 2-12(a),

$$i_R = \frac{v}{R_v + R}$$

and from Fig. 2-12(b),

$$i_R = \frac{iR_i}{R_i + R}$$

Then, for equivalence,

$$\frac{v}{R_v + R} = \frac{iR_i}{R_i + R}$$

which will hold if we take $i = v/R_v$ and $R_i = R_v$.

2.27 Find the current in, and the voltage across, the 2 Ω resistance in Fig. 2-13(a).

Using the results of Problem 2.26, we transform the 5-A current source to a voltage source; the circuit then becomes as shown in Fig. 2-13(b). For the two loops, KVL gives

$$25 - 15I_1 - 10 + 3I_2 - 3I_1 = 0$$
$$20 - 3I_2 + 3I_1 - 2I_2 = 0$$

which when solved yield $I_2 = 5$ A and $V_{ab} = (5)(2) = 10$ V.

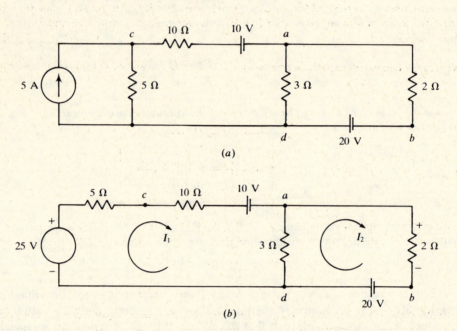

(a)

(b)

Fig. 2-13

Supplementary Problems

2.28 (a) What is the ohmic value of a 1000-W/110-V resistor? (b) Determine the current rating of this resistor.   Ans.   (a) 12.1 Ω; (b) 9.09 A

2.29 A conductor has resistance 5.4 Ω at 20 °C and 7.0 Ω at 100 °C. Determine (a) the resistance at 0 °C, and (b) the temperature at which the resistance will be 6.0 Ω. Assume that the resistance varies linearly with temperature.   Ans.   (a) 5.0 Ω; (b) 50 °C

2.30 Temperature coefficients of resistance for copper and platinum are respectively 0.00427 °C^{-1} and 0.00357 °C^{-1}. Resistors made of copper and platinum have respective resistances 17 Ω and 18 Ω at 20 °C. Calculate the temperature at which the two resistances will be equal.   Ans.   165.4 °C

2.31 The resistance of a certain coil is 4.0 Ω at 20 °C and is 4.25 Ω at 80 °C. At what temperature will the resistance be 4.67 Ω? Assume a constant temperature coefficient of resistance.   Ans.   180.8 °C

2.32 Calculate the powers dissipated in the three resistances of Problem 2.3, and thereby verify the result of Problem 2.3(b).   Ans.   2000 W, 1000 W, 500 W

2.33 If a source of 200 V is connected across the circuit of Problem 2.4, find (a) the voltage across the 8 Ω resistance, and (b) the power dissipated in the 1 Ω resistance.   Ans.   (a) 40 V; (b) 25 W

2.34 Determine the ratio of the currents, I_1/I_2, at 60 °C in the resistors of Problem 2.9.   Ans.   1.0384

2.35 A battery has internal resistance R_i and terminal voltage V_t. Show that the power supplied to a resistive load cannot exceed $V_t^2/2R_i$.

2.36 For the battery of Problem 2.35, $V_t = 96$ V and $R_i = 50$ mΩ. Discrete loads of 150, 100, 50, 30, and 20 mΩ are connected, one at a time, across the battery. Plot the curve of power supplied versus the ohmic value of the load. Hence verify that the maximum power transfer occurs when $R_i = R_{load} = 50$ mΩ.

2.37 Two inductances, $L_1 = 30$ mH and $L_2 = 60$ mH, are connected in parallel, and the combination is connected in series with an inductance $L_3 = 10$ mH. Determine the equivalent inductance.
Ans. 30 mH

2.38 What is the equivalent capacitance between the terminals a and b of the capacitive system shown in Fig. 2-14?   *Ans.* 14C

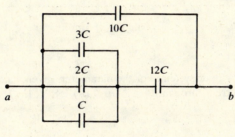

Fig. 2-14

2.39 A 40-μF capacitor is connected in parallel with a 60-μF capacitor and across a time-varying voltage source. At a certain instant, the total current supplied by the source is 10 A. Determine the instantaneous currents through the individual capacitors.   *Ans.* $i_{40\mu F} = 4$ A, $i_{60\mu F} = 6$ A

2.40 A 50-μF capacitor is charged to 300 μC. An uncharged 100-μF capacitor is then connected in parallel with the first capacitor. Evaluate the charge transferred to the 100-μF capacitor.   *Ans.* 200 μC

2.41 A rectangular block of copper measures 2.54 cm by 5.08 cm by 7.62 cm. If the conductivity of copper is 58 MS/m, determine the resistances between pairs of parallel faces.   *Ans.* 0.113 $\mu\Omega$, 0.452 $\mu\Omega$, 1.018 $\mu\Omega$

2.42 A 3-Ω resistor carries the current waveform shown in Fig. 2-15. Calculate the average power dissipated in the resistor.   *Ans.* 100 W

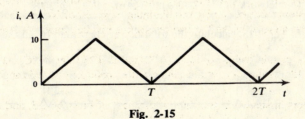

Fig. 2-15

2.43 The v-i characteristic of a nonlinear circuit element is shown in Fig. 2-16(a), together with a linear approximation (dashed). Obtain the equivalent circuit, using a resistor and a voltage source, for this element if it were to operate at (a) 10 V, (b) 30 V.   *Ans.* (a) see Fig. 2-16(b); (b) see Fig. 2-16(c)

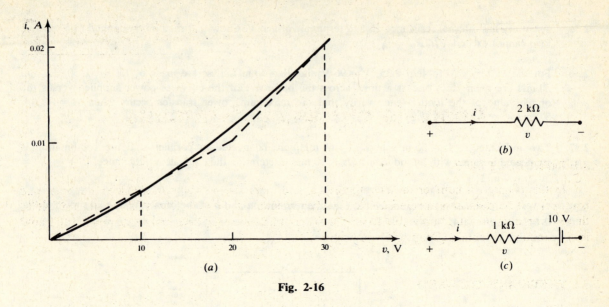

Fig. 2-16

2.44 Figure 2-17 shows a 3-wire dc system. For the parameters given, calculate (*a*) the voltage across the 10-Ω resistor, and (*b*) the current through it.   *Ans.*   (*a*) 158 V; (*b*) 15.8 A

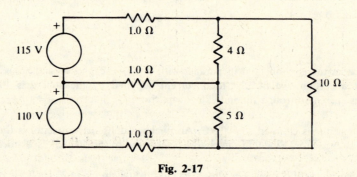

Fig. 2-17

2.45 For the circuit shown in Fig. 2-13(*a*), determine the power supplied to the network by the (*a*) 5-A current source, (*b*) 10-V battery, and (*c*) 20-V battery.   *Ans.*   (*a*) 83.33 W; (*b*) −16.7 W; (*c*) 100 W

2.46 Show that the two circuits of Fig. 2-12, as related in Problem 2.26, are duals.

2.47 Obtain a dual of the network of Fig. 2-13(*a*).   *Ans.*   See Fig. 2-18.

2.48 For the network of Fig. 2-18, calculate the total power consumed by all the resistances.
Ans.   1131 W

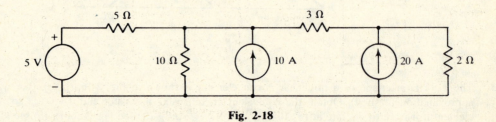

Fig. 2-18

Chapter 3

Analysis of Resistive Circuits

3.1 INTRODUCTION

In this chapter, a number of laws, theorems, and methods (not necessarily independent of one another) will be discussed in reference to resistive circuits. Along with Ohm's law, KVL, and KCL, these techniques are also applicable to ac circuits containing inductances and capacitances, and hold in the frequency domain (Chapter 5), as well.

3.2 THEVENIN'S THEOREM

Thévenin's theorem may be stated as follows: At the terminals *12* in Fig. 3-1(*a*), the arbitrary linear network *A*, containing resistances and energy sources, can be replaced by an equivalent circuit consisting of a voltage source V_{Th} in series with a resistance R_{Th}.

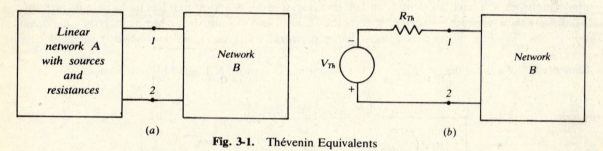

Fig. 3-1. Thévenin Equivalents

The voltage V_{Th} is the open-circuit voltage across *12*, and R_{Th} is the ratio of the open-circuit voltage to the short-circuit current. Alternatively, R_{Th} is the equivalent resistance at the terminals *12* when all independent sources are suppressed. Figure 3-1(*a*) shows the arbitrary network, and its Thévenin equivalent is pictured in Fig. 3-1(*b*).

Thévenin's theorem is useful in network reduction, especially in determining the current through a specified branch of a network.

3.3 NORTON'S THEOREM

Norton's equivalent circuit is the dual of Thévenin's equivalent circuit, and may be obtained from the latter by source transformation (Problem 2.26). Explicitly, *Norton's theorem* may be stated as follows: At the terminals *12* in Fig. 3-2(*a*), the arbitrary linear network *A*, containing resistances

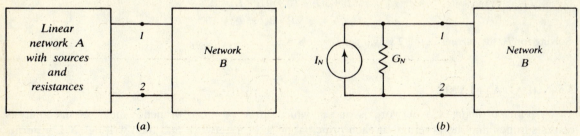

Fig. 3-2. Norton Equivalents

and energy sources, can be replaced by an equivalent circuit consisting of a current source I_N in parallel with a conductance G_N, shown in Fig. 3-2(b).

The current I_N is the short-circuit current (through 12, when these terminals are short-circuited) and G_N is the ratio of the short-circuit current to the open-circuit voltage. It is to be noted that $G_N = 1/R_{Th}$ where R_{Th} is the Thévenin equivalent resistance.

3.4 SUPERPOSITION THEOREM

The *superposition theorem* states that for a linear resistive circuit, containing voltage and/or current sources, the current through (or the voltage across) any element can be obtained by the algebraic addition of the individual currents (or voltages) due to each independent source acting alone. Nonacting independent voltage sources are replaced by their internal resistances (or short circuits, for ideal voltage sources), and nonacting independent current sources are replaced by their internal conductances (or open circuits, for ideal current sources).

3.5 MESH ANALYSIS

We have already used mesh analysis, e.g. in Problem 2.27, without explicitly defining it. We now consider the technique more formally. First, we define a *flat* or *planar* network as one that can be represented by a diagram (*graph*) on a plane surface such that no two branches cross. Any loop of a planar network will have an interior and an exterior; a *mesh* is a loop that has no loop in its interior. (If the graph is viewed as a map, the meshes bound the "countries.") Mesh analysis consists in applying KVL, (*2.6*), to the meshes of a planar network; KCL, (*2.7*), will be implicitly satisfied.

Example 3.1   Solve for the current in the 2-Ω resistor of the network of Fig. 3-3 by mesh analysis.

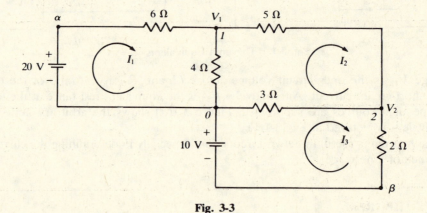

Fig. 3-3

There are three meshes, and we define I_k as the current flowing in mesh k ($k = 1, 2, 3$). Then, KVL gives

$$6I_1 + 4(I_1 - I_2) = 20$$
$$5I_2 + 3(I_2 - I_3) + 4(I_2 - I_1) = 0$$
$$2I_3 + 3(I_3 - I_2) = 10$$

Solving for I_3, the current in the 2 Ω, we get $I_3 = 2.98$ A.

3.6 NODAL ANALYSIS

A *principal node* of a network is one at which three or more branches meet. Nodal analysis consists in defining the potential at each principal node of a network and then writing KCL in terms of these potentials. Since only potential differences are important, one of the potentials may be taken as zero and the method applied at the remaining principal nodes only.

Example 3.2   By nodal analysis, obtain the current in the 2-Ω resistor of the network of Fig. 3-3.

In Fig. 3-3 we identify the principal nodes 0, 1, and 2, and choose node 0 as the reference node. Next, we define the voltages of nodes 1 and 2 with respect to node 0: $V_{10} \equiv V_1$, $V_{20} \equiv V_2$. We now apply KCL at nodes 1 and 2 to obtain:

$$\frac{20 - V_1}{6} + \frac{V_2 - V_1}{5} - \frac{V_1}{4} = 0$$

$$\frac{V_1 - V_2}{5} - \frac{V_2 + 10}{2} - \frac{V_2}{3} = 0$$

Solving for V_2, we get $V_2 = -4.046$ V. Hence, the current in 2-Ω resistor is

$$I_3 = \frac{V_2 + 10}{2} = \frac{-4.046 + 10}{2} = 2.98 \text{ A}$$

and is directed from node 2 to node β. This result is in agreement with Example 3.1.

3.7 NETWORK REDUCTION AND SOURCE TRANSFORMATION

We have referred to the process of network reduction in Chapter 2, while discussing the laws of series and parallel resistances, (*2.8*) and (*2.9*). We may obtain an equivalent source for several voltage sources in series, as illustrated in Fig. 3-4; or for several current sources in parallel, as shown in Fig. 3-5. Combining resistances to obtain an equivalent resistance and sources to get an equivalent source considerably simplifies computations. However, there exist resistance configurations that cannot be simplified by the series/parallel laws of resistances. Such cases may be handled by the *wye-delta* (*Y-Δ*) *transformation*. A wye circuit is shown in Fig. 3-6(*a*) and a delta circuit in Fig. 3-6(*b*). The two circuits are equivalent if the resistance between any two terminals of the one is equal

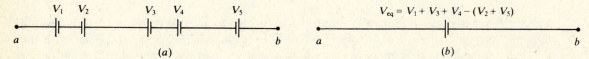

Fig. 3-4

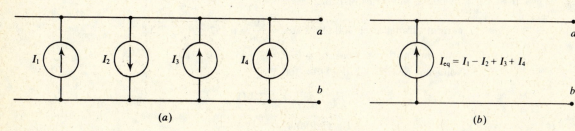

Fig. 3-5

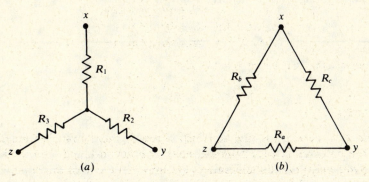

Fig. 3-6

to the resistance between the same two terminals of the other. The criterion for equivalence may be shown to be:

$$R_1 = \frac{R_b R_c}{R_a + R_b + R_c} \qquad R_2 = \frac{R_a R_c}{R_a + R_b + R_c} \qquad R_3 = \frac{R_a R_b}{R_a + R_b + R_c} \qquad (3.1)$$

or, reciprocally,

$$R_a = \frac{R_1 R_2 + R_2 R_3 + R_3 R_1}{R_1} \qquad R_b = \frac{R_1 R_2 + R_2 R_3 + R_3 R_1}{R_2} \qquad R_c = \frac{R_1 R_2 + R_2 R_3 + R_3 R_1}{R_3} \quad (3.2)$$

Solved Problems

3.1 Three 2100-Ω resistors, all in parallel, are connected across a 210-V dc source. Calculate (a) the total current, (b) the current through each resistor, and (c) the total power dissipated in the resistors.

From the law of parallel resistances,

$$\frac{1}{R_{ep}} = \frac{1}{2100} + \frac{1}{2100} + \frac{1}{2100} \qquad \text{or} \qquad R_{ep} = 700 \; \Omega$$

(a)
$$I = \frac{V}{R_{ep}} = \frac{210}{700} = 0.3 \text{ A}$$

(b)
$$\frac{1}{3} I = 0.1 \text{ A}$$

(c)
$$VI = (210)(0.3) = 63 \text{ W}$$

3.2 Three 700-Ω resistors, all in parallel, are to be connected to a 210-V dc source. It is desired to limit the voltage across these resistors to 110 V by connecting a resistor in series with the parallel combination. Determine (a) the value of the series resistor, and (b) the total power drawn from the 210-V source.

(a) Let R_x be the value of the series resistor.

$$\text{current through each 700-}\Omega\text{ resistor} = \frac{110}{700} = 0.157 \text{ A}$$

$$\text{total current drawn from the source} = 3(0.157) = 0.471 \text{ A}$$

$$\text{voltage across } R_x = 210 - 110 = 100 \text{ V}$$

Hence
$$R_x = \frac{100}{0.471} = 212.3 \; \Omega$$

(b)
$$VI = (210)(0.471) = 98.9 \text{ W}$$

3.3 Obtain the source voltage V for the circuit shown in Fig. 3-7.

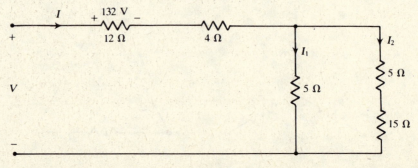

Fig. 3-7

From Fig. 3-7, $I = 132/12 = 11$ A. From the laws of series and parallel resistances, the total circuit resistance is

$$R = 12 + 4 + \frac{5(5 + 15)}{5 + 5 + 15} = 20 \ \Omega$$

Hence
$$V = IR = (11)(20) = 220 \ V$$

3.4   Determine I_1 and I_2 shown in Fig. 3-7.

Using current division, (*2.11*),

$$I_1 = \frac{5 + 15}{5 + 5 + 15}(11) = 8.8 \ A$$

$$I_2 = \frac{5}{5 + 5 + 15}(11) = 2.2 \ A$$

3.5   Replace the network of Fig. 3-8(*a*) to the left of terminals *ab* by its Thévenin equivalent circuit. Hence, determine *I*.

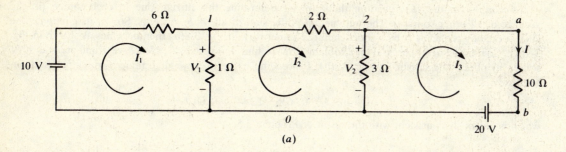

(a)

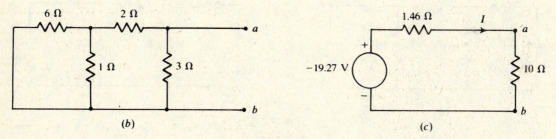

(b)   (c)

Fig. 3-8

To determine R_{Th}, short-circuit the voltage sources to obtain the network of Fig. 3-8(*b*). The 6 Ω and 1 Ω in parallel are equivalent to

$$\frac{(6)(1)}{6 + 1} = \frac{6}{7} \ \Omega$$

which, in series with the 2 Ω, is equivalent to

$$\frac{6}{7} + 2 = \frac{20}{7} \ \Omega$$

Then, the parallel combination of (20/7) Ω and 3 Ω gives

$$R_{Th} = R_{ab} = \frac{(20/7)(3)}{(20/7) + 3} = \frac{60}{41} = 1.46 \ \Omega$$

To find V_{Th}, open-circuit the terminals *ab*. The 10-V battery then sees a resistance

$$6 + \frac{(1)(2 + 3)}{1 + 2 + 3} = \frac{41}{6} \ \Omega$$

and so a current of

$$\frac{10}{41/6} = \frac{60}{41} = 1.46 \text{ A}$$

is drawn from it, of which

$$\frac{1}{1+2+3}(1.46) = 0.243 \text{ A}$$

passes through the 3-Ω resistor. Then,

$$V_{Th} = (0.243)(3) - 20 = -19.27 \text{ V}$$

Thévenin's equivalent circuit becomes as shown in Fig. 3-8(c), whence

$$I = \frac{-19.27}{1.46 + 10} = -1.68 \text{ A}$$

The negative sign indicates that the current actually flows in the 10 Ω from b to a.

3.6   Solve for the current I of Fig. 3-8(a) by applying the superposition theorem.

According to the superposition theorem, we determine the current due to each source (in the absence of all other sources). The net current is the sum of all these currents. So, let us eliminate the 20-V source by replacing it by a short circuit. The network of Fig. 3-8(a) then becomes as shown in Fig. 3-9(a). Combining resistances as in Problem 3.5, we find $I_1 = 0.0636$ A. Next, we eliminate the 10-V source to obtain the circuit of Fig. 3-9(b). Proceeding as before, we determine $I_2 = -1.744$ A. Consequently,

$$I = I_1 + I_2 = 0.0636 - 1.744 = -1.68 \text{ A}$$

which is obviously consistent with the result of Problem 3.5.

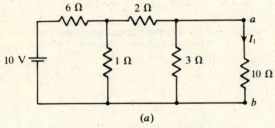

(a) (b)

Fig. 3-9

3.7   Determine I of Fig. 3-8(a) by mesh analysis.

In terms of the three mesh currents I_1, I_2, and I_3 ($=I$) indicated in Fig. 3-8(a), we have:

$$7I_1 - I_2 = 10$$
$$-I_1 + 6I_2 - 3I_3 = 0$$
$$-3I_2 + 13I_3 = -20$$

Solving for I_3 yields $I_3 = I = -1.68$ A, which is a further check of our previous results.

3.8   Evaluate I of Fig. 3-8(a) by nodal analysis.

Figure 3-8(a) can be redrawn so that the two principal nodes labeled 0 coincide. Choosing this single node 0 as the reference, we have the nodal equations

$$\frac{10 - V_1}{6} - \frac{V_1}{1} - \frac{V_1 - V_2}{2} = 0$$

$$\frac{V_1 - V_2}{2} - \frac{V_2}{3} - \frac{V_2 - 20}{10} = 0$$

from which $V_2 = 3.2$ V. Hence,

$$I = \frac{V_2 - 20}{10} = -1.68 \text{ A}$$

In Problems 3.5 through 3.8 we have applied four different methods of network analysis. Nodal analysis ordinarily requires the least amount of work.

3.9   A resistive network is shown in Fig. 3-10(a). Determine the value of the equivalent resistance at the terminals AB.

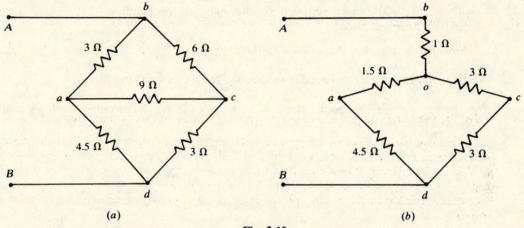

(a)   (b)

Fig. 3-10

We transform the delta network abc of Fig. 3-10(a) to wye by use of (3.1):

$$R_{ao} = \frac{(9)(3)}{3+6+9} = 1.5 \ \Omega \qquad R_{bo} = \frac{(6)(3)}{3+6+9} = 1 \ \Omega \qquad R_{co} = \frac{(6)(9)}{3+6+9} = 3 \ \Omega$$

The resulting network, Fig. 3-10(b), has resistances in a series-parallel combination, and we find:

$$R_{od} = \frac{(1.5+4.5)(3+3)}{1.5+4.5+3+3} = 3 \ \Omega$$

Hence, $R_{AB} = R_{bo} + R_{od} = 1 + 3 = 4 \ \Omega$.

3.10   In the network of Fig. 3-10(a), the 4.5 Ω is replaced by a variable resistance R_{ad}, and a 110-V source is connected across AB, terminal A being positive. What must be the value of the variable resistance for it to draw maximum power from the source?

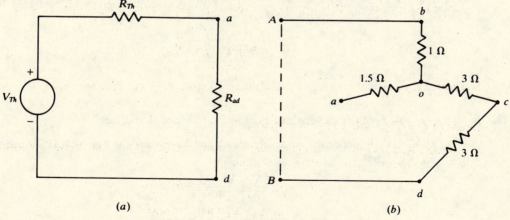

(a)   (b)

Fig. 3-11

The calculations will be simplified if we work with the equivalent network, Fig. 3-10(b), wherein the 4.5 Ω is replaced by R_{ad}. The Thévenin equivalent of that network at ad is shown in Fig. 3-11(a); for maximum power transfer we must have $R_{ad} = R_{Th}$ (see Problems 2.35 and 2.36).

To find R_{Th}, remove R_{ad} and replace the 110-V source by a short circuit [dashed line in Fig. 3-11(b)]. Then R_{Th} is the resistance measured at ad. Now, the short-circuiting has put the 1.5 Ω in series with a parallel combination of 1 Ω and $3 + 3 = 6$ Ω; hence,

$$R_{Th} = R_{ad} = 1.5 + \frac{(1)(6)}{1+6} = \frac{33}{14} \, \Omega$$

3.11 Calculate the total power supplied to the network of Problem 3.10 under the derived maximum-power condition.

From Fig. 3-11(a), with $R_{ad} = R_{Th} = (33/14) \, \Omega$,

$$\text{total power} = \frac{V_{Th}^2}{R_{Th} + R_{ad}} = \frac{7}{33} V_{Th}^2 \quad \text{(W)}$$

The Thévenin voltage V_{Th} appears across ad or across od in Fig. 3-11(b) when the 110-V source is put across AB. Thus, by voltage division,

$$V_{Th} = V_{od} = \frac{6}{7}(110) = \frac{660}{7} \, \text{V}$$

and

$$\text{total power} = \frac{7}{33}\left(\frac{660}{7}\right)^2 = \frac{13\,200}{7} = 1886 \, \text{W}$$

3.12 A resistive network with voltage and current sources is shown in Fig. 3-12(a). Determine the currents I_1 and I_2 by mesh analysis.

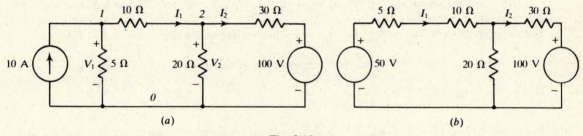

(a)   (b)

Fig. 3-12

In order to apply mesh analysis, we first transform the 10-A current source in parallel with the 5-Ω resistor to a $10 \times 5 = 50$ V voltage source in series with a 5-Ω resistor. Thus, we obtain the network of Fig. 3-12(b), for which mesh equations are

$$35I_1 - 20I_2 = 50$$
$$-20I_1 + 50I_2 = -100$$

Solving, $I_1 = 0.37$ A and $I_2 = -1.85$ A.

3.13 Solve for I_1 and I_2 of the network of Fig. 3-12(a) by nodal analysis.

For nodal analysis, define the node voltages V_1 and V_2 as shown in Fig. 3-12(a). At nodes 1 and 2, respectively, KCL gives

$$\frac{V_1}{5} + \frac{V_1 - V_2}{10} = 10$$

$$\frac{V_2}{20} + \frac{V_2 - 100}{30} = \frac{V_1 - V_2}{10}$$

which have the solution

$$V_1 = \frac{1300}{27} \text{ V} \qquad V_2 = \frac{400}{9} \text{ V}$$

Thus, $\qquad I_1 = \frac{V_1 - V_2}{10} = \frac{10}{27} = 0.37 \text{ A} \qquad I_2 = \frac{V_2 - 100}{30} = \frac{-50}{27} = -1.85 \text{ A}$

3.14 For the network of Fig. 3-12(a), calculate the power supplied by the current source and by the voltage source. Verify that the sum of the powers from the two sources is the total power dissipated in all the resistances.

power supplied by the 10-A source $= 10 \times V_1 = \frac{10(1300)}{27} = 481.48$ W

power supplied by the 100-V source $= 100 \times (-I_2) = (100)(1.85) = 185$ W

total power supplied by the two sources $= 481.48 + 185 = 666.48$ W

power dissipated in the 5 $\Omega = \frac{V_1^2}{5} = \frac{1}{5}\left(\frac{1300}{27}\right)^2 = 463.65$ W

power dissipated in the 10 $\Omega = I_1^2 \times 10 = (0.37)^2(10) = 1.37$ W

power dissipated in the 20 $\Omega = \frac{V_2^2}{20} = \frac{1}{20}\left(\frac{400}{9}\right)^2 = 98.76$ W

power dissipated in the 30 $\Omega = I_2^2 \times 30 = (1.85)^2(30) = 102.67$ W

total power dissipated $= 463.65 + 1.37 + 98.76 + 102.67 = 666.45$ W

3.15 Solve for the current in the 20-Ω resistor of the network of Fig. 3-12(a) by applying Norton's theorem.

The Norton equivalent circuit is shown in Fig. 3-13(a). Recall from Section 3.3 that R_N is the resistance at *20* in the absence of all sources; voltage sources are removed by short-circuiting and current sources by open-circuiting. Thus, from Fig. 3-13(b),

$$R_{20} = R_N = \frac{(10+5)(30)}{10+5+30} = 10 \ \Omega$$

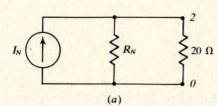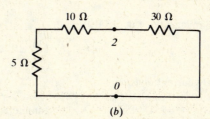

(a)   (b)

Fig. 3-13

The Norton current I_N is the short-circuit current through *20* (when *20* is short-circuited in the presence of all sources). Notice from Fig. 3-12(a) that when *20* is short-circuited, the 100-V source and the 30-Ω series resistor are short-circuited through *20*, and so the current contribution due to the 100-V source is

$$I_N' = \frac{100}{30} = \frac{10}{3} \text{ A}$$

The short-circuit current through *20* due to the 10-A source is, by the current-division rule,

$$I_N'' = \frac{5}{5+10}(10) = \frac{10}{3} \text{ A}$$

Hence, by superposition,

$$I_N = I_N' + I_N'' = \frac{10}{3} + \frac{10}{3} = \frac{20}{3} \text{ A}$$

The Norton current divides between $R_N = 10\ \Omega$ and the 20 Ω; the current in the latter is therefore

$$I_{20} = \frac{10}{10 + 20}\left(\frac{20}{3}\right) = 2.22\ \text{A}$$

Check: From Problem 3.13, $I_{20} = I_1 - I_2 = 0.37 - (-1.85) = 2.22$ A.

3.16   Two networks are shown in Fig. 3-14; the ammeters in both circuits have negligible resistances. Calculate the ammeter readings in the two networks, and comment on the results. Notice that in the two networks the positions of the ammeter and the voltage source are interchanged.

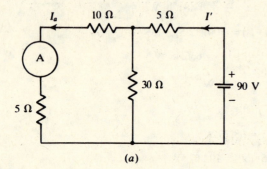 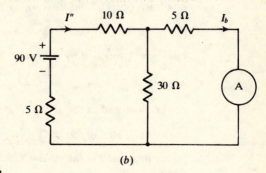

(a)   (b)

Fig. 3-14

Referring to Fig. 3-14(a), the resistance seen by the voltage source is

$$R_e' = 5 + \frac{(10 + 5)(30)}{10 + 5 + 30} = 15\ \Omega$$

so that

$$I' = \frac{90}{15} = 6\ \text{A}$$

Hence, by current division,

$$\text{ammeter reading} \equiv I_a = \frac{30}{45}(6) = 4\ \text{A}$$

Similarly, for Fig. 3-14(b),

$$R_e'' = 15 + \frac{(30)(5)}{30 + 5} = \frac{135}{7}\ \Omega \qquad \text{and} \qquad I'' = \frac{90}{135/7} = \frac{14}{3}\ \text{A}$$

By current division,

$$\text{ammeter reading} \equiv I_b = \frac{30}{35}\left(\frac{14}{3}\right) = 4\ \text{A}$$

The equality of the ammeter readings constitutes a special case of the *reciprocity theorem*:   In a linear bilateral network, if a voltage V in some branch produces a current I in some other branch, then the same voltage V in the second branch will produce the same current I in the first branch.

3.17   Determine the power supplied to the network of Fig. 3-15(a).

We transform the wye network into a delta network via (3.2); the resulting network is shown in Fig. 3-15(b). The total resistance across the voltage source is

$$R = \frac{\left[\dfrac{(6)(16.5)}{6 + 16.5} + \dfrac{(9)(11)}{9 + 11}\right]\left[\dfrac{(3)(33)}{3 + 33}\right]}{\dfrac{(6)(16.5)}{6 + 16.5} + \dfrac{(9)(11)}{9 + 11} + \dfrac{(3)(33)}{3 + 33}} = \frac{17}{8} = 2.125\ \Omega$$

Hence,

$$\text{power supplied} = \frac{(100)^2}{2.125} = 4705.9\ \text{W}$$

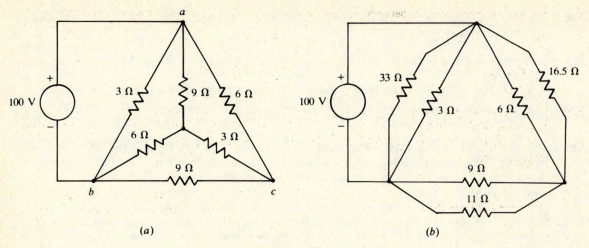

Fig. 3-15

Supplementary Problems

3.18 In the circuit of Fig. 3-16, an ammeter of 0.1 Ω resistance is used to measure the line current I. The range of the ammeter is 15 A. (a) With the switch S open, is the range of the ammeter adequate? (b) With the switch S closed, a shunt resistance R_x is connected across the ammeter. Determine the value of R_x such that the ammeter reads full scale.   *Ans.* (a) no; (b) $R_x = 0.0584$ Ω

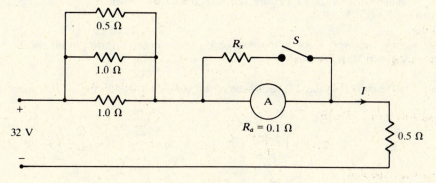

Fig. 3-16

3.19 Using mesh analysis, determine the currents I_1 and I_2 in Fig. 3-17. First transform the current sources to voltage sources.   *Ans.* $I_1 = 5$ A, $I_2 = 10$ A

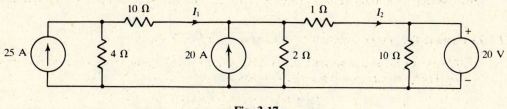

Fig. 3-17

3.20 Solve for I_1 and I_2 of the network of Fig. 3-17 by using nodal equations only.

3.21 Determine the power dissipated in the 1 Ω and 4 Ω resistances of the network of Fig. 3-17.
     Ans. $P_{1\Omega} = 100$ W, $P_{4\Omega} = 1600$ W

3.22 Calculate the current in the 2 Ω resistance of the network of Fig. 3-17 by applying the superposition theorem.   *Ans.*   15 A

3.23 (*a*) Determine the power supplied by each source to the entire network of Fig. 3-17. (*b*) Do all sources supply power?
Ans.   (*a*) $P_{25A} = 2000$ W, $P_{20A} = 600$ W, $P_{20V} = -160$ W. (*b*) No; the 20-V source absorbs power.

3.24 Calculate the total power dissipated in all the resistances of the network of Fig. 3-17.   *Ans.*   2440 W

3.25 The 3-Ω resistor in branch *cd* of Fig. 3-10(*a*) is replaced by a 9-Ω resistor, the other resistances remaining unchanged. If a 110-V source is connected at the terminals *AB*, (*a*) how much power will be drawn from the source? (*b*) how much power will be dissipated in the new 9-Ω resistor?
Ans.   (*a*) 2420 W; (*b*) 484 W

3.26 For the circuit shown in Fig. 3-18, determine the value of R_x such that the same amount of power is supplied to the 10 Ω resistance by the current and by the voltage source.   *Ans.*   20 Ω

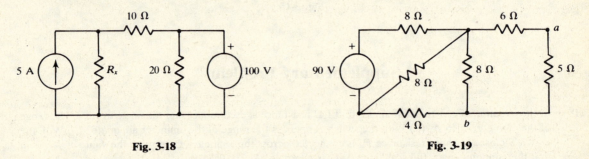

Fig. 3-18   Fig. 3-19

3.27 The terminals *20* of Fig. 3-13(*a*) are open-circuited. How much voltage will appear across the open circuit?   *Ans.*   33.33 V

3.28 Calculate by Thévinin's theorem the voltage V_{ab} in the network shown in Fig. 3-19.
Ans. $V_{Th} = 30$ V, $R_{Th} = 10$ Ω, $V_{ab} = 10$ V

3.29 Calculate the current in the 2 Ω resistance of the network shown in Fig. 3-20.   *Ans.*   5 A

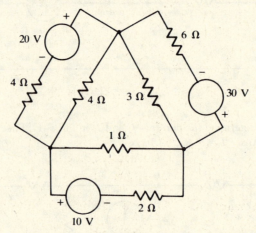

Fig. 3-20

3.30 Solve for the current in the 1 Ω resistance of Fig. 3-20 by (*a*) Thévenin's theorem, (*b*) the superposition theorem.   *Ans.*   (*a*) 0 A; (*b*) 0 A

3.31 Open-circuit the voltage source of Fig. 3-15(*a*) and determine the circuit resistance at the terminals *ac*.   *Ans.*   2.8 Ω

Chapter 4

Transients in DC Circuits

4.1 INTRODUCTION

Voltages and currents in an electrical network are *transient phenomena* insofar as they are dependent upon initial conditions and change with time. Only in the limit of infinite time do these quantities become independent of the initial conditions (the *steady state*). In dc circuits, steady-state values are constants, independent of time.

Because Ohm's law is time-independent, purely resistive circuits experience no transients; at each instant the voltage or current has its steady-state value. However, a capacitive or an inductive element has a time-dependent *v-i* relationship; from (*2.4*) and (*2.5*),

$$v(t) = v(t_0) + \frac{1}{C} \int_{t_0}^{t} i(u)\, du \tag{4.1}$$

for C, and

$$i(t) = i(t_0) + \frac{1}{L} \int_{t_0}^{t} v(u)\, du \tag{4.2}$$

for L. Consequently, a circuit involving L or C will, in response to some switching operation, exhibit transient behavior.

4.2 RESPONSE OF SOURCE-FREE CIRCUITS

Let us consider the source-free RL circuit of Fig. 4-1. As the excitation (input) is zero, the response (output) i will be determined solely by the initial current I_0. By KVL,

$$L\frac{di}{dt} + Ri = 0 \tag{4.3}$$

Integrating,

$$\int_{I_0}^{i} \frac{di}{i} = -\int_{0}^{t} \frac{R}{L}\, dt \qquad \text{or} \qquad i = I_0 e^{-Rt/L} \tag{4.4}$$

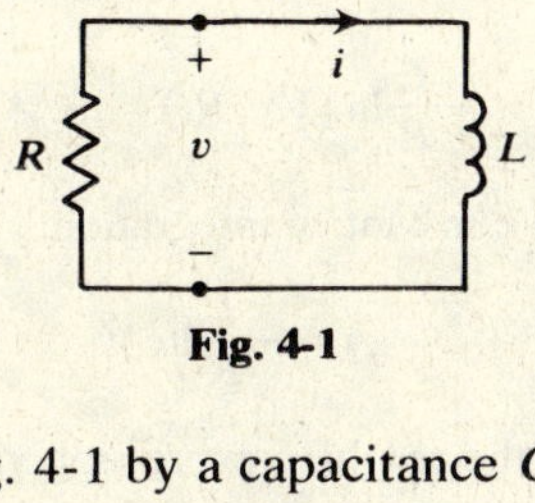

Fig. 4-1

Now replace the inductance of Fig. 4-1 by a capacitance C. The voltage response, v, satisfies

$$C\frac{dv}{dt} + \frac{v}{R} = 0 \tag{4.5}$$

The solution to (*4.5*) is

$$v = V_0 e^{-t/RC} \tag{4.6}$$

where V_0 is the initial voltage across the capacitance.

35

The source-free response of a circuit is more often called its *natural response*. As shown by (4.4) and (4.6), the natural responses of *RL* and *RC* circuits are purely transient, with steady-state value zero. Mathematically, this is due to the fact that the governing differential equations, (4.3) and (4.5), are *homogeneous*. The quantity L/R in (4.4) has the dimension [T]; it is known as the *time constant* (τ) of the *RL* circuit. The smaller the time constant, the more rapid the decay of the current *i*. Similarly, the voltage decay in the *RC* circuit is controlled by the time constant $\tau \equiv RC$.

4.3 COMPLETE RESPONSE OF CIRCUITS

When sources are present, the response of a circuit will be made up of two components: the natural response, which gives the transient behavior, and the *forced response*, which is the steady-state component. In mathematical terms, the solution to the governing *nonhomogeneous* differential equation is the sum of the *complementary function* (the solution of the corresponding homogeneous equation) and a *particular integral*.

Example 4.1 Solve for the current in the *RL* circuit of Fig. 4-2(*a*), where the dc voltage, *V*, is applied at $t = 0$ by closing the switch. Assume zero initial conditions ($i = 0$ at $t = 0$).

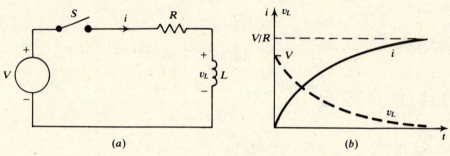

(*a*)   (*b*)

Fig. 4-2

For this case (4.3) is replaced by

$$L\frac{di}{dt} + Ri = V \tag{4.7}$$

which may be solved by separating the variables as

$$\frac{L}{V - Ri}\, di = dt$$

By integration we obtain

$$-\frac{L}{R}\ln\,(V - Ri) = t + A \tag{4.8}$$

Using the initial condition to evaluate the constant of integration,

$$A = -\frac{L}{R}\ln\,V \tag{4.9}$$

Substituting (4.9) in (4.8) and simplifying the resulting expression yields

$$i = \frac{V}{R}(1 - e^{-Rt/L}) \tag{4.10}$$

Observe that the complete current response, (4.10), has the predicted form:

$$i = i_n + i_f = -\frac{V}{R}e^{-Rt/L} + \frac{V}{R} \tag{4.11}$$

where i_n is given by (4.4) together with the initial condition, and where i_f is a constant solution to (4.7). Current *i*, as well as inductor voltage v_L, is graphed in Fig. 4-2(*b*).

Example 4.2   In Fig. 4-3(a), the switch is closed at $t = 0$, with no initial charge on C. Solve for i and v_C.

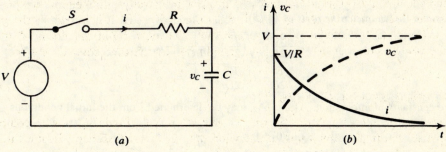

(a)   (b)

Fig. 4-3

For $t > 0$, KVL gives

$$Ri + \frac{1}{C}\int_0^t i\, dt = V$$

which (since $i = dq/dt$) may also be written as

$$R\frac{dq}{dt} + \frac{1}{C}q = V \tag{4.12}$$

Notice that (4.12) and (4.7) are of the same form. The solution becomes, as in (4.11),

$$q = q_n + q_f = Q_0 e^{-t/RC} + CV \tag{4.13}$$

where, from the initial condition, $Q_0 = -CV$. Thus

$$q = CV(1 - e^{-t/RC}) \tag{4.14}$$

and from (4.14),

$$i = \frac{dq}{dt} = \frac{V}{R}e^{-t/RC} \qquad v_C = \frac{q}{C} = V(1 - e^{-t/RC})$$

The two functions are plotted in Fig. 4-3(b).

Example 4.3   Determine the current response of the series RLC circuit, Fig. 4-4(a), when a voltage V is suddenly applied by closing the switch at $t = 0$. Initial current and initial charge on C are zero.

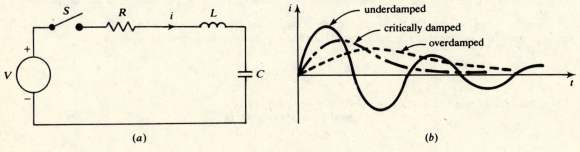

(a)   (b)

Fig. 4-4

The voltage equation is

$$L\frac{di}{dt} + Ri + \frac{1}{C}\int_0^t i\, dt = V$$

or, differentiating with respect to time,

$$L\frac{d^2i}{dt^2} + R\frac{di}{dt} + \frac{1}{C}i = 0 \tag{4.15}$$

Since (4.15) is homogeneous, $i_f = 0$ and

$$i = i_n = A_1 e^{s_1 t} + A_2 e^{s_2 t} \tag{4.16}$$

where s_1 and s_2 are the *characteristic roots* of (4.15):

$$s_1 = -\frac{R}{2L} + \sqrt{\left(\frac{R}{2L}\right)^2 - \frac{1}{LC}} \equiv -\alpha + \beta$$

$$s_2 = -\frac{R}{2L} - \sqrt{\left(\frac{R}{2L}\right)^2 - \frac{1}{LC}} \equiv -\alpha - \beta \tag{4.17}$$

and where the constants of integration, A_1 and A_2, may be determined from the initial conditions

$$i(0) = 0 \qquad L\frac{di}{dt}\bigg|_0 = V$$

(See Problems 4.7 and 4.8.)

Because α is always a positive real number, the transient current (4.16) eventually decays in magnitude like an exponential function. The more exact features of this decay depend on the circuit parameters R, L, and C as they enter in the constant β. We define $\omega_0 \equiv 1/\sqrt{LC}$, the *resonant frequency* (in rad/s) of the circuit, so that

$$\beta = \sqrt{\alpha^2 - \omega_0^2}$$

Case 1: $\alpha > \omega_0$.

Here, β is real and positive, and $\beta < \alpha$. The solution takes the form

$$i = A_1 e^{-(\alpha - \beta)t} + A_2 e^{-(\alpha + \beta)t} \tag{4.18}$$

i.e. the sum of two decaying exponentials. In this case the circuit is said to be *overdamped*.

Case 2: $\alpha = \omega_0$.

It can be shown that as $\beta \to 0$, (4.16) goes over into

$$i = (A_1 + A_2 t)e^{-\alpha t} \tag{4.19}$$

The circuit is said to be *critically damped*.

Case 3: $\alpha < \omega_0$.

Now β is a pure imaginary, $\beta = j|\beta|$, and (4.16) becomes

$$i = e^{-\alpha t}(A_1 e^{j|\beta|t} + A_2 e^{-j|\beta|t}) \tag{4.20}$$

or, equivalently (see Problem 4.21),

$$i = A e^{-\alpha t} \sin(|\beta|t + \psi) \tag{4.21}$$

As given by (4.21), the response is a damped sine wave, of frequency $|\beta|$ (rad/s); the circuit is *underdamped*.

Figure 4-4(b) illustrates the three kinds of damping. In the critically damped and underdamped cases, the response goes to zero essentially as $e^{-\alpha t}$, and α is called the *damping coefficient*.

Solved Problems

4.1   The switch in the circuit shown in Fig. 4-5 has been in position *1* for a long time. At $t = 0$, the switch is suddenly brought to position *2*. Determine the current $i(t)$ through the capacitance for $t > 0$.

At the instant of switching, the capacitor was charged to the battery voltage, 20 V. Hence, for $t > 0$, the capacitor voltage must be given by (4.6), with

$$V_0 = 20 \text{ V} \qquad RC = (500 \times 10^3)(500 \times 10^{-6}) = 250 \text{ s}$$

that is, $v = 20e^{-t/250}$ (V). Hence the current through the capacitor is

$$i(t) = C\frac{dv}{dt} = (500 \times 10^{-6})\left[20\left(-\frac{1}{250}\right)e^{-t/250}\right] \quad \text{(A)} = -40e^{-t/250} \quad (\mu\text{A})$$

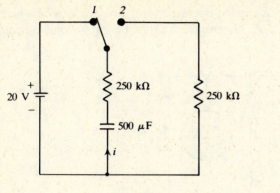

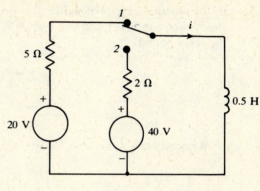

Fig. 4-5   **Fig. 4-6**

4.2 In the circuit of Fig. 4-6, the switch has been in position *1* for a long time. At $t = 0$, it is switched to position *2*. Solve for the inductor current $i(t)$ for $t > 0$.

Using the concept of forced and natural responses (Section 4.3), we have, for $t > 0$,

$$i(t) = i_f + i_n = \frac{V}{R} + I_0 e^{-Rt/L}$$

where $V = 40$ V, $R = 2\ \Omega$, $L = 0.5$ H, and $I_0 = 20/5 = 4$ A. Thus, $i(t) = 20 + 4e^{-4t}$ (A).

4.3 Set up a differential equation with initial conditions for i in the circuit shown in Fig. 4-7, where the sources V and I are simultaneously turned on at $t = 0$. At $t = 0-$, the capacitor is uncharged and the inductor is unenergized.

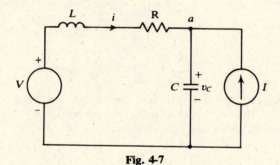

Fig. 4-7

For $t > 0$, KCL at node a yields

$$i - C\frac{dv_C}{dt} + I = 0 \qquad (1)$$

and KVL around the left mesh yields

$$L\frac{di}{dt} + Ri + v_C - V = 0 \qquad (2)$$

Now eliminate dv_C/dt between (*1*) and the derivative of (*2*), to obtain

$$L\frac{d^2 i}{dt^2} + R\frac{di}{dt} + \frac{1}{C}i = -\frac{I}{C} \qquad (3)$$

The capacitor voltage and inductor current must be continuous at $t = 0$. Therefore, the initial conditions to be imposed on (*3*) read:

$$i(0) = 0 \qquad L\frac{di}{dt}\bigg|_0 = V \qquad (4)$$

We could solve (*3*) by writing $i = i_n + i_f$, where i_n is given by (*4.16*) and (*4.17*), and where $i_f = -I$. Then, conditions (*4*) could be used to determine the integration constants.

4.4   In the circuit of Fig. 4-8, switch S represents a current-operated relay, the contacts of which close when $i_L = 0.9$ A and open when $i_L = 0.25$ A. Determine the time period for one cycle of the relay operation.

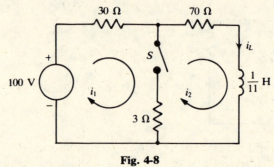

Fig. 4-8

Let us measure time from a moment at which the switch opens. Then the inductor current builds up from 0.25 A, with time constant

$$\frac{L}{R} = \frac{1/11}{70 + 30} = \frac{1}{1100} \text{ s}$$

towards a final value of $100/(70 + 30) = 1$ A. That is,

$$i_L(t) = 1 - 0.75 \, e^{-1100t} \quad \text{(A)}$$

This current reaches 0.9 A when $1100t_1 = \ln 7.5$, or $t_1 = 1.83$ ms. At this instant, S closes, producing a two-mesh circuit with mesh currents i_1 and $i_2 = i_L$. KVL gives

$$33i_1 - 3i_2 = 100$$

$$73i_2 - 3i_1 + \frac{1}{11}\frac{di_2}{dt} = 0$$

or, eliminating i_1,

$$\frac{di_2}{dt} + 800i_2 = 100 \tag{1}$$

The solution of (1) that satisfies the initial condition $i_2(t_1) = 0.9$ A is $i_2(t) = i_L(t) = 0.125 + 0.775e^{-800(t-t_1)}$. This current decays to 0.25 A when $800(t_2 - t_1) = \ln 6.2$ or $t_2 - t_1 = 2.28$ ms. Hence the relay period is $t_2 = t_1 + 2.28 = 1.83 + 2.28 = 4.11$ ms.

4.5   An inductance L, a capacitance C, and a resistance R are all connected in parallel. If the initial current through the inductance is I_0 and the initial charge on the capacitance is Q_0, determine the natural behavior of the circuit in terms of the common voltage $v(t)$ across the elements.

For $t > 0$, KCL gives

$$\frac{v}{R} + \left[\frac{1}{L}\int_0^t v(u)\,du + I_0\right] + C\frac{dv}{dt} = 0$$

or, taking the time derivative,

$$C\frac{d^2v}{dt^2} + \frac{1}{R}\frac{dv}{dt} + \frac{1}{L}v = 0 \tag{1}$$

with initial conditions

$$v(0) = \frac{Q_0}{C} \qquad C\frac{dv}{dt}\Big|_0 = -\left(\frac{Q_0}{RC} + I_0\right) \tag{2}$$

Equation (1) has the same form as (4.15), the equation for the natural current in a series RLC circuit; indeed, (1) and (4.15) are duals (Section 2.5). Thus we have the solution

$$v(t) = A_1 e^{s_1 t} + A_2 e^{s_2 t}$$

where
$$s_1 = -\alpha + \sqrt{\alpha^2 - \omega_0^2}$$
$$s_2 = -\alpha - \sqrt{\alpha^2 - \omega_0^2}$$

$$\alpha = \frac{1}{2RC} \equiv \text{damping coefficient}$$

$$\omega_0 = \frac{1}{\sqrt{LC}} \equiv \text{resonant frequency}$$

Observe that, under the duality, the resonant frequency does not change. The constants A_1 and A_2 may be evaluated from (2).

4.6   Consider the natural behavior of a critically damped, series RLC circuit (see Fig. 4-4) in which the initial current is zero. Determine the time at which the current reaches its maximum value, if $R = 5 \, \Omega$ and $L = 10$ mH.

For a critically damped circuit the natural current is given by (4.19):

$$i = (A_1 + A_2 t)e^{-\alpha t}$$

Since $i = 0$ at $t = 0$, we have $A_1 = 0$, and $i = A_2 t e^{-\alpha t}$. For a maximum i,

$$\frac{di}{dt} = A_2 e^{-\alpha t}(1 - \alpha t) = 0$$

from which

$$t = \frac{1}{\alpha} = \frac{2L}{R} = \frac{2(10 \times 10^{-3})}{5} = 4 \text{ ms}$$

4.7   The series RLC circuit of Fig. 4-4 is critically damped. If $R = 200 \, \Omega$ and $L = 100$ mH, (a) determine C. (b) With zero initial conditions, the switch is closed at $t = 0$. Solve for the voltage across the capacitor, $v_C(t)$, if $V = 100$ V.

(a)   The condition for critical damping is $\alpha^2 = \omega_0^2$, or

$$\left(\frac{R}{2L}\right)^2 = \frac{1}{LC}$$

Thus
$$C = \frac{4L}{R^2} = \frac{4(100 \times 10^{-3})}{(200)^2} = 10 \ \mu\text{F}$$

(b)   By (4.19), the current in the circuit is

$$i(t) = (A_1 + A_2 t)e^{-\alpha t} \qquad \text{with} \qquad \alpha = \frac{R}{2L} = 1000 \text{ s}^{-1}$$

At $t = 0$, $i = 0$ and $di/dt = V/L = 1000$ A/s; thus,

$$A_1 = 0 \qquad A_2 = 1000 \text{ A/s}$$

and $i(t) = 1000t e^{-1000t}$ (A). Then,

$$v_C(t) = \frac{1}{C}\int_0^t i(u)\, du = \frac{1}{10 \times 10^{-6}}\int_0^t 1000 u e^{-1000u}\, du = 100[1 - e^{-1000t}(1 + 1000t)] \quad \text{(V)}$$

4.8   The parameters of the circuit of Fig. 4-4 are such that the circuit is underdamped. With zero initial conditions, the switch is closed at $t = 0$. (a) Obtain an expression for the voltage across the capacitor. (b) Because the circuit is underdamped, the voltage across C will undergo (damped) oscillations. Determine the time at which the first maximum of the capacitor voltage occurs. What is the value of the voltage at this instant?

(a)   Applying the initial conditions to (4.21), we obtain

$$i(0) = 0 = A \sin \psi$$

$$\left. \frac{di}{dt} \right|_0 = \frac{V}{L} = A(-\alpha \sin \psi + |\beta| \cos \psi)$$

which imply that $\sin \psi = 0$ and $A \cos \psi = V/L|\beta|$. Thus, (4.21) becomes

$$i(t) = \frac{V}{L|\beta|} e^{-\alpha t} \sin |\beta| t$$

From this, using

$$\alpha^2 + |\beta|^2 = \omega_0^2 \equiv \frac{1}{LC}$$

we obtain

$$v_C(t) = \frac{1}{C} \int_0^t i(u) \, du = V \left[1 - \frac{e^{-\alpha t}}{|\beta|} (\alpha \sin |\beta| t + |\beta| \cos |\beta| t) \right]$$

$$= V \left[1 - \frac{e^{-\alpha t}}{\sin \phi} \sin (|\beta| t + \phi) \right]$$

in which the new phase angle, ϕ, is defined by

$$\cos \phi \equiv \frac{\alpha}{\omega_0}$$

(b)   From the expression for $i = C(dv_C/dt)$ obtained in (a), the first maximum of $v_C(t)$ occurs at $t = \pi/|\beta|$. We have

$$v_C(\pi/|\beta|) = V \left[1 - \frac{e^{-\alpha \pi/|\beta|}}{\sin \phi} \sin (\pi + \phi) \right] = V(1 + e^{-\pi \cot \phi})$$

4.9   Calculate the *percent overshoot* of the capacitor voltage in Problem 4.8.

The function $v_C(t)$ is graphed in Fig. 4-9. It is seen that the greatest deviation of the voltage from its steady-state value occurs at $t = \pi/|\beta|$. Expressing this peak overshoot as a fraction,

$$\text{percent overshoot} \equiv \frac{v_C(\pi/|\beta|) - V}{V} \times 100\% \equiv 100 e^{-\pi \cot \phi}$$

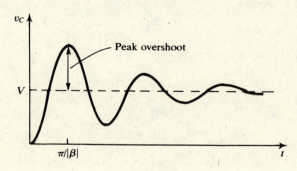

Fig. 4-9

4.10   For the circuit of Problem 4.8, $V = 100$ V, $R = 200 \ \Omega$, and $L = 100$ mH. Calculate C such that the peak overshoot is 5 percent.

By Problem 4.9,

$$100 e^{-\pi \cot \phi} = 5 \qquad \text{or} \qquad \cot \phi = \frac{\ln 20}{\pi} = 0.954$$

whence $\cos \phi = \alpha/\omega_0 = 0.690$. For this circuit,

$$\alpha = \frac{R}{2L} = \frac{200}{2(100 \times 10^{-3})} = 1000 \text{ s}^{-1}$$

and so

$$\frac{1}{\omega_0^2} = \left(\frac{0.690}{1000}\right)^2 = 4.76 \times 10^{-7} = LC$$

from which

$$C = \frac{4.76 \times 10^{-7}}{100 \times 10^{-3}} = 4.76 \ \mu\text{F}$$

4.11 At $t = 0$, when the capacitor charge and the inductor current are zero in Fig. 4-10, switch S is suddenly opened. Set up a differential equation for the node voltage v and state the corresponding initial conditions.

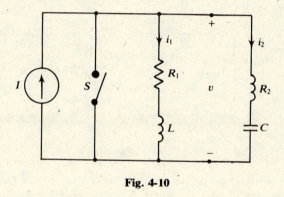

Fig. 4-10

For $t > 0$, we have

$$i_1 + i_2 = I \tag{1}$$

$$v = R_1 i_1 + L \frac{di_1}{dt} \tag{2}$$

$$v = R_2 i_2 + \frac{1}{C} \int_0^t i_2 \, dt \tag{3}$$

and their first and second derivatives

$$\frac{di_1}{dt} + \frac{di_2}{dt} = 0 \qquad \frac{d^2 i_1}{dt^2} + \frac{d^2 i_2}{dt^2} = 0$$

$$\frac{dv}{dt} = R_1 \frac{di_1}{dt} + L \frac{d^2 i_1}{dt^2} \tag{4}$$

$$\frac{dv}{dt} = R_2 \frac{di_2}{dt} + \frac{1}{C} i_2 \qquad \frac{d^2 v}{dt^2} = R_2 \frac{d^2 i_2}{dt^2} + \frac{1}{C} \frac{di_2}{dt}$$

Equations (1) and (2) and the five equations (4) compose a system from which i_1 and i_2 and their first and second derivatives may be eliminated, giving

$$L \frac{d^2 v}{dt^2} + (R_1 + R_2) \frac{dv}{dt} + \frac{1}{C} v = \frac{R_1 I}{C} \tag{5}$$

which is the sought equation.

To find the initial conditions on (5), note that at $t = 0+$, the capacitor voltage is zero and $i_2 = I$; hence,

$$v(0+) = R_2 I \tag{6}$$

Moreover, (2) becomes

$$R_2 I = L \left.\frac{di_1}{dt}\right|_{0+}$$

and (1) and the fourth equation (4) give

$$\left.\frac{dv}{dt}\right|_{0+} = -R_2 \left.\frac{di_1}{dt}\right|_{0+} + \frac{I}{C}$$

These two relations imply

$$\left.\frac{dv}{dt}\right|_{0+} = \left(\frac{1}{C} - \frac{R_2^2}{L}\right) I \tag{7}$$

For a numerical application, see Problem 4.29.

4.12   Assuming that Thévenin's and Norton's theorems are applicable to circuits containing inductances and/or capacitances, represent (a) a capacitance C, with an initial charge Q_0, by a Thévenin equivalent circuit; (b) an inductance L, with an initial current I_0, by a Norton equivalent circuit.

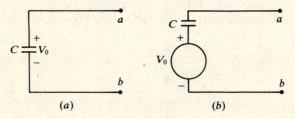

Fig. 4-11

(a)   The open-circuit voltage across the capacitor, Fig. 4-11(a), is $V_0 = Q_0/C$. Hence, the Thévenin circuit becomes as shown in Fig. 4-11(b), where the capacitor is initially uncharged.

(b)   See Fig. 4-12; the initial current in L in Fig. 4-12(b) is zero.

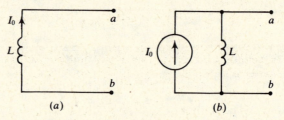

Fig. 4-12

4.13   Switch S of the network shown in Fig. 4-13(a) has been closed for a long time. At $t = 0$, the switch is suddenly opened. Evaluate the voltage across the 200-Ω resistor.

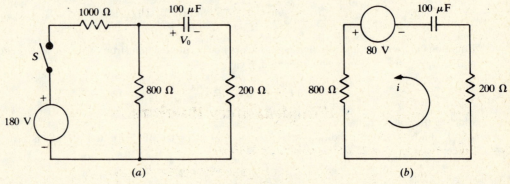

Fig. 4-13

Before S was opened, the network was under steady state and the capacitance was charged to a voltage V_0. This voltage is the same as the voltage across the 800 Ω resistance; thus, by voltage division,

$$V_0 = \frac{800}{1800}(180) = 80 \text{ V}$$

In view of Problem 4.12(a), the circuit for $t > 0$ may be represented as in Fig. 4-13(b). We have:

$$800i + 200i + \frac{1}{100 \times 10^{-6}} \int_0^t i\,dt = 80$$

This equation has been solved earlier, in Example 4.2. Hence,

$$i = \frac{80}{1000}e^{-t/(1000)(100 \times 10^{-6})} \quad \text{(A)}$$

and the voltage across the 200 Ω becomes

$$v = 200i = 16e^{-10t} \quad \text{(V)}$$

4.14 The switch of the network of Fig. 4-14(a) is suddenly opened at $t = 0$; prior to this, the network was under steady state. Determine the voltage across the 6 Ω resistance for $t > 0$.

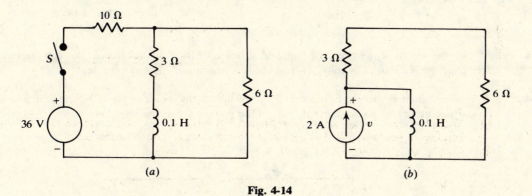

Fig. 4-14

The current in the inductance for $t < 0$ can be verified to be 2 A. Applying Problem 4.12(b), the circuit for $t > 0$ becomes as shown in Fig. 4-14(b), for which we have

$$2 = \frac{v}{9} + \frac{1}{0.1} \int_0^t v\,dt \tag{1}$$

The solution to (1) is

$$v = 18e^{-90t} \quad \text{(V)}$$

Hence the voltage across the 6 Ω resistance becomes

$$v_{6\Omega} = \frac{6}{9}(18e^{-90t}) = 12e^{-90t} \quad \text{(V)}$$

Supplementary Problems

4.15 The voltage across a 20 μF capacitance is given by

$$v_C = 10.75 - 1.5e^{-1000t} + 0.75e^{-2000t} \quad \text{(V)}$$

What is the current through the capacitance?   *Ans.* $0.03(e^{-1000t} - e^{-2000t}) \quad \text{(A)}$

4.16   For the circuit of Fig. 4-3(a), $V = 100$ V, $R = 5000$ Ω, and $C = 400$ μF; S is closed at $t = 0$. (a) Obtain
an expression for the instantaneous power dissipated in the resistance. (b) What is the instantaneous power
absorbed by the capacitance? (c) When is the power delivered to the capacitance a maximum? (d)
Determine the voltage across the capacitance at the instant found in (c).
Ans.   (a) $2e^{-t}$ (W); (b) $2(e^{-0.5t} - e^{-t})$ (W); (c) $t = 1.39$ s; (d) 50 V

4.17   Refer to the circuit of Fig. 4-6. Assume that the switch was at *2* for a long time, then switched to *1* at
$t = 0$. Determine the voltage across the inductance.   *Ans.* $120e^{-10t}$ (V)

4.18   The capacitor of the circuit shown in Fig. 4-15 has no charge when the switch is closed at $t = 0$. Calculate
(a) the current through, and (b) the voltage across, the capacitor.
Ans.   (a) $5e^{-400t}$ (A); (b) $250(1 - e^{-400t})$ (V)

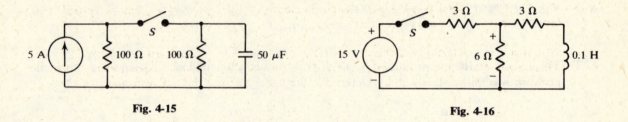

Fig. 4-15   **Fig. 4-16**

4.19   At $t = 0$ the switch of the circuit of Fig. 4-16 is closed. With no initial current through the inductance,
determine the instantaneous voltage across the 6 Ω resistance.   *Ans.* $6 + 4e^{-50t}$ (V)

4.20   With no initial energy stored in the inductance of the circuit shown in Fig. 4-17, the switch is closed at
$t = 0$. Evaluate (a) the source current, and (b) the voltage across the inductance.
Ans.   (a) $2.5(2 - e^{-15t})$ (A); (b) $120e^{-15t}$ (V)

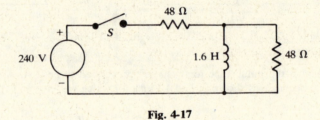

Fig. 4-17

4.21   Put (4.20) in the form (4.21).
Ans.   In order that i be real, A_1 and A_2 in (4.20) must be complex conjugates. Then,

$$A \equiv 2|A_1| = 2|A_2| \qquad \psi \equiv \frac{\pi}{2} + \arg A_1 = \frac{\pi}{2} - \arg A_2$$

4.22   A resistance R is connected in parallel with a parallel combination of a 20 mH inductance and a 50 μF
capacitance. For what value of R will the circuit be critically damped?   *Ans.*   10 Ω

4.23   The value of R in the circuit of Problem 4.8 is changed to 100 Ω. Verify that the circuit is now
underdamped. Determine (a) the damping coefficient, and (b) the resonant frequency.
Ans.   (a) 500 s^{-1}; (b) 1000 rad/s

4.24   For the data and circuit of Problem 4.7, determine how long it takes for the voltage across the capacitor
to come within 5% of its final value.   *Ans.*   5 ms (approx.)

4.25   For the circuit and data of Problem 4.10, determine the time it takes the voltage across the capacitor to
come within 5% of its final value.   *Ans.*   2 ms (approx.)

4.26   Show that the circuit of Fig. 4-18 cannot have damped oscillations, regardless of the values of R_1, R_2, C_1, and C_2. In particular, show that i_2 can never have the form of a damped oscillation.

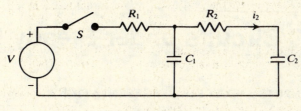

Fig. 4-18

4.27   In the circuit of Fig. 4-18 the capacitances are replaced by inductances. Show that the current i_2 can never have the form of a damped oscillation.

4.28   The circuit shown in Fig. 4-10 has $I = 1.0$ A, $L = 100$ mH, $C = 17.78\ \mu$F, $R_1 = 90\ \Omega$, and $R_2 = 0\ \Omega$. Determine i_1 when the switch is suddenly opened and the initial conditions are zero. (*Hint*:   Derive the differential equation

$$L\frac{d^2 i_1}{dt^2} + (R_1 + R_2)\frac{di_1}{dt} + \frac{1}{C}i_1 = \frac{I}{C}$$

for $t > 0$.)   *Ans.* $1.0 + 1.25 e^{-450t}\sin(600t - 127°)$ (A)

4.29   The values of the parameters of the circuit shown in Fig. 4-10 are: $R_1 = 60\ \Omega$, $R_2 = 90\ \Omega$, $L = 100$ mH, $C = 17.78\ \mu$F. A current source of 1.0 A is suddenly applied (as shown) with zero initial conditions. Determine the node voltage v.   *Ans.* $v = 30[2 + (1 - 75t)e^{-750t}]$ (V)

4.30   In the circuit of Fig. 4-14(*a*), the 3 Ω resistance is replaced by a short circuit. The circuit was under steady state for $t < 0$; the switch is opened at $t = 0$. Determine the current through the 6 Ω resistance for $t > 0$.   *Ans.* $3.6 e^{-0.6t}$ (A)

4.31   The switch in the network of Fig. 4-19 has been closed for a long time; at $t = 0$ it is suddenly opened. Solve for the voltage v for $t > 0$.   *Ans.* $4 e^{-0.1t}$ (V)

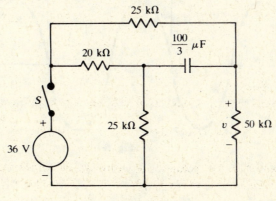

Fig. 4-19

Chapter 5

AC Circuits under Steady State

5.1 SINUSOIDAL INPUTS AND THEIR REPRESENTATIONS

In preceding chapters, the voltage and current sources exciting the networks were dc (or unidirectional). In contrast, many electrical networks of practical importance are excited by ac sources. The most common example of an ac network is the electric power network, containing power generating stations, transmission lines, substations, etc. Whereas *ac* is an abbreviation for *a*lternating *c*urrent, implying a periodic reversal of direction (of current or voltage), we will restrict attention to quantities varying *sinusoidally*; e.g., the voltage

$$v = V_m \sin \omega t \tag{5.1}$$

A cosine function might be used instead of the sine function, but we shall accord preference to the form *(5.1)*.

Sinusoidal variation may be depicted graphically as in Fig. 5-1, which is a plot of *(5.1)*. In *(5.1)* and in Fig. 5-1, V_m is known as the *amplitude* or *maximum value* and ω is called the *angular frequency* (in rad/s). Notice that the function v is periodic, repeating itself at regular intervals of 2π rad or $T = 2\pi/\omega$ (s). The rate of repetition is called the *frequency*, *f*. Since a function having period T must undergo $1/T$ periods per second,

$$f = \frac{1}{T} = \frac{\omega}{2\pi} \tag{5.2}$$

Frequency is measured in *hertz* (Hz), where 1 Hz = 1 cycle/second.

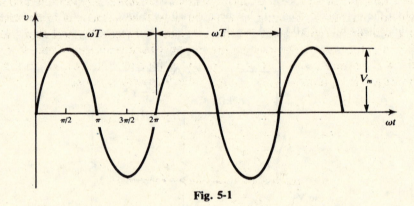

Fig. 5-1

If we have another quantity, say a current i, in addition to the voltage of Fig. 5-1, such that the current and voltage may be sketched as in Fig. 5-2, then, with reference to v of *(5.1)*, i may be written as

$$i = I_m \sin (\omega t + \theta) \tag{5.3}$$

where θ is known as the *phase angle* between v and i; the two sinusoids are *out of phase* by the amount θ. Notice from *(5.3)* and Fig. 5-2 that i is ahead of v, or *i leads v*, by θ. Conversely, *v lags i* by θ. Whereas ωt must be given in radians, it is customary to specify θ in degrees.

48

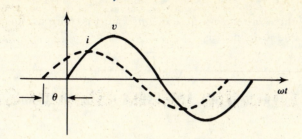

Fig. 5-2

Another way of representing sinusoids is provided by Euler's formula,

$$\cos \omega t + j \sin \omega t = e^{j\omega t} \tag{5.4}$$

$$\cos \omega t = \text{Re}\,(e^{j\omega t}) \qquad \sin \omega t = \text{Im}\,(e^{j\omega t}) \tag{5.5}$$

Often, in practice, we suppress Re or Im and write simply

$$v = V_m \sin \omega t = V_m e^{j\omega t} \tag{5.6}$$

or, instead of (5.3),

$$i = I_m e^{j(\omega t + \theta)} \tag{5.7}$$

Expressions (5.6) and (5.7) are examples of the *complex-exponential representation* of sinusoids.

A third way to indicate a sinusoid is the *polar* or *phasor* representation. Thus, (5.6) and (5.7) are replaced by

$$\mathbf{V} = V_m \underline{/0^\circ} \tag{5.8}$$

$$\mathbf{I} = I_m \underline{/\theta} \tag{5.9}$$

Boldface symbols are used to emphasize that phasors are complex vectors; for instance, $\mathbf{I}$ has magnitude $|\mathbf{I}| = I_m$ (a real number) and direction $\arg \mathbf{I} = \theta$. Phasors (5.8) and (5.9) are depicted in Fig. 5-3. Although the time factor $e^{j\omega t}$ is not explicit in the phasor representation, the two vectors in Fig. 5-3 are imagined to rotate as a rigid unit about O, with angular velocity ω in the counterclockwise direction. That $\mathbf{I}$ will hit the fixed point P *before* $\mathbf{V}$ does so reflects the fact that i *leads* v (by θ). Indeed, i and v are the respective projections of the rotating vectors $\mathbf{I}$ and $\mathbf{V}$ on the vertical (imaginary) axis. We say that i and v, which are explicit functions of time, are in the *time domain*, whereas their representative phasors $\mathbf{I}$ and $\mathbf{V}$ are in the *frequency domain*.

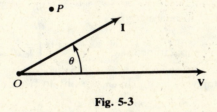

Fig. 5-3

In Section 5.2 we shall find it convenient to rescale the magnitudes of our phasors; this will have no effect on the basic properties of these quantities.

5.2 INSTANTANEOUS, AVERAGE, AND RMS VALUES

By the *instantaneous value* of a quantity (in the time domain) we mean the value of the quantity at a given instant. Often, we will be interested in the average values of time-varying quantities. Obviously, the average value of a sine (or cosine) function over a period is zero. For sinusoids, then, another concept, that of the *root-mean-square* or *rms* value, is more useful. For an arbitrary periodic function $v(t)$, the rms value is defined as

$$V \equiv \sqrt{\frac{1}{T} \int_{t_0}^{t_0+T} v^2(t)\, dt} \tag{5.10}$$

where, of course, V is independent of t_0. The motive for such a definition may be gathered from Example 5.1.

Example 5.1   Consider a resistance, R, connected across a dc voltage source, V_{dc}. The power absorbed by R is

$$P_{dc} = \frac{V_{dc}^2}{R} \tag{5.11}$$

Now, let an ac voltage source, $v = V_m \sin \omega t$, be connected across R. In this case, the *instantaneous power* is given by

$$p = \frac{v^2}{R} = \frac{V_m^2 \sin^2 \omega t}{R}$$

Hence the *average power* over a period, P_{ac}, is

$$P_{ac} = \frac{1}{T} \int_0^T p\, dt = \frac{V_m^2}{2\pi R} \int_0^{2\pi} \sin^2 \omega t\, d(\omega t) = \frac{V_m^2}{2R} \tag{5.12}$$

Comparing (*5.12*) to (*5.11*), we see that, in respect to power dissipation, an ac source of amplitude V_m is equivalent to a dc source of magnitude

$$V_{dc} = \frac{V_m}{\sqrt{2}} = \sqrt{\frac{1}{T} \int_0^T v^2\, dt} \equiv V \tag{5.13}$$

For this reason, the rms value of a sinusoid, $V = V_m/\sqrt{2}$, is also called the *effective* value.

   From this point on, unless an explicit statement is made to the contrary, all phasors will reflect rms values rather than amplitudes. Thus, a time-domain voltage $v = V_m \sin(\omega t + \phi)$ will be indicated in the frequency domain as $\mathbf{V} = V\underline{/\phi}$, where $V = V_m/\sqrt{2}$.

5.3   STEADY-STATE RESPONSES OF R, L, AND C TO SINUSOIDAL INPUTS

   From Chapters 1 and 2, the instantaneous v-i relationships for the three circuit elements are:

$$\text{resistance, } R: \qquad v_R = Ri_R \tag{5.14}$$

$$\text{inductance, } L: \qquad v_L = L\frac{di_L}{dt} \tag{5.15}$$

$$\text{capacitance, } C: \qquad v_C = \frac{1}{C}\int i_C\, dt \tag{5.16}$$

For a sinusoidal input voltage, $v_R = V_m \sin \omega t$, we obtain from (*5.14*)

$$i_R = \frac{V_m}{R} \sin \omega t \equiv I_m \sin \omega t$$

Thus the current is *in phase* with the applied voltage, and its amplitude is $1/R$ times the amplitude of the voltage. Dividing both amplitudes by $\sqrt{2}$ to obtain rms values, we may rewrite (*5.14*) in the frequency domain as

$$\mathbf{V}_R = R\mathbf{I}_R \tag{5.14a}$$

where $$\mathbf{V}_R = \frac{V_m}{\sqrt{2}}\underline{/0°} = V_R\underline{/0°}$$

$$\mathbf{I}_R = \frac{I_m}{\sqrt{2}}\underline{/0°} = I_R\underline{/0°}$$

The phasor relationship (*5.14a*) is diagrammed in Fig. 5-4(*a*). Of course, if the input had been $v_R = V_m \sin(\omega t + \phi)$, then both phasors in (*5.14a*) would have angle ϕ instead of $0°$.

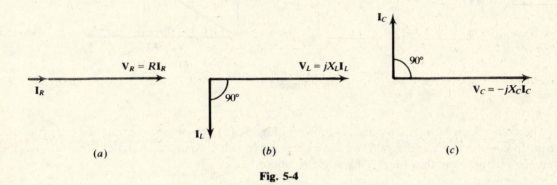

(*a*)   (*b*)   (*c*)

Fig. 5-4

For an inductance L, with applied voltage $v_L = V_m \sin \omega t$, the steady-state solution of (*5.15*) is

$$i_L = -\frac{V_m}{\omega L} \cos \omega t \equiv I_m \sin(\omega t - 90°)$$

Thus, the current *lags* the voltage by $90°$ and its rms value is $1/\omega L$ times the rms voltage. In phasor form,

$$\mathbf{V}_L = j\omega L \mathbf{I}_L \tag{5.15a}$$

We define $X_L \equiv \omega L$, the *inductive reactance*, measured in ohms. Figure 5-4(*b*) is the diagram of (*5.15a*); observe that multiplying a phasor (here $\mathbf{I}_L$) by the imaginary unit j is equivalent to rotating the phasor counterclockwise through $90°$. Indeed, if $z = re^{j\theta}$ is an arbitrary complex number, then

$$jz = e^{j\pi/2}(re^{j\theta}) = re^{j(\theta + \pi/2)}$$

Similarly, $1/j = -j$ may be considered an operator that rotates a phasor clockwise through $90°$.

Finally, for a capacitance C, it can be readily shown that the frequency-domain version of (*5.16*) is

$$\mathbf{V}_C = \frac{1}{j\omega C}\mathbf{I}_C \tag{5.16a}$$

in which the quantity $1/\omega C$ is the *capacitive reactance*, X_C. The phasor diagram, Fig. 5-4(*c*) shows that the current *leads* the voltage by $90°$.

Comparing the time-domain equations (*5.15*) and (*5.16*) with their frequency-domain counterparts (*5.15a*) and (*5.16a*), we see that the inverse operations $d(\ )/dt$ and $\int (\ )\,dt$ go over into the inverse *arithmetic* operations $j\omega(\ )$ and $(1/j\omega)(\ )$. It is this fact that makes the phasor representation so useful. (The Laplace transform has similar properties; see Chapter 6.)

5.4 IMPEDANCE AND SERIES *RLC* CIRCUITS

Let us now consider steady-state ac circuits with more than one element. For the series *RL* circuit, represented in the frequency domain in Fig. 5-5(*a*), $\mathbf{I}_R = \mathbf{I}_L = \mathbf{I}$. Therefore, we have (in terms of rms values) from (*5.14a*) and (*5.15a*),

$$\mathbf{V} = \mathbf{V}_R + \mathbf{V}_L = (R + j\omega L)\mathbf{I} \equiv \mathbf{Z}\mathbf{I} \tag{5.17}$$

where

$$\mathbf{Z} = R + j\omega L = R + jX_L \tag{5.18}$$

is called the *impedance* of the circuit, and carries the units Ω. While $\mathbf{V}$ and $\mathbf{I}$ are phasors, $\mathbf{Z} = \mathbf{V}/\mathbf{I}$ is not a phasor (it has no implicit time factor $e^{j\omega t}$), but is merely a complex number. Thus, when we express the impedance in polar form as $\mathbf{Z} = Z\underline{/\theta}$, Z is no rms value but is just the modulus of $\mathbf{Z}$.

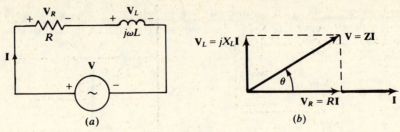

Fig. 5-5

The voltage-current relationships are shown in Fig. 5-5(*b*), the phasor diagram for the series *RL* circuit. In constructing this diagram, we have chosen **I** (which is common to the entire circuit) as the reference phasor. Note that **I** lags **V** by angle θ, where

$$\theta = \arg \mathbf{Z} = \arctan \frac{X_L}{R} \tag{5.19}$$

For reasons that will become apparent in Section 5.6, we define

$$power\ factor \equiv \cos\theta = \frac{R}{Z} \tag{5.20}$$

and say that the series *RL* circuit has a *lagging* power factor.

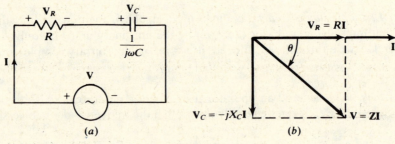

Fig. 5-6

Next, considering the series *RC* circuit of Fig. 5-6(*a*), we have the phasor diagram of Fig. 5-6(*b*) and

$$\mathbf{V} = \mathbf{V}_R + \mathbf{V}_C = \left(R + \frac{1}{j\omega C}\right)\mathbf{I} \equiv \mathbf{ZI} \tag{5.21}$$

where

$$\mathbf{Z} = R - \frac{j}{\omega C} = R - jX_C \tag{5.22}$$

This circuit has a *leading* power factor (θ is negative and so **I** *leads* **V**).

Finally, for the series *RLC* circuit, as shown in Fig. 5-7(*a*), we obtain

$$\mathbf{Z} \equiv \frac{\mathbf{V}}{\mathbf{I}} = R + j\left(\omega L - \frac{1}{\omega C}\right) \tag{5.23}$$

$$= R + j(X_L - X_C)$$

The phasor diagram, Fig. 5-7(*b*) is drawn for leading power factor ($X_L < X_C$). There are also the possibilities of lagging power factor ($X_L > X_C$) and unity power factor ($X_L = X_C$).

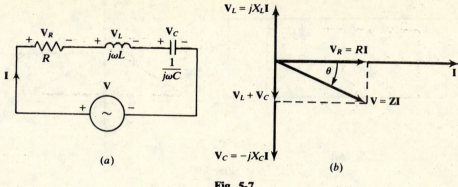

Fig. 5-7

5.5 ADMITTANCE AND PARALLEL *RLC* CIRCUITS

A parallel *RL* circuit, under sinusoidal steady state, is shown in Fig. 5-8(*a*). For this circuit, $\mathbf{V}_R = \mathbf{V}_L = \mathbf{V}$, and (5.14a) and (5.15a) give

$$\mathbf{I} = \mathbf{I}_R + \mathbf{I}_L = \frac{\mathbf{V}}{R} + \frac{\mathbf{V}}{j\omega L} = (G - jB_L)\mathbf{V} = \mathbf{Y}\mathbf{V} \qquad (5.24)$$

where
$$\mathbf{Y} \equiv G - jB_L \equiv admittance,\ S$$
$$G \equiv 1/R \equiv conductance,\ S$$
$$B_L \equiv 1/X_L \equiv inductive\ susceptance,\ S$$

By its definition,

$$\mathbf{Y} = \frac{1}{\mathbf{Z}} \qquad (5.25)$$

and so the angle, ϕ, of $\mathbf{Y}$ is the negative of the angle, θ, of $\mathbf{Z}$. See the phasor diagram, Fig. 5-8(*b*).

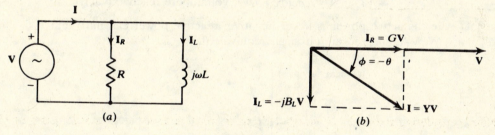

Fig. 5-8

Analogously, for a parallel *RC* circuit we will have

$$\mathbf{Y} \equiv \frac{\mathbf{I}}{\mathbf{V}} = G + jB_C \qquad (5.26)$$

where $B_C \equiv \omega C = 1/X_C$ (S) is the *capacitive susceptance*.

For the general parallel *RLC* circuit, Fig. 5-9(*a*), we have $\mathbf{I} = \mathbf{I}_R + \mathbf{I}_L + \mathbf{I}_C$, which leads to

$$\mathbf{Y} \equiv \frac{\mathbf{I}}{\mathbf{V}} = G + j\left(\omega C - \frac{1}{\omega L}\right) = G + j(B_C - B_L) \qquad (5.27)$$

Figure 5-9(*b*) is the phasor diagram for the case $B_C < B_L$ (lagging power factor). Notice that in drawing phasor diagrams for parallel circuits we select $\mathbf{V}$ as the reference phasor, because $\mathbf{V}$ is common to all the elements.

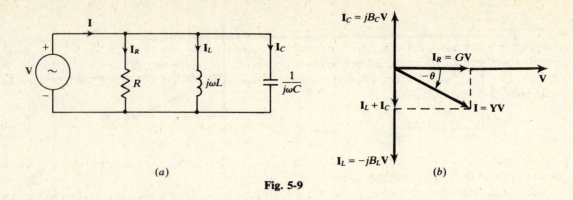

(a)   (b)

Fig. 5-9

5.6 POWER IN AC CIRCUITS

Let $v = V_m \sin \omega t$ and $i = I_m \sin(\omega t - \theta)$ be the steady-state voltage and current in an ac circuit; here θ is the angle of the impedance of the circuit. The instantaneous power, p, is then

$$p = vi = V_m I_m \sin \omega t \sin(\omega t - \theta)$$

If T is the period of the voltage, or current, waveform, then the average power, P, is given by

$$P = \frac{1}{T} \int_0^T p \, dt = \frac{1}{2} V_m I_m \cos \theta = VI \cos \theta \qquad (5.28)$$

where V and I are the rms voltage and rms current, respectively. The definition (5.20) now becomes meaningful: the *apparent power*, VI, must be multiplied by the factor $\cos \theta$ to obtain the actual average power. Sometimes, P is referred to as the *true power*, *real power*, *in-phase power*, or *active power*. This concept becomes easier to understand if we refer to Fig. 5-10 (drawn for the case $\theta > 0$), from which the following definitions of (average) power emerge:

$$VI \equiv \textit{apparent power}, \text{ voltampere (VA)}$$
$$P \equiv VI \cos \theta \equiv \textit{(active or real) power}, \text{ W}$$
$$Q \equiv VI \sin \theta \equiv \textit{reactive power}, \text{ voltampere reactive (var)}$$

Note that whereas active power is never negative, reactive power can have either sign. The three power units, W, VA, and var, are physically equivalent.

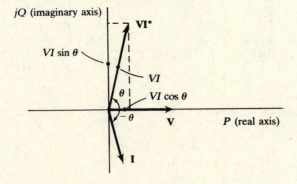

Fig. 5-10

All three kinds of power may be derived from a single *complex power*, defined as $\mathbf{VI}^*$ (VA). In fact, since $\mathbf{V} = V\underline{/0°}$ and $\mathbf{I} = I\underline{/-\theta}$,

$$\mathbf{VI}^* = (V\underline{/0°})(I\underline{/+\theta}) = VI\underline{/\theta} = VI \cos \theta + jVI \sin \theta = P + jQ$$

and so

$$VI = |\mathbf{VI}^*| \qquad P = \text{Re}(\mathbf{VI}^*) \qquad Q = \text{Im}(\mathbf{VI}^*) \qquad (5.29)$$

5.7 RESONANCE

Under a certain condition, series or parallel circuits containing R's, L's, *and* C's behave as purely resistive circuits. This condition is that of *resonance*. Consider first the series RLC circuit, Fig. 5-7(a). If the angular frequency of the input voltage is such that the inductive and capacitive reactances become equal—i.e., if

$$\omega_0 L = \frac{1}{\omega_0 C} \qquad \text{or} \qquad \omega_0 = \frac{1}{\sqrt{LC}} \equiv \text{resonant frequency} \qquad (5.30)$$

—then, by (5.23), $Z = R = $ minimum. Thus, for a given voltage amplitude, the current amplitude is a maximum and the circuit operates at unity power factor.

Similarly, the parallel circuit of Fig. 5-9(a) behaves as a purely resistive circuit at its resonant frequency, also given by (5.30). However, unlike the series circuit, the parallel RLC circuit has *maximum* impedance at the resonant frequency.

(The resonant frequency was shown in Example 4.3 and Problem 4.5 to be of importance in the transient response of RLC circuits. Now it is seen to be important in the sinusoidal steady state as well.)

A general normalized resonance curve is shown in Fig. 5-11, which illustrates certain concepts relating to resonance phenomena. The frequencies ω_1 and ω_2 whose ordinates are $1/\sqrt{2}$ of the peak ordinate are called the *half-power frequencies*. (At these frequencies, the power into the series RLC circuit actually is one-half the maximum power possible at the given voltage. For the parallel RLC circuit, the power is frequency-independent, but the name is used anyhow.) The separation of the half-power frequencies is a measure of the sharpness of the resonance peak; we define

$$\text{bandwidth} \equiv \omega_2 - \omega_1 \qquad (5.31)$$

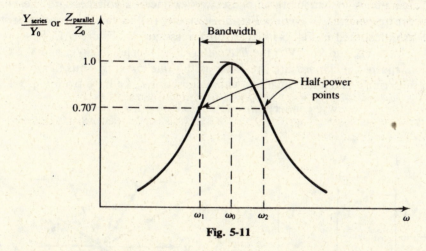

Fig. 5-11

This geometrical notion of bandwidth may be given physical significance by introducing the *quality factor* (a dimensionless number),

$$Q \equiv \frac{2\pi(\text{maximum energy stored in circuit})}{\text{total energy lost per period}} \qquad (5.32)$$

As shown in Problems 5.35 and 5.15, the quality factors for a series resonant circuit and a parallel resonant circuit are

$$Q_s = \frac{\omega_0 L}{R} \qquad Q_p = \omega_0 CR \qquad (5.33)$$

These expressions correspond under the duality transformations $R \to 1/R$, $L \to C$, $C \to L$. Then (Problem 5.36), for either circuit,

$$\text{bandwidth} = \frac{\omega_0}{Q} \qquad (5.34)$$

5.8 THREE-PHASE CIRCUITS

Most electric power systems are *three-phase* in that they involve three voltage sources having the same amplitude and frequency but displaced from each other by 120° (in time).

$$v_a = V_m \sin \omega t \qquad\qquad\qquad V_a = V\underline{/0°}$$
$$v_b = V_m \sin (\omega t - 120°) \qquad\qquad \text{or} \qquad V_b = V\underline{/-120°}$$
$$v_c = V_m \sin (\omega t - 240°) = V_m \sin (\omega t + 120°) \qquad V_c = V\underline{/120°}$$

where, as usual, $V = V_m/\sqrt{2}$. See Fig. 5-12.

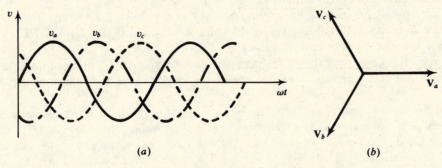

Fig. 5-12

The three voltage sources are mutually connected either in wye or in delta, as defined (for resistances) in Fig. 3-6. A system of three-phase, wye-connected voltages and the corresponding phasor diagram are respectively shown in Fig. 5-13(a) and (b). A delta-connected system and its phasor diagram are illustrated in Fig. 5-14(a) and (b), respectively. Figure 5-13(b) shows the *phase voltages* $V_a = V_{an}$, $V_b = V_{bn}$, and $V_c = V_{cn}$; the *line voltages* are labeled V_{ab}, V_{bc}, and V_{ca}. Obviously, in a wye connection, the *phase currents* are the same as the *line currents*. Thus, if we denote a phase quantity with a subscript p and a line quantity with a subscript ℓ, we have from Fig. 5-13:

$$\text{wye connection:} \quad V_\ell = \sqrt{3}V_p \qquad I_\ell = I_p \qquad\qquad (5.35)$$

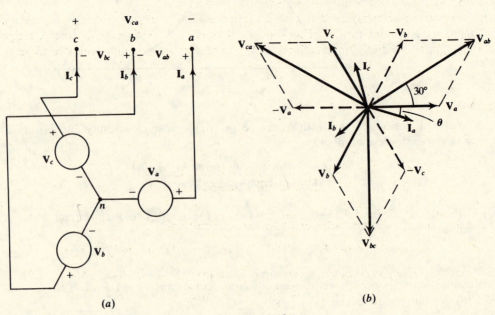

Fig. 5-13

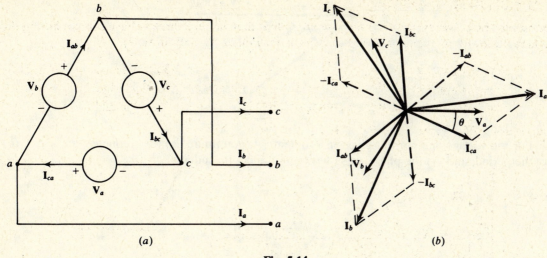

Fig. 5-14

From Fig. 5-14, we obtain

$$\text{delta connection:} \quad V_\ell = V_p \qquad I_\ell = \sqrt{3}I_p \tag{5.36}$$

In Figs. 5-13(b) and 5-14(b), the currents are arbitrarily drawn lagging the voltage.

From (5.35) and (5.36) it is clear that, for either connection, the apparent power in the three-phase circuit is given by

$$\text{apparent power} = \sqrt{3}V_\ell I_\ell = 3V_p I_p \quad \text{(VA)} \tag{5.37}$$

and the active and reactive powers are

$$P = \sqrt{3}V_\ell I_\ell \cos\theta \quad \text{(W)} \qquad Q = \sqrt{3}V_\ell I_\ell \sin\theta \quad \text{(var)} \tag{5.38}$$

where θ, the angle by which $\mathbf{V}_p$ leads $\mathbf{I}_p$, is as indicated in Figs. 5-13(b) and 5-14(b).

Just as resistances can be connected in wye or delta, so impedances can be connected to constitute three-phase loads. For three-phase circuit analysis, we generally reduce the circuit to per phase, and then study the per phase circuit in the usual manner.

Finally, we wish to emphasize that because impedance, **Z**, of an ac circuit is analogous to resistance, *R*, of a dc circuit, most of the rules and laws (such as Thévenin's theorem, reciprocity, Kirchhoff's laws) developed in Chapters 2 and 3 for dc circuits are equally applicable to ac circuits.

Solved Problems

5.1   The instantaneous voltage and current for an ac circuit are

$$v = 155.6\sin 377t \quad \text{(V)} \qquad i = 7.07\sin(377t - 36.87°) \quad \text{(A)}$$

Represent these (a) as complex exponentials, and (b) in a phasor diagram.

(a) $$v = 155.6\,e^{j377t} \quad \text{(V)} \qquad i = 7.07\,e^{j(377t-36.87°)} \quad \text{(A)}$$

(b)   See Fig. 5-15.

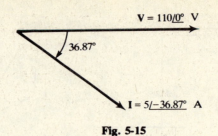

Fig. 5-15

5.2   For the voltage and current given in Problem 5.1, determine (*a*) the frequency (in hertz), (*b*) the period, and (*c*) the phase angle between v and i (in radians).

(*a*) $\omega = 377 = 2\pi f$ or $f = \dfrac{377}{2\pi} = 60$ Hz

(*b*) $T = \dfrac{1}{f} = \dfrac{1}{60} = 0.0167$ s

(*c*)   phase angle $= 36.87° = 36.87°\left(\dfrac{\pi \text{ rad}}{180°}\right) = 0.64$ rad

5.3   The voltage wave of Fig. 5-16 is applied to a 20-Ω resistor. If electrical energy costs \$0.06 per kWh, how much would it cost to operate the circuit for 24 hours?

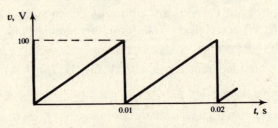

Fig. 5-16

From Fig. 5-16, $v = 10^4 t$ (V) for $0 < t < 0.01$ s, and v is periodic of period 0.01 s. Then:

$$i = \frac{v}{R} = \frac{10^4}{20}t = 500t \quad \text{(A)}$$

$$p = vi = (10^4 t)(500t) = 5 \times 10^6 t^2 \quad \text{(W)}$$

$$P = \frac{1}{T}\int_0^T p\, dt = \frac{1}{0.01}\int_0^{0.01} 5 \times 10^6 t^2\, dt = \frac{500}{3} \text{ W} = \frac{1}{6} \text{ kW}$$

$$\text{energy} = P \times \text{time} = \frac{1}{6} \times 24 = 4 \text{ kWh}$$

$$\text{cost} = \$0.06 \times 4 = \$0.24$$

5.4   Determine the rms value of the voltage waveform of Problem 5.3.

Using (*5.10*),

$$V = \left[\frac{1}{0.01}\int_0^{0.01}(10^4 t)^2\, dt\right]^{1/2} = \left[\frac{10^8}{0.01}\frac{t^3}{3}\Big|_0^{0.01}\right]^{1/2} = \frac{100}{\sqrt{3}} \text{ V}$$

5.5   Given $v = 200 \sin 377t$ (V) and $i = 8 \sin(377t - 30°)$ (A) for an ac circuit. Determine (*a*) the power factor, (*b*) true power, (*c*) apparent power, and (*d*) reactive power.

(*a*)   The current lags the voltage by $\theta = 30°$.

power factor $= \cos 30° = 0.866$ lagging

(b)   From the data, $V = (200/\sqrt{2})$ V and $I = (8/\sqrt{2})$ A. Therefore,

$$\text{true power} = VI \cos \theta = \left(\frac{200}{\sqrt{2}}\right)\left(\frac{8}{\sqrt{2}}\right)(0.866) = 692.8 \text{ W}$$

(c)

$$\text{apparent power} = VI = \left(\frac{200}{\sqrt{2}}\right)\left(\frac{8}{\sqrt{2}}\right) = 800 \text{ VA}$$

(d)

$$\text{reactive power} = VI \sin \theta = \left(\frac{200}{\sqrt{2}}\right)\left(\frac{8}{\sqrt{2}}\right)(0.5) = 400 \text{ var}$$

5.6   A coil has a resistance of 10 Ω and draws a current of 5 A when connected across a 100-V, 60-Hz source. Determine (a) the inductance of the coil, (b) the power factor of the circuit, and (c) the reactive power.

(a)
$$Z = \frac{100}{5} = \sqrt{R^2 + (\omega L)^2} = \sqrt{10^2 + (377L)^2} \qquad \text{or} \qquad L = 45.94 \text{ mH}$$

(b)
$$\cos \theta = \frac{R}{Z} = \frac{1}{2} \qquad \text{or} \qquad \theta = 60°$$

(c)
$$\text{reactive power} = VI \sin \theta = (100)(5)(\sin 60°) = 433 \text{ var}$$

5.7   A series RLC circuit is excited by a 100-V, 79.6-Hz source and has the following data: $R = 100$ Ω, $L = 1$ H, $C = 5$ μF. Calculate (a) the input current, and (b) the voltages across the elements.

(a)
$$\omega = 2\pi f = 2\pi(79.6) = 500 \text{ rad/s}$$
$$X_L = \omega L = (500)(1) = 500 \text{ } \Omega$$
$$X_C = \frac{1}{\omega C} = \frac{10^6}{(500)(5)} = 400 \text{ } \Omega$$
$$\mathbf{Z} = R + j(X_L - X_C) = 100 + j(500 - 400) = 100 + j100 = 141.4\underline{/45°} \text{ } \Omega$$
$$\mathbf{I} = \frac{\mathbf{V}}{\mathbf{Z}} = \frac{100\underline{/0°}}{141.4\underline{/45°}} = 0.707\underline{/-45°} \text{ A}$$

(b)
$$\mathbf{V}_R = R\mathbf{I} = (100)(0.707\underline{/-45°}) = 70.7\underline{/-45°} \text{ V}$$
$$\mathbf{V}_L = jX_L\mathbf{I} = (500\underline{/90°})(0.707\underline{/-45°}) = 353.5\underline{/45°} \text{ V}$$
$$\mathbf{V}_C = -jX_C\mathbf{I} = (400\underline{/-90°})(0.707\underline{/-45°}) = 282.8\underline{/-135°} \text{ V}$$

5.8   For the circuit shown in Fig. 5-17(a), evaluate the current through, and the voltage across, each element. Then draw a phasor diagram showing all the voltages and currents.

Applying nodal analysis at node 1, with $\mathbf{V}_1 = \mathbf{V}_{10}$,

$$\frac{\mathbf{V}_1 - 173.2}{10} + \frac{\mathbf{V}_1}{20} + \frac{\mathbf{V}_1}{-j11.55} = 0 \qquad \text{whence} \qquad \mathbf{V}_1 = 100\underline{/-30°} \text{ V}$$

From this,

$$\mathbf{V}_{10\Omega} = 173.2\underline{/0°} - \mathbf{V}_1 = 100\underline{/30°} \text{ V}$$

$$\mathbf{I} = \frac{\mathbf{V}_{10\Omega}}{10} = \frac{100\underline{/30°}}{10} = 10\underline{/30°} \text{ A}$$

$$\mathbf{I}_1 = \frac{\mathbf{V}_1}{20} = \frac{100\underline{/-30°}}{20} = 5\underline{/-30°} \text{ A}$$

$$\mathbf{I}_2 = \frac{\mathbf{V}_1}{-j11.55} = \frac{100\underline{/-30°}}{11.55\underline{/-90°}} = 8.66\underline{/60°} \text{ A}$$

It can be readily verified that $\mathbf{I} = \mathbf{I}_1 + \mathbf{I}_2$. The phasor diagram of Fig. 5-17(b) shows all voltages and currents.

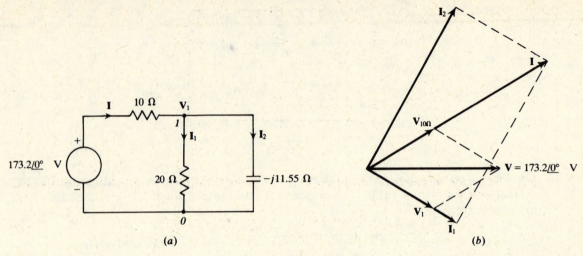

Fig. 5-17

5.9   For the circuit shown in Fig. 5-18, calculate the current supplied by the voltage source and the voltage across the current source.

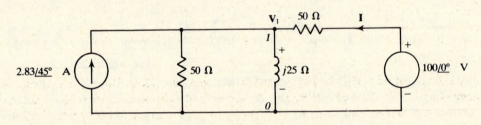

Fig. 5-18

At node *1* we have, with $\mathbf{V}_1 \equiv \mathbf{V}_{10}$ = voltage across current source,

$$\text{current in} = \text{current out}$$

$$2.83\underline{/45°} - \frac{\mathbf{V}_1}{50} = \frac{\mathbf{V}_1}{j25} + \frac{\mathbf{V}_1 - 100}{50}$$

from which $\mathbf{V}_1 = 25 + j75$ V. Then the current from the voltage source is

$$\mathbf{I} = \frac{100 - \mathbf{V}_1}{50} = 1.5 - j1.5 \quad \text{A}$$

5.10   For the circuit of Fig. 5-18, using the relationship $P = \text{Re}(\mathbf{VI}^*)$, determine the power delivered by (*a*) the voltage source, (*b*) the current source. (*c*) Verify that the sum of these two powers is the power dissipated in the two resistances.

(*a*)   Since, from Problem 5.9, $\mathbf{I} = 1.5 - j1.5$ A,

$$P_V = \text{Re}\left[(100 + j0)(1.5 + j1.5)\right] = 150 \text{ W}$$

(*b*)   Again, from Problem 5.9, $\mathbf{V}_1 = 25 + j75$ V and $2.83\underline{/45°} = 2 + j2$ A. Thus,

$$P_I = \text{Re}\left[(25 + j75)(2 - j2)\right] = 50 + 150 = 200 \text{ W}$$

(*c*)   power dissipated in the two resistances

$$= I^2(50) + \frac{V_1^2}{50} = [(1.5)^2 + (1.5)^2](50) + \frac{(25)^2 + (75)^2}{50} = 225 + 125 = 350 \text{ W} = P_V + P_I$$

5.11   (*a*) Show that the circuit of Fig. 5-19 is resonant *at all frequencies* if $R = \sqrt{L/C}$. (*b*) Calculate input current and power under universal resonance, given $V = 100$ V and $R = 40$ Ω.

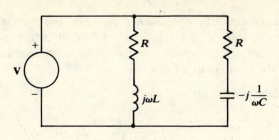

Fig. 5-19

(a) The voltage source "sees" the impedances $R + j\omega L$ and $R - j(1/\omega C)$ in parallel; hence, the *input impedance* is

$$\mathbf{Z}_{in} = \frac{[R + j\omega L][R - j(1/\omega C)]}{2R + j[\omega L - (1/\omega C)]} = \frac{R^2 + (L/C) + jRX}{2R + jX} \qquad (1)$$

where $X \equiv \omega L - (1/\omega C)$. Subtracting R from both sides of (1),

$$\mathbf{Z}_{in} - R = \frac{(L/C) - R^2}{2R + jX} \qquad (2)$$

and (2) shows that $\mathbf{Z}_{in} = R$, for all ω, if $L/C = R^2$.

(b)

$$I_{in} = \frac{V}{Z_{in}} = \frac{100}{40} = 2.5 \text{ A}$$

$$P = VI_{in} \cos 0° = (100)(2.5)(1) = 250 \text{ W}$$

5.12 A voltage source $V\underline{/0°}$ having an internal impedance $\mathbf{Z}_S = R_S + jX_S$ supplies a load having impedance $\mathbf{Z}_L = R_L + jX_L$. If R_L and X_L are individually variable, show that maximum power is transferred to the load when $R_L = R_S$ and $X_L = -X_S$; that is, when $\mathbf{Z}_L = \mathbf{Z}_S{}^*$. (This is known as the *maximum-power-transfer condition*.) What is the power transferred to the load under this condition?

Because $\mathbf{Z}_S$ and $\mathbf{Z}_L$ are in series, we have the load current

$$I_L = \frac{V}{\sqrt{(R_S + R_L)^2 + (X_S + X_L)^2}}$$

Power transferred to the load is

$$P = I_L^2 R_L = \frac{V^2 R_L}{(R_S + R_L)^2 + (X_S + X_L)^2} \leq \frac{V^2 R_L}{(R_S + R_L)^2}$$

$$= \frac{V^2}{4R_S}\left[1 - \left(\frac{R_S - R_L}{R_S + R_L}\right)^2\right] \leq \frac{V^2}{4R_S}$$

Equality in the first line is attained for $X_L = -X_S$; in the second, for $R_L = R_S$. Hence, P attains its absolute maximum, $V^2/4R_S$, for $\mathbf{Z}_L = \mathbf{Z}_S^*$.

5.13 A 20 Ω resistance is connected in series with a parallel combination of a capacitance C and a 15 mH pure inductance. At angular frequency $\omega = 1000$ rad/s, find C such that the line current is 45° out of phase with the line voltage.

For **V** and **I** to be 45° out of phase, the net reactance of the parallel LC combination must be ± 20 Ω, since $R = 20$ Ω. Hence,

$$\frac{1}{\pm j20} = \frac{1}{j(1000)(0.015)} + j1000C$$

or

$$C = \frac{1}{1000}\left(\frac{1}{15} \mp \frac{1}{20}\right) = 16.67 \ \mu\text{F}, \ 116.7 \ \mu\text{F}$$

For the smaller (larger) capacitance, **I** lags (leads) **V** by 45°.

5.14 A 46-mH inductive coil has a resistance of 10 Ω. (*a*) How much current will it draw if connected across a 100-V, 60-Hz source? (*b*) What is the power factor of the coil? (*c*) Determine the value of the capacitance that must be connected across the coil to make the power factor of the overall circuit unity.

(*a*) $\qquad \omega L = (2\pi \times 60)(46 \times 10^{-3}) = 17.34\ \Omega \qquad$ and $\qquad \mathbf{Z}_L = 10 + j17.34 = 20.0\underline{/60°}\ \Omega$

Then,

$$\mathbf{I}_L = \frac{100\underline{/0°}}{20.0\underline{/60°}} = 5.0\underline{/-60°}\ \text{A}$$

(*b*) $\qquad\qquad\qquad\qquad\qquad\qquad$ power factor $= \cos 60° = 0.5$ lagging

(*c*) The admittance of the parallel combination will be

$$\frac{1}{\mathbf{Z}_L} + j\omega C = \frac{1}{10 + j17.34} + j377C = \frac{1}{40} + j(377C - 0.0434)$$

For unity power factor, the imaginary part must vanish, yielding

$$C = \frac{0.0434}{377} = 115\ \mu\text{F}$$

5.15 Derive the quality factor Q_p of the parallel RLC circuit at resonance.

Assuming the steady state and resonance, the applied voltage and inductor current have the forms

$$v = V_m \sin \omega_0 t \qquad\qquad i_L = \frac{V_m}{\omega_0 L} \sin (\omega_0 t - 90°) = -\frac{V_m}{\omega_0 L} \cos \omega_0 t$$

where $\omega_0 = 1/\sqrt{LC}$ is the resonant frequency. Then the instantaneous energy in the circuit is

$$w(t) = \frac{1}{2} L i_L^2 + \frac{1}{2} C v^2 = \frac{1}{2} C V_m^2 \cos^2 \omega_0 t + \frac{1}{2} C V_m^2 \sin^2 \omega_0 t$$

$$= \frac{1}{2} C V_m^2 = \text{constant}$$

Hence, the numerator in (*5.32*), the definition of Q_p, has the value $\pi C V_m^2$.

As for the denominator, the average power loss in the circuit is

$$P = \frac{V^2}{R} = \frac{(V_m/\sqrt{2})^2}{R} = \frac{V_m^2}{2R}$$

and so the energy lost per period is

$$\left(\frac{V_m^2}{2R}\right)\left(\frac{2\pi}{\omega_0}\right) = \frac{\pi V_m^2}{R\omega_0}$$

Consequently,

$$Q_p = \frac{\pi C V_m^2}{\pi V_m^2/R\omega_0} = \omega_0 CR$$

5.16 For the circuit of Fig. 5-19, show that the locus of $\mathbf{Z}_{\text{in}}(\omega)$ (input impedance as a function of frequency) is a circle in the complex plane.

By Problem 5.11(*a*),

$$\mathbf{Z}_{\text{in}} - R = \frac{(L/C) - R^2}{2R + jX}$$

in which X varies with ω. Here we suppose that $L/C \neq R^2$. Figure 5-20(*a*) constructs the locus of $1/(2R + jX)$ as ω runs from 0 to ∞; it is seen to be a circle of diameter $1/2R$, centered on the real axis. This circle, uniformly magnified by the factor $(L/C) - R^2$ and then translated $+ R$ units along the real axis, gives the locus of $\mathbf{Z}_{\text{in}}$, Fig. 5-20(*b*). Note that $|\mathbf{Z}_{\text{in}}|$ is maximized at $\omega = \omega_0$ (parallel resonance).

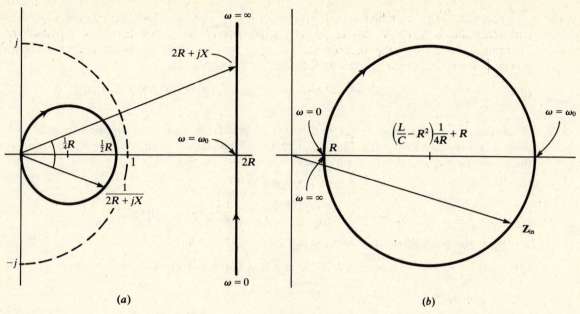

Fig. 5-20

5.17 A power P is to be transmitted over a power transmission line at a line-to-line voltage V and power factor $\cos\theta$. This power may be transmitted by a 1-phase, 2-wire line or by a 3-phase, 3-wire line, the choice depending only upon which line requires the least copper for a given copper (resistive) loss. Which system will be chosen and what is the approximate saving in copper?

Let I_1 and I_3 be the respective line currents for the 1-phase and 3-phase lines. For equal power transmission, we have, by (5.38),

$$P = VI_1\cos\theta = \sqrt{3}\,VI_3\cos\theta \qquad (1)$$

For equal power loss,

$$2I_1^2R_1 = 3I_3^2R_3 \qquad (2)$$

where R_1 and R_3 are the respective resistances of each wire of the 1-phase and 3-phase lines.

From (1) and (2) we obtain

$$R_3 = 2R_1 \qquad (3)$$

Since $R = \rho\ell/A$, with A_1 and A_3 as the respective cross-sectional areas of the wires of the 1-phase and 3-phase lines, (3) implies

$$A_1 = 2A_3 \qquad (4)$$

But the respective volumes of copper are

$$(\text{volume})_1 = 2(\ell A_1) \qquad (\text{volume})_3 = 3(\ell A_3) \qquad (5)$$

Together, (4) and (5) give

$$(\text{volume})_1 = \frac{4}{3}(\text{volume})_3$$

Consequently, the 3-phase system is preferred, and results in a 25% saving in the volume of copper.

5.18 A 3-phase voltage source has a phase voltage of 120 V and supplies a wye-connected load having impedance $36 + j48$ Ω per phase. Calculate (a) the line voltage, (b) the line current, (c) the power factor, and (d) the total 3-phase power supplied to the load.

(a) $$V_\ell = \sqrt{3}\,V_p = \sqrt{3}\times 120 = 207.8 \text{ V}$$

(b)
$$I_\ell = I_p = \frac{V_p}{Z} = \frac{120}{\sqrt{(36)^2 + (48)^2}} = 2 \text{ A}$$

(c)
$$\cos\theta = \frac{R}{Z} = \frac{36}{60} = 0.6 \text{ lagging}$$

(d)
$$\text{power} = \sqrt{3}V_\ell I_\ell \cos\theta = \sqrt{3}(207.8)(2)(0.6) = 432 \text{ W}$$

5.19 If the phase impedances of Problem 5.18 are connected in delta and supplied by a 207.8-V, 3-phase source [see Problem 5.18(a)], calculate (a) the phase current, (b) the line current, (c) the power factor, and (d) the total power.

(a)
$$I_p = \frac{207.8}{60} = 3.46 \text{ A}$$

(b)
$$I_\ell = \sqrt{3}I_p = \sqrt{3} \times 3.46 = 6 \text{ A}$$

(c) From Problem 5.18(c), $\cos\theta = 0.6$ lagging.

(d)
$$\text{total power} = \sqrt{3}V_\ell I_\ell \cos\theta = \sqrt{3}(207.8)(6)(0.6) = 1296 \text{ W}$$

Supplementary Problems

5.20 Given: $v = 70.7 \sin(314t - 20°)$ (V), $i = 2.12 \cos(314t + 10°)$ (A). (a) What is the phase angle between v and i? (b) What is the frequency, in Hz? (c) Express v and i as phasors (rms).
Ans.   (a) 120°; (b) 50 Hz; (c) $\mathbf{V} = 50.0\underline{/-20°}$ V, $\mathbf{I} = 1.5\underline{/100°}$ A

5.21 Determine the average power from the data of Problem 5.20. Does the circuit absorb power?
Ans. −37.5 W; no

5.22 For the data of Problem 5.1, calculate (a) instantaneous power; (b) average power; (c) reactive power; (d) apparent power; and (e) power factor.
Ans.   (a) $1100 \sin 377t \sin(377t - 36.87°)$ (W); (b) 440 W; (c) 330 var; (d) 550 VA; (e) 0.8 lagging

5.23 Find the rms value of the current obtained in Problem 5.3.   *Ans.* 2.89 A $= \dfrac{100/\sqrt{3}}{20}$ A

5.24 For the circuit of Fig. 5-5(a), if $\mathbf{V} = V\underline{/0°}$ (V), write an expression for the instantaneous steady-state current.
Ans. $i(t) = \dfrac{\sqrt{2}V}{\sqrt{R^2 + \omega^2 L^2}} \sin\left(\omega t - \tan^{-1}\dfrac{\omega L}{R}\right)$ (A)

5.25 A series *RLC* circuit carries a current of $10\underline{/0°}$ A. Its parameters are: $R = 8\ \Omega$, $X_L = 12\ \Omega$, $X_C = 6\ \Omega$. Determine (a) the voltages across the three elements, and (b) the voltage across the entire circuit.
Ans.   (a) $\mathbf{V}_R = 80\underline{/0°}$ V, $\mathbf{V}_L = 120\underline{/90°}$ V, $\mathbf{V}_C = 60\underline{/-90°}$ V; (b) $\mathbf{V} = 100\underline{/36.9°}$ V

5.26 In the ac bridge circuit of Fig. 5-21,
$$\mathbf{Z}_1 = \mathbf{Z}_2 = \mathbf{Z}_D = 10 + j0\ \ \Omega \qquad \mathbf{Z}_3 = -\mathbf{Z}_4 = j10\ \Omega$$
and $\mathbf{V} = 40\underline{/0°}$ V. Find the current (a) in the capacitor, (b) in the inductor.
Ans.   (a) $|1 + j3| = 3.162$ A; (b) $|1 - j3| = 3.162$ A

5.27 For the bridge circuit of Problem 5.26, determine the power drawn from the source.   *Ans.* 80 W

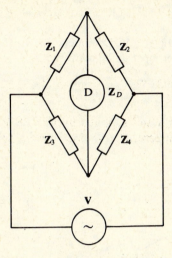

Fig. 5-21   **Fig. 5-22**

5.28   An ac circuit with a current excitation is shown in Fig. 5-22. Determine (a) the voltage across the inductance, and (b) the power dissipated in the two resistances.   *Ans.*   (a) $20\underline{/0°}$ V; (b) 2 W

5.29   An impedance $15 + j20$ Ω is connected across a 125-V, 60-Hz source. Find (a) the instantaneous current through the load, (b) the instantaneous power, and (c) the average active and reactive powers.
Ans.   (a) $7.07 \sin(377t - 53.1°)$ (A); (b) $1250 \sin 377t \sin(377t - 53.1°)$ (W); (c) 375 W, 500 var

5.30   A voltage source of 100 V has internal impedance $0.1 + j0.1$ Ω and supplies a load having that same impedance. Calculate the power absorbed by the load.   *Ans.*   125 W

5.31   A 75 Ω resistance is connected in parallel with a 10 μF capacitance. Determine an equivalent series RC circuit such that the two circuits have the same impedance at an angular frequency of 1000 rad/s.   *Ans.* $R = 48$ Ω, $C = 27.8$ μF

5.32   An inductive coil has resistance 30 Ω, but the inductance is unknown. The coil is connected in parallel with a 100-Ω resistor, and the combination, when connected across a 100-V, 60-Hz source, draws 400 W. Determine the value of the inductance.   *Ans.*   26.53 mH

5.33   A parallel RL circuit consumes 480 W at 120 V and has a lagging power factor of 0.8. It is desired to make the power factor unity by connecting a capacitance in parallel with the RL circuit. If the source frequency is 60 Hz, find the value of the required capacitance.   *Ans.*   66.3 μF

5.34   Determine the relationship between the impedances $Z_1, \ldots, Z_4$ in the ac bridge of Fig. 5-21 such that the current through the detector D is zero. (Under this condition, the bridge is *balanced*.)
Ans. $Z_1 Z_4 = Z_2 Z_3$

5.35   In a resonant series RLC circuit show that the total stored energy is a constant (cf. Problem 5.15). From this, calculate Q_s.
Ans. $w(t) = \dfrac{1}{2} L I_m^2$, $Q_s = \dfrac{\omega_0 L}{R}$

5.36   For either the series or the parallel RLC circuit, evaluate the half-power frequencies ω_1 and ω_2 (see Fig. 5-11) in terms of the quality factor from Problem 5.35 or Problem 5.15, and the resonant frequency ω_0. Hence prove (*5.34*). [*Hint.* Consider only the series circuit: the results will also hold for the parallel circuit, by duality.]
Ans. $\omega_1 = \omega_0 \left[\sqrt{1 + \left(\dfrac{1}{2Q}\right)^2} - \dfrac{1}{2Q} \right]$, $\omega_2 = \omega_0 \left[\sqrt{1 + \left(\dfrac{1}{2Q}\right)^2} + \dfrac{1}{2Q} \right]$

5.37 From Problem 5.36, deduce that $\omega_0 = \sqrt{\omega_1 \omega_2}$.

5.38 A 3-phase network is shown in Fig. 5-23, with phase voltage $\mathbf{V}_{an} = 120\underline{/0°}$ V. Determine: (a) $\mathbf{I}_a$, (b) $\mathbf{V}_{a'n}$, (c) the total power loss in the line resistances, and (d) total power input to the network.
Ans. (a) $2\underline{/-53.1°}$ A; (b) $124.96\underline{/-2.9°}$ V; (c) 48 W; (d) 480 W

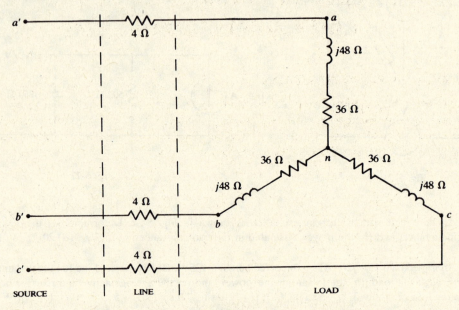

Fig. 5-23

5.39 Three-phase wye-connected capacitances are connected in parallel with the load of Fig. 5-23 such that the combination becomes purely resistive. (a) What is the value of capacitance per phase, if the system operates at 60 Hz? (b) If voltage $\mathbf{V}_{a'n}$ remains at 124.96 V, as in Problem 5.38, determine the current $\mathbf{I}_a$.   Ans.   (a) 35.37 μF; (b) 1.2 A

Chapter 6

Special Forcing Functions and Laplace Transforms

6.1 SPECIAL FORCING FUNCTIONS

In the preceding chapters we have encountered the two most common forcing functions—the dc source and the sinusoidal ac source (which was considered only under steady state). Here we shall examine three special types of forcing functions which, though not often physically realized, do provide an insight into the behavior of circuits and systems.

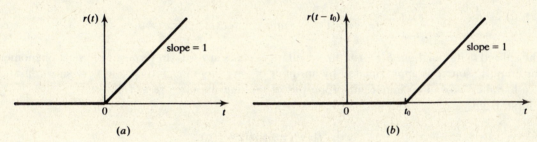

Fig. 6-1

The *unit ramp* source, $r(t)$, is plotted in Fig. 6-1(a); analytically,

$$r(t) = \begin{cases} t & (t \geq 0) \\ 0 & (t \leq 0) \end{cases} \qquad (6.1)$$

If the ramp is shifted in time, as in Fig. 6-1(b), the analytical form becomes

$$r(t - t_0) = \begin{cases} t - t_0 & (t \geq t_0) \\ 0 & (t \leq t_0) \end{cases} \qquad (6.2)$$

Notice that the unit ramp has a unit slope.

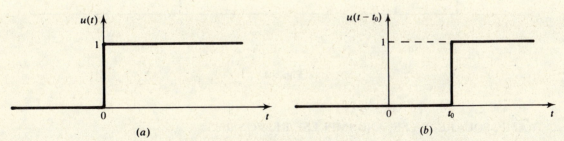

Fig. 6-2

The *unit step* source, $u(t)$, which is plotted in Fig. 6-2(a), is given by

$$u(t) = \begin{cases} 1 & (t > 0) \\ 0 & (t < 0) \end{cases} \qquad (6.3)$$

67

Figure 6-2(b) shows the shifted unit step

$$u(t - t_0) = \begin{cases} 1 & (t > t_0) \\ 0 & (t < t_0) \end{cases} \tag{6.4}$$

Comparing Figs. 6-1 and 6-2, we observe that for all values of t except $t = 0$,

$$u(t) = \frac{d}{dt}[r(t)] \tag{6.5}$$

and, for all values of t,

$$r(t) = \int_{-\infty}^{t} u(\tau)\, d\tau \tag{6.6}$$

To understand the concept of unit impulse, consider a pulse of duration T and amplitude $1/T$. The "area" within the pulse is unity. We may maintain this unit area of the pulse if we decrease the duration of the pulse and reciprocally increase the amplitude [Fig. 6-3(a)]. In the limit as the pulse duration tends to zero, we obtain the *unit impulse* source, $\delta(t)$, indicated by the "spike" in Fig. 6-3(b). The function $\delta(t)$ is also known as the *delta function*. Mathematically, $\delta(t)$ is defined by

$$\int_{-\infty}^{\infty} \delta(\tau)\, d\tau = 1 \qquad \text{and} \qquad \delta(t) = 0 \text{ for } t \neq 0 \tag{6.7}$$

In the shifted delta function, $\delta(t - t_0)$, the singularity occurs at $t = t_0$. In terms of physical phenomena, a unit impulse source is approximated by an input of very large magnitude and very short duration; e.g., a hammer blow. Two important properties of the unit impulse are as follows.

First, from

$$\int_{-\infty}^{\infty} f(\tau)\delta(\tau - t_0)\, d\tau = f(t_0) \tag{6.8}$$

we observe that $\delta(\tau - t_0)$ picks out the value of $f(\tau)$ at $\tau = t_0$. This is the *sifting property* of the delta function. Second, the unit impulse forcing function yields the natural response of a system, as will be seen in Example 6.3.

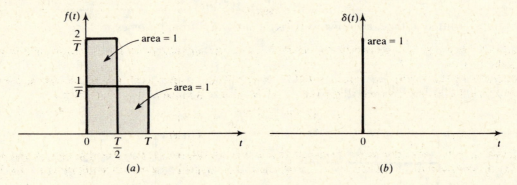

Fig. 6-3

6.2   RAMP, SQUARE-PULSE, AND IMPULSE RESPONSES

We know that the complete response of a circuit consists of two parts, the transient response and the steady-state response. The transient response is the natural response of the circuit to an initial energization; it is independent of the source. Therefore, in order to find the response of a circuit to a given source, we need only the steady-state response to that source. Once the steady-state solution is obtained, we add to it the general form of the transient solution and then evaluate the unknown constants in the transient solution from the initial conditions (applied to the sum of the two parts).

Example 6.1 A resistance of $1 \, M\Omega$ is connected in series with a $5 \, \mu F$ capacitance. A ramp voltage, $r(t) = \max\{0, 2t\}$ (V) is applied to the circuit. Determine the circuit current, $i(t)$.

From Section 4.2, the circuit equation,

$$R\frac{di}{dt} + \frac{1}{C}i = \frac{dr(t)}{dt}$$

has the transient solution

$$i_n = Ae^{-t/RC} = Ae^{-t/5} \quad (A)$$

For $t > 0$, $dr/dt = 2$ V/s, so that, by inspection, the steady-state solution is

$$i_f = 2C = 10^{-5} \, A$$

Applying the initial condition $i(0+) = 0$ to the complete solution, $i = i_n + i_f$, we get $A = -10^{-5}$ A; therefore,

$$i(t) = 10^{-5}(1 - e^{-t/5}) \quad (A)$$

Ramp signals are especially useful in characterizing certain types of feedback control systems (see Chap. 17).

A common signal encountered in the study of electric circuits is the *square pulse*, shown in Fig. 6-3(a) and redrawn in Fig. 6-4(a). Notice that this pulse can be generated by a positive step and a delayed negative step, as illustrated in Fig. 6-4(b). The response of a circuit to a pulse then becomes the superposition of the responses to the two step functions.

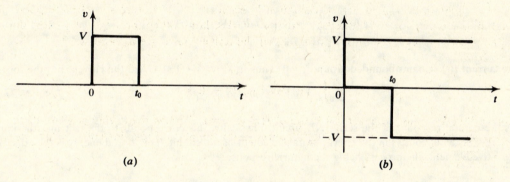

(a)   (b)

Fig. 6-4

Example 6.2 Consider the RC circuit of Example 6.1. Find the current produced by a voltage pulse of 100 V and 1 s duration.

Referring to Fig. 6-4(b), we have $V = 100$ V, $t_0 = 1$ s. Either voltage step produces a zero steady-state current; hence the complete current response for $t > 0$ is the superposition of the two transients:

$$i = Ae^{-t/RC} + Be^{-(t-t_0)/RC}u(t - t_0)$$
$$= Ae^{-t/5} + Be^{-(t-1)/5}u(t - 1) \quad (A)$$

[The factor $u(t - 1)$ makes the second transient zero in the interval $0 < t < 1$ s.] The constants A and B are determined from the initial condition, $i(0+) = V/R$, and the continuity of the capacitor voltage at $t = 1$ s. Thus,

$$A = \frac{V}{R} = 10^{-4} \, A$$

and
$$v = Ri + v_C$$
$$\Delta v|_{t=1} = R \, \Delta i|_{t=1} + 0$$
$$-V = R \cdot B$$
$$B = -\frac{V}{R} = -10^{-4} \, A$$

Our solution now takes the form

$$i = 10^{-4}e^{-t/5}[1 - e^{1/5}u(t - 1)] \quad (A) \qquad (t > 0)$$

This function is sketched in Fig. 6-5.

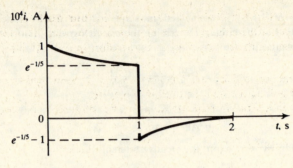

Fig. 6-5

Example 6.3 An impulsive voltage, $V\delta(t)$ (V), is applied to a series RL circuit. Determine the circuit current, $i(t)$.

The circuit equation is

$$L\frac{di}{dt} + Ri = V\delta(t)$$

from which

$$L\int_{0-}^{0+}\frac{di}{dt}\,dt + R\int_{0-}^{0+} i\,dt = V\int_{0-}^{0+}\delta(t)\,dt$$
$$L\,i(0+) + \qquad 0 \qquad = V\cdot(1\text{ s})$$

since the current $i(t)$ remains bounded around $t = 0$. Thus, $i(0+) = V\cdot(1\text{ s})/L$ and the current has the form

$$i(t) = \frac{V\cdot(1\text{ s})}{L}\,e^{-Rt/L} \qquad (t > 0)$$

which is the natural response of the series RL circuit. Physically, this corresponds to the fact that the impulse stores energy (instantaneously) in the energy-storage element of the circuit. This stored energy is then dissipated in the resistive element, the process being the natural response.

6.3 LAPLACE TRANSFORMS

Heretofore we have determined circuit quantities in the time domain by solving differential, or integro-differential, equations, using classical techniques. Laplace transforms offer another solution method. For our purposes, it is sufficient to define the Laplace transform of a function $f(t)$, where $f(t) \equiv 0$ for $t \leq 0$, as

$$\mathcal{L}\{f(t)\} \equiv F(s) = \int_0^\infty e^{-st}f(t)\,dt \qquad (6.9)$$

In (6.9), s is a complex variable, $s = \sigma + j\omega$, with σ chosen large enough to make the infinite integral converge.

Example 6.4 Given $f(t) = e^{-at}$ ($a > 0$), find $F(s)$.

Substituting in (6.9) yields

$$F(s) = \int_0^\infty e^{-st}e^{-at}dt = \frac{e^{-(s+a)t}}{-(s+a)}\bigg|_0^\infty = \frac{1}{s+a}$$

where, to make the exponential vanish at the upper limit, it has been assumed that $\sigma > -a$. However, this restriction may be dropped, and the result

$$F(s) = \frac{1}{s+a}$$

may be considered valid over the entire complex s-plane.

The table of Laplace transforms below should suffice for our work. Lines 10–16 express general properties of the Laplace transform; line 14, the *convolution theorem*, is often useful in determining the time-function that corresponds to a composite expression in the *s*-domain.

Example 6.5   At $t = 0$, an unenergized series *RL* circuit is suddenly connected across a dc voltage source, V. Find the circuit current (for $t > 0$) by the Laplace transform method.

We develop the solution in a series of steps; these are to be followed in any circuit problem to which the method applies.

Step 1:   Write the governing equation

$$L\frac{di}{dt} + Ri = V \qquad (t > 0) \tag{1}$$

Step 2:   Take the Laplace transform of (*1*)

Using line 10 of Table 6-1, with $i(0+) = 0$, and also line 2, we obtain

$$LsI(s) + RI(s) = \frac{V}{s} \tag{2}$$

Table 6-1.   Laplace Transform Pairs

	$f(t)$	$F(s)$
1	$\delta(t)$	1
2	a	$\dfrac{a}{s}$
3	t	$\dfrac{1}{s^2}$
4	e^{-at}	$\dfrac{1}{s+a}$
5	te^{-at}	$\dfrac{1}{(s+a)^2}$
6	$\sin \omega t$	$\dfrac{\omega}{s^2 + \omega^2}$
7	$\cos \omega t$	$\dfrac{s}{s^2 + \omega^2}$
8	$e^{-at} \sin \omega t$	$\dfrac{\omega}{(s+a)^2 + \omega^2}$
9	$e^{-at} \cos \omega t$	$\dfrac{s+a}{(s+a)^2 + \omega^2}$
10	$\dfrac{d}{dt}[f(t)]$	$sF(s) - f(0+)$
11	$\dfrac{d^2}{dt^2}[f(t)]$	$s^2F(s) - sf(0+) - f'(0+)$
12	$\displaystyle\int_0^t f(\tau)\,d\tau$	$\dfrac{1}{s}F(s)$
13	$af(t) + bg(t)$	$aF(s) + bG(s)$
14	$\displaystyle\int_0^t f(\tau)g(t - \tau)\,d\tau$	$F(s)\,G(s)$
15	$f(\infty)$ (final value)	$\displaystyle\lim_{s \to 0} sF(s)$
16	$f(0+)$ (initial value)	$\displaystyle\lim_{\substack{s \to \infty \\ s\ \text{real}}} sF(s)$

Step 3: Solve (2) for the unknown transform

$$I(s) = \frac{V}{s(Ls + R)} \tag{3}$$

Step 4: Take the inverse Laplace transform of (3)
Writing

$$I(s) = \frac{1}{s} \cdot \frac{V/L}{s + (R/L)}$$

we have by line 4 of Table 6-1,

$$\mathcal{L}^{-1}\left\{\frac{V/L}{s + (R/L)}\right\} = \frac{V}{L} e^{-Rt/L}$$

Hence, by line 12,

$$i(t) = \int_0^t \frac{V}{L} e^{-R\tau/L}\, d\tau = \frac{V}{R}(1 - e^{-Rt/L})$$

The same result for $i(t)$ would have been obtained if we had used line 14 [with $G(s) = 1/s$] or the method of partial fractions to invert $I(s)$.

Solved Problems

6.1　Express the sinusoidal pulse of Fig. 6-6 in terms of unit step functions.

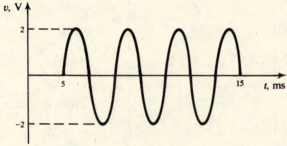

Fig. 6-6

The sinusoid has 7 half-periods in 10 ms; i.e.

$$\frac{7T}{2} = 10^{-2} \qquad \text{or} \qquad \omega = \frac{2\pi}{T} = 700\pi \text{ rad/s}$$

and the amplitude is 2 V. Therefore, it is the function

$$2 \sin 700\pi(t - 5 \times 10^{-3})$$

that must be cut off to zero for $t < 5 \times 10^{-3}$ s and for $t > 15 \times 10^{-3}$ s. Thus,

$$v(t) = [2 \sin 700\pi(t - 5 \times 10^{-3})][u(t - 5 \times 10^{-3}) - u(t - 15 \times 10^{-3})] \quad \text{(V)}$$

6.2　A series RL circuit experiences a pulse of voltage, V, occurring during the interval $t_0 < t < t_1$. Determine the circuit current, $i(t)$.

The pulse is indicated in Fig. 6-7(a), and the electric circuit, with step voltage sources simulating the pulse, is shown in Fig. 6-7(b). Let $i_0(t)$ and $i_1(t)$ be the currents due to sources $Vu(t - t_0)$ and $-Vu(t - t_1)$, respectively, each acting alone. Then, from Chapter 4 (Example 4.1), we may directly write

$$i_0(t) = \frac{V}{R}[1 - e^{-R(t - t_0)/L}]\, u(t - t_0) \qquad i_1(t) = -\frac{V}{R}[1 - e^{-R(t - t_1)/L}]u(t - t_1)$$

and, by superposition, $i(t) = i_0(t) + i_1(t)$. The current is plotted in Fig. 6-7(a).

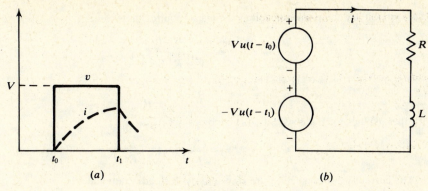

Fig. 6-7

In this case, the initial conditions $i_0(t_0) = 0$ and $i_1(t_1) = 0$, separately imposed on the two currents, automatically ensure that $i(t_0) = 0$ and that $i(t)$ is continuous at $t = t_1$ (the current in an inductor cannot change abruptly).

6.3   A single sawtooth voltage waveform is shown in Fig. 6-8(a). Synthesize this waveform by a superposition of step and ramp functions.

The three components needed in the synthesis are shown in Fig. 6-8(b). Analytically,

$$v(t) = \frac{V}{t_0} r(t) - Vu(t - t_0) - \frac{V}{t_0} r(t - t_0)$$

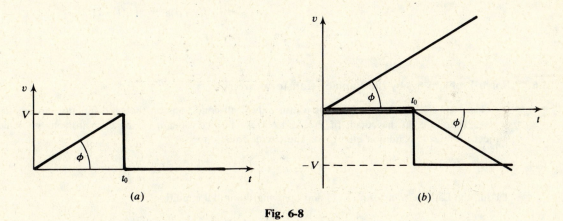

Fig. 6-8

6.4   Find the current in a series RC circuit if the capacitor is initially uncharged and the applied voltage is of the form shown in Fig. 6-8(a).

The differential equation for the current is

$$R\frac{di}{dt} + \frac{1}{C} i = \frac{dv}{dt} = \frac{V}{t_0} u(t) - \frac{V}{t_0} u(t - t_0) \qquad (1)$$

The solution proceeds as in Example 6.2, except that now the two steps composing the forcing function give rise to currents with steady-state values CV/t_0 and $-CV/t_0$, respectively. Thus, for $t > 0$,

$$i(t) = \left[Ae^{-t/RC} + \frac{CV}{t_0} \right] + \left[Be^{-(t-t_0)/RC} - \frac{CV}{t_0} \right] u(t - t_0) \qquad (2)$$

The initial condition, $i(0+) = 0$ [because $v(0) = 0$], determines

$$A = -\frac{CV}{t_0}$$

The continuity of the capacitor voltage at $t = t_0$ requires that

$$\Delta v|_{t=t_0} = R \, \Delta i|_{t=t_0}$$

$$-V = R\left[B - \frac{CV}{t_0}\right]$$

$$B = \frac{CV}{t_0} - \frac{V}{R}$$

With these values for the constants, (2) becomes

$$i(t) = \frac{CV}{t_0}(1 - e^{-t/RC}) - \frac{CV}{t_0}\left[1 - \left(1 - \frac{t_0}{RC}\right)e^{-(t-t_0)/RC}\right]u(t - t_0) \qquad (3)$$

The current is, of course, a pure transient, with steady-state value

$$\frac{CV}{t_0} - \frac{CV}{t_0} = 0$$

6.5   Establish the following *shifting property* of the Laplace transform: If $F(s) = \mathcal{L}\{f(t)\}$, then

$$e^{-as}F(s) = \mathcal{L}\{f(t - a) \, u(t - a)\}$$

We have

$$\mathcal{L}\{f(t - a) \, u(t - a)\} \equiv \int_0^\infty f(t - a)u(t - a)e^{-st} \, dt$$

$$= \int_a^\infty f(t - a)e^{-st} \, dt$$

$$= e^{-as}\int_0^\infty f(t - a)e^{-s(t-a)} \, d(t - a)$$

$$\equiv e^{-as}F(s)$$

6.6   Solve Problem 6.4 by the Laplace transform method.

The Laplace transform method is useful only if all conditions on the time function are applied at $t = 0+$. Therefore, in the circuit of Problem 6.4, we first consider $q(t)$, the capacitor charge, which is subject to the sole condition $q(0+) = 0$. The circuit equation is

$$R\frac{dq}{dt} + \frac{1}{C}q = v = \begin{cases} Vt/t_0 & (0 < t < t_0) \\ 0 & (t > t_0) \end{cases} \qquad (1)$$

Taking the Laplace transform of (1) and applying the initial condition,

$$RsQ(s) + \frac{1}{C}Q(s) = \frac{V}{t_0}\int_0^{t_0} te^{-st} \, dt$$

$$= \frac{V}{t_0}\left(\frac{1}{s^2} - \frac{1}{s^2}e^{-t_0 s} - \frac{t_0}{s}e^{-t_0 s}\right) \qquad (2)$$

Now, because $i = dq/dt$ and $q(0+) = 0$, $I(s) = sQ(s)$. Solving (2), with $\alpha \equiv 1/RC$, then gives

$$I(s) = sQ(s) = \frac{V}{Rt_0}\frac{1}{s(s + \alpha)} - \frac{V}{Rt_0}\frac{e^{-t_0 s}}{s(s + \alpha)} - \frac{V}{R}\frac{e^{-t_0 s}}{s + \alpha}$$

$$\equiv I_1(s) + I_2(s) + I_3(s) \qquad (3)$$

Inverting (3) termwise, we have, as in Example 6.5,

$$i_1(t) = \frac{V}{Rt_0\alpha}(1 - e^{-\alpha t})$$

Then, by the shifting theorem (Problem 6.5),

$$i_2(t) = -\frac{V}{Rt_0\alpha}[1 - e^{-\alpha(t-t_0)}] \, u(t - t_0)$$

Finally, by line 4 of Table 6-1 and the shifting theorem,

$$i_3(t) = -\frac{V}{R} e^{-\alpha(t - t_0)} u(t - t_0)$$

The reader should verify that the solution just found,

$$i(t) = i_1(t) + i_2(t) + i_3(t)$$

coincides with (3) of Problem 6.4. He should also identify currents i_1, i_2, and i_3 as the separate responses to the three voltage components shown in Fig. 6-8(b).

6.7   A series RLC circuit has $R = 250\ \Omega$, $L = 10$ mH, and $C = 100\ \mu$F. Assuming zero initial conditions, use Laplace transforms to determine the (natural) current for a unit impulse voltage input.

The circuit voltage equation is, for an initially uncharged capacitor,

$$\frac{1}{C} \int_0^t i(\tau)\, d\tau + Ri + L\frac{di}{dt} = v(t) \tag{1}$$

With the given numerical values, (1) becomes

$$10^4 \int_0^t i(\tau)\, d\tau + 250i + 10^{-2}\frac{di}{dt} = \delta(t) \tag{2}$$

Taking the Laplace transform of (2) yields

$$\frac{10^4}{s} I(s) + 250 I(s) + 10^{-2} s I(s) = 1 \tag{3}$$

whence, using partial fractions,

$$I(s) = \frac{100s}{s^2 + 25\,000s + 10^6} = \frac{100s}{(s + 24\,960)(s + 40)} = \frac{100.16}{s + 24\,960} - \frac{0.16}{s + 40}$$

Inverting, $i(t) = 100.16\, e^{-24960t} - 0.16\, e^{-40t}$ (A) $\tag{4}$

The handling of the delta function in the above solution was quite different from that in Example 6.3, where it was used merely to derive the initial condition, $i(0+) = V \cdot 1/L$, to be applied to a homogeneous equation. Here, on the other hand, we have extended the Laplace transform back to $t = 0-$, with the result that di/dt in (2) gives rise to no initial-value term in (3). The correct initial value is built into the solution (4):

$$i(0+) = 100.16 - 0.16 = 100\ \text{A} = \frac{1\ \text{V} \cdot \text{s}}{10\ \text{mH}}$$

6.8   The voltage across a 2-μF capacitor is given, in the transform domain, by

$$V_C(s) = \frac{60s + (8 \times 10^4)}{s^2 + 10^6}$$

What is the current through the capacitor?
By the initial-value theorem, line 16 of Table 16-1,

$$v_C(0+) = \lim_{\substack{s \to \infty \\ s\ \text{real}}} s V_C(s) = 60\ \text{V}$$

Then, from $i = C(dv_C/dt)$, it follows that

$$I(s) = C[s V_C(s) - v_C(0+)] = (2 \times 10^{-6}) \left[\frac{60s^2 + (8 \times 10^4)s}{s^2 + 10^6} - 60 \right]$$

$$= (0.16)\frac{s}{s^2 + 10^6} - (0.12)\frac{10^3}{s^2 + 10^6}$$

Inverting by use of lines 7 and 6 of Table 6-1,

$$i(t) = 0.16 \cos 1000t - 0.12 \sin 1000t = 0.20 \sin (1000t + 127°)\quad \text{(A)}$$

6.9   Refer to Fig. 6-9. The circuit was under steady state with the switch at position 1. At $t = 0$ the switch is moved to position 2. Solve for the circuit current for $t > 0$.

For $t > 0$,

$$4\frac{di}{dt} + 2i = 12$$

which transforms to

$$4sI(s) - 4i(0+) + 2I(s) = \frac{12}{s} \qquad (1)$$

As $i(0+) = i(0-)$ is the steady-state current with the switch in position 1,

$$i(0+) = \frac{6}{2} = 3 \text{ A} \qquad (2)$$

From (1) and (2),

$$I(s) = \frac{6(s+1)}{s(2s+1)} = 6\left(\frac{1}{s} - \frac{1}{2s+1}\right) = 6\left(\frac{1}{s} - \frac{0.5}{s+0.5}\right) \qquad (3)$$

Hence $i(t) = 6(1 - 0.5e^{-0.5t})$ (A)

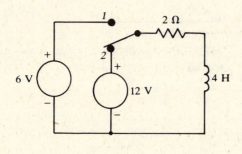

Fig. 6-9

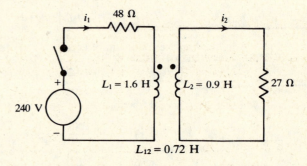

$L_{12} = 0.72$ H

Fig. 6-10

6.10   Without solving for $i(t)$, find the steady-state current in the circuit of Fig. 6-9.

Applying the final-value theorem to (3) of Problem 6.8, we have

$$\lim_{t \to \infty} i(t) = \lim_{s \to 0} s\left[\frac{6(s+1)}{s(2s+1)}\right] = 6 \text{ A}$$

6.11   For the circuit shown in Fig. 6-10, solve for i_1 and i_2 if the switch is closed at $t = 0$. The mutual inductance is of additive polarity.

The circuit equations are, for $t > 0$,

$$1.6\frac{di_1}{dt} + 48i_1 - 0.72\frac{di_2}{dt} = 240$$

$$-0.72\frac{di_1}{dt} + 0.9\frac{di_2}{dt} + 27i_2 = 0$$

These equations in the transform domain become (zero initial values)

$$(1.6s + 48)I_1(s) - 0.72sI_2(s) = \frac{240}{s}$$

$$-0.72sI_1(s) + (0.9s + 27)I_2(s) = 0$$

Solving for $I_1(s)$ and $I_2(s)$ and taking their inverse transforms,

$$i_1(t) = 5 - 2.5e^{-75t} - 2.5e^{-75t/4} \quad \text{(A)} \qquad i_2(t) = -3.33e^{-75t} + 3.33e^{-75t/4} \quad \text{(A)}$$

6.12 Refer to Problem 4.11. When the circuit parameters are such that

$$\sqrt{\frac{L}{C}} = R_1 = R_2 \equiv R$$

then the solution to (5) that obeys the initial conditions (6) and (7) is obviously $v(t) = RI$; i.e. the circuit becomes purely resistive and experiences no transient. Verify this phenomenon of *universal resonance* (already treated in Problem 5.11) by a Laplace-transform solution of (1), (2), and (3).

In view of the initial conditions $i_1(0+) = 0$ and $i_2(0+) = I$, we obtain the transformed equations

$$I_1(s) + I_2(s) = \frac{I}{s}$$

$$V(s) = RI_1(s) + LsI_1(s)$$

$$V(s) = RI_2(s) + \frac{I_2(s)}{Cs}$$

Solve the second and third equations respectively for $I_1(s)$ and $I_2(s)$, substitute in the first equation, and use $R^2 = L/C$ to obtain

$$V(s) = \frac{RI}{s} \qquad \text{whence} \qquad v(t) = RI \qquad (t > 0)$$

6.13 An inductively coupled circuit is shown in Fig. 6-11. The switch is closed at $t = 0$. Assuming zero initial conditions, determine the current $i_2(t)$.

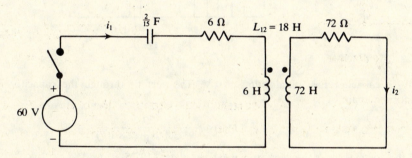

Fig. 6-11

The circuit equations in the transform domain become

$$\left(6s + 6 + \frac{15}{2s}\right)I_1(s) - 18sI_2(s) = \frac{60}{s}$$

$$-18sI_1(s) + (72s + 72)I_2(s) = 0$$

Solving for $I_2(s)$ gives:

$$I_2(s) = \frac{10s}{s^3 + 8s^2 + 9s + 5} = \frac{10s}{(s + s_0)(s + s_1)(s + s_2)}$$

where $s_0 = 6.782$, $s_1 = s_2^* = 0.609 + j0.607$. The method of partial fractions then gives

$$I_2(s) = \left[\frac{-10s_0}{(s_1 - s_0)(s_2 - s_0)}\right]\frac{1}{s + s_0} + \left[\frac{-10s_1}{(s_0 - s_1)(s_2 - s_1)}\right]\frac{1}{s + s_1} + \left[\frac{-10s_2}{(s_0 - s_2)(s_1 - s_2)}\right]\frac{1}{s + s_2}$$

$$= \frac{-1.763}{s + 6.782} + \left\{\frac{0.882 - j0.726}{s + s_1} + (\text{complex conjugate})\right\}$$

Inverting,

$$i_2(t) = -1.763\,e^{-6.782t} + \{(0.882 - j0.726)e^{-(0.609+j0.607)t} + \text{(complex conjugate)}\}$$

$$= -1.763\,e^{-6.782t} + \{1.142\,e^{-0.609t}e^{-j(0.607t+39.5°)} + \text{(complex conjugate)}\}$$

$$= -1.763e^{-6.782t} + 2.284e^{-0.609t}\cos(0.607t + 39.5°) \quad \text{(A)}$$

Supplementary Problems

6.14 Show graphically that *any* waveform may be approximated, as closely as desired, by a superposition of (shifted) step functions.

    Ans. Figure 6-12 shows an arbitrary waveform approximated to the desired precision by square pulses. Each square pulse may, in turn, be expressed as the difference of two step functions (as in Fig. 6-4).

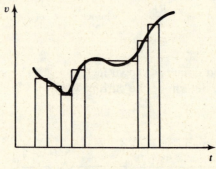

Fig. 6-12

6.15 A 20-μF capacitor has initial voltage 10 V. The current through the capacitor is expressed in the transform domain by

$$I_C(s) = \frac{30}{s^2 + 3000s + (2 \times 10^6)}$$

    Find the voltage across the capacitor, $v_C(t)$.   *Ans.* $10.75 - 1.5e^{-1000t} + 0.75e^{-2000t}$ (V)

6.16 For the circuit shown in Fig. 4-10 (Chapter 4): $I = 5$ A, $R_1 = R_2 = 1000\ \Omega$, $C = 500\ \mu$F, and $L = 5$ mH. The switch is opened at $t = 0$. Find the transformed current in the inductive branch.

    Ans. $I_1(s) = \dfrac{10^6(s+2)}{s[s^2 + (4 \times 10^5)s + (4 \times 10^5)]}$

6.17 Given $I = 2$ A, $R_1 = R_2 = 10\ \Omega$, $L = 0.5$ mH, and $C = 10\ \mu$F in the circuit of Fig. 4-10 (Chapter 4). The switch is opened at $t = 0$. Find (a) an expression for $V(s)$, the transformed voltage across the circuit, and (b) the steady-state value of v.

    Ans. (a) $\dfrac{20[s^2 + (3 \times 10^4)s + (2 \times 10^8)]}{s[s^2 + (4 \times 10^4)s + (2 \times 10^8)]}$; (b) 20 V

6.18 The following are the values of the elements of the circuit of Fig. 4-10 (Chapter 4): $I = 1.0$ A, $L = 100$ mH, $C = (160/9)\ \mu$F, $R_1 = 90\ \Omega$, and $R_2 = 0\ \Omega$. The switch is opened at $t = 0$. Determine (a) the current in the inductive branch, $i_1(t)$; and (b) the voltage, $v(t)$, across the circuit.

    Ans. (a) $i_1(t) = 1 + 1.25\,e^{-450t}\sin(600t - 126.9°)$ (A); (b) $v(t) = 90 + 93.75\,e^{-450t}\sin(600t - 73.8°)$ (V)

6.19 Repeat Problem 6.18 for the case when $R_2 = 60\ \Omega$, other values remaining unchanged.

    Ans. (a) $i_1(t) = 1 - (1 + 150t)e^{-750t}$ (A); (b) $v(t) = 30[3 - (1 + 75t)e^{-750t}]$ (V)

6.20 Refer to the bridge circuit of Fig. 5-21 (Chapter 5). Arms 1 and 4 are pure capacitances, C, and arms 2 and 3 are pure inductances, L. The detector D is a pure resistance, R. The values of R, L, and C are such that $R = \sqrt{L/C}$. Instead of the ac source shown in Fig. 5-21, let a dc voltage V be suddenly applied. Determine the voltage across the detector.   *Ans.* $V(1 - 2e^{-Rt/L})$

6.21 Solve Problems 4.14, 4.18, 4.19, 4.20, and 4.28 using Laplace transforms.

Chapter 7

Diodes

7.1 INTRODUCTION

The diode is among the oldest and most widely used electronic devices. It may be defined as a near-unilateral conductor for which the state of conductivity is determined by the polarity of its terminal voltage. This chapter will deal mainly with the *semiconductor diode*, formed by the metallurgical junction of a *p*-type and an *n*-type semiconductor material.

7.2 TERMINAL CHARACTERISTICS OF SEMICONDUCTOR DIODES

The schematic symbol of the common or *rectifier diode* is shown in Fig. 7-1. The two terminals have been labeled *anode* (*p*-type) and *cathode* (*n*-type), which makes understandable the choice of *diode* as the name for this device. When terminal voltage is positive ($v_D \geq 0$), the diode is said to be *forward-biased*, or "on"; the positive current that flows ($i_D \geq 0$) is called *forward current*. For a negative value of v_D, the diode is *reverse-biased*, or "off"; the corresponding small negative current is referred to as *reverse current*.

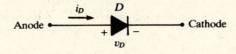

Fig. 7-1

Use of the Fermi-Dirac probability function to predict charge neutralization gives the following equation for the diode current:

$$i_D = I_o(e^{v_D/V_T} - 1) \quad \text{(A)} \tag{7.1}$$

where
$$V_T \equiv kT/q \quad \text{(V)}$$
$$v_D \equiv \text{diode terminal voltage, V}$$
$$I_o \equiv \text{saturation current, A}$$
$$T \equiv \text{absolute temperature, K}$$
$$k \equiv \text{Boltzmann's constant } (1.38 \times 10^{-23} \text{ J/K})$$
$$q \equiv \text{charge of an electron } (1.6 \times 10^{-19} \text{ C})$$

While (*7.1*) serves as a useful model of the junction diode insofar as temperature effects and dynamic resistance are concerned, it is seen from Fig. 7-2 to have regions of inaccuracy:

1. Actual forward voltage drop is greater than predicted (due to ohmic resistance of metal contacts and semiconductor material).
2. Actual reverse current at small negative values of v_D is greater than predicted (due to leakage current (I_s) along the surface of the semiconductor material).
3. Actual reverse current increases to significantly larger values than predicted for large negative values of v_D (due to a complex phenomenon called *avalanche breakdown*).

All circuits studied in this book will be assumed to operate above reverse breakdown; i.e., in the range $v_D > -V_R$.

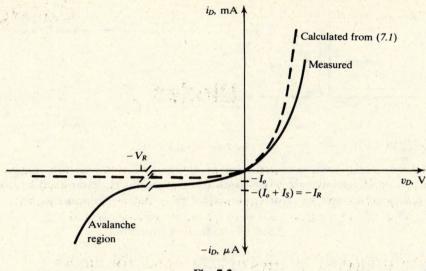

Fig. 7-2

Commercially available diodes are manufactured by proper *p*- and *n*-doping (impurity addition) of either intrinsic Si or Ge. A semiconductor diode processed with Si base material has terminal characteristics distinct from those of a Ge device; a comparison is made in Fig. 7-3. For a reverse bias $(-V_R < v_D < -0.1$ V), either diode will exhibit a near-constant reverse current I_R. Typically, $1 < I_R < 500$ μA for Ge, while $10^{-3} < I_R < 1$ μA for Si. For a forward bias, the onset of low-resistance conduction is between 0.2 and 0.3 V for Ge, while 0.6 to 0.7 V is required for a Si diode.

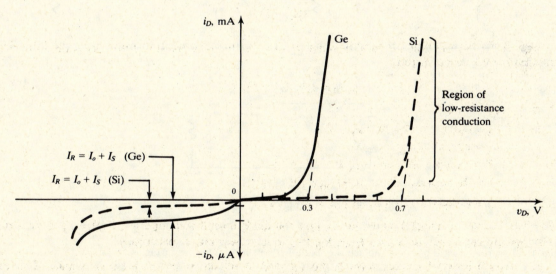

Fig. 7-3

7.3 CIRCUIT ANALYSIS FOR AN IDEAL DIODE

The *perfect* or *ideal diode* would be a two-state device that exhibits zero impedance when forward-biased and infinite impedance when reverse-biased (Fig. 7-4). Note that since either current or voltage is zero at any instant, no power is dissipated by an ideal diode.

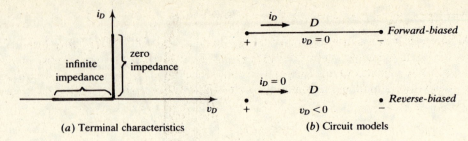

(a) Terminal characteristics   (b) Circuit models

Fig. 7-4. Ideal Diode

In many circuit applications where forward voltage drops and reverse currents are small, sufficiently accurate answers result when the actual diode is modeled as ideal.

Ideal Diode Analysis Procedure

Step 1.   Assume forward bias and replace the ideal diode with a short circuit.

Step 2.   Evaluate diode current i_D by any linear circuit analysis technique.

Step 3.   If $i_D \geq 0$, the analysis is valid, the diode is actually forward-biased, and Step 4 is omitted.

Step 4.   If $i_D < 0$, the analysis is invalid. Replace the diode with an open circuit, forcing $i_D = 0$, and solve for desired circuit quantities by any method of linear circuit analysis. Voltage v_D will be found to have a negative value.

Example 7.1   Find voltage v_L in the circuit of Fig. 7-5(a).

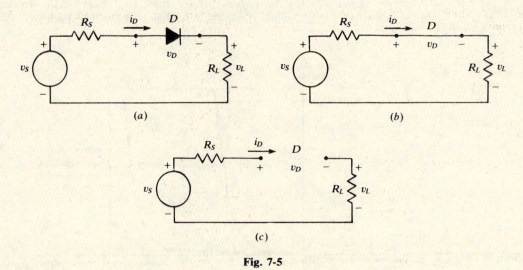

Fig. 7-5

Step 1.   Assume forward bias and replace the diode with a short circuit, as in Fig. 7-5(b).

Step 2.   By Ohm's law,

$$i_D = \frac{v_S}{R_S + R_L}$$

Step 3.   If $v_S \geq 0$, then $i_D \geq 0$ and

$$v_L = i_D R_L = \frac{R_L}{R_S + R_L} v_S$$

Step 4.   If $v_S < 0$, then $i_D < 0$, and the result of Step 3 is invalid. The diode must be replaced by an open circuit, as illustrated in Fig. 7-5(c), and all analysis performed again. The result is $v_L = i_D R_L = 0$.

7.4 GRAPHICAL CIRCUIT ANALYSIS

A graphical solution necessarily assumes that the diode is resistive, and therefore instantaneously characterized by its i_D-versus-v_D curve. The balance of the network under study must be linear, so that a Thévenin's equivalent looking back into the network from the terminals of the diode exists (Fig. 7-6). Thus, the two simultaneous equations to be solved graphically for i_D and v_D are the diode characteristic

$$i_D = f_1(v_D) \tag{7.2}$$

and the *load line*

$$i_D = f_2(v_D) \equiv -\frac{1}{R_{Th}} v_D + \frac{v_{Th}}{R_{Th}} \tag{7.3}$$

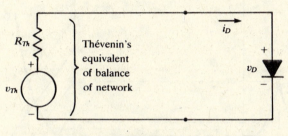

Fig. 7-6

Example 7.2 If all sources in the original linear portion of a network are dc, v_{Th} is also a dc source. Figure 7-7(a) represents the reduction of such a network; determine i_D and v_D.

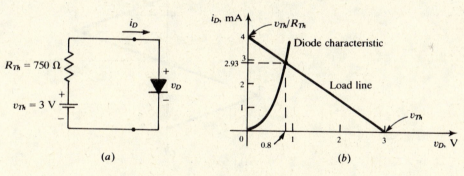

(a)   (b)

Fig. 7-7

Equation (7.3) for the Thévenin's equivalent network is superimposed on the diode characteristic curve in Fig. 7-7(b). The desired solution, $i_D = 2.93$ mA and $v_D = 0.8$ V, is given by the point of intersection of the two plots.

Example 7.3 If all sources in the original linear portion of a network are time-varying, then v_{Th} is also a time-varying source. The reduction of such a network, Fig. 7-8(a), has a Thévenin's source that is a 2-V-peak triangular wave. Find i_D and v_D for this network.

No longer is there a unique value of i_D that satisfies the simultaneous equations; rather, there exists a value of i_D corresponding to each value that v_{Th} takes on as time progresses. An acceptable solution for i_D may be found by considering a finite number of values of v_{Th} over a cycle. Since v_{Th} is repetitive, i_D will be repetitive (with the same period) and only one cycle need be considered.

In Fig. 7-8(b), a scaled plot of v_{Th} is laid out with the v_{Th}-axis parallel to the v_D-axis of the diode characteristic curve. Select a point on the v_{Th} plot, such as $v_{Th} = 0.5$ V at $t = t_1$. Consider time stopped at $t = t_1$ and construct a load line on the diode characteristic, intersecting the v_D-axis at $v_{Th} = 0.5$ V and the i_D-axis at $v_{Th}/R_{Th} = 0.5/50 = 10$ mA. Determine the value of i_D corresponding to $t = t_1$, and plot the point (t_1, i_D) in the

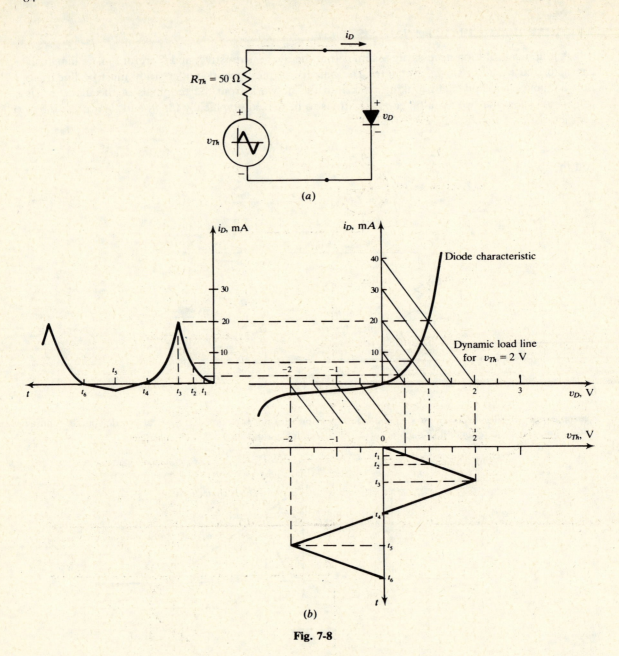

Fig. 7-8

coordinate system constructed to the left of the diode characteristic curve. Let time progress to some new value, $t = t_2$, and repeat the entire process. Continue until one cycle of v_{Th} is complete. Since the load line is continually changing, it is referred to as a *dynamic load line*. The solution i_D differs drastically in form from v_{Th} because of the nonlinearity of the diode.

Example 7.4 If both dc and time-varying sources are present in the original linear portion of a network, then v_{Th} is a series combination of a dc and a time-varying source. Suppose that the Thévenin's source can be modeled by a 0.7-V battery and 0.1-V-peak sinusoidal source, as indicated in Fig. 7-9(a). Find i_D and v_D for the network.

A scaled plot of v_{Th} is laid out with the v_{Th}-axis parallel to the v_D-axis of the diode characteristic curve. Consider v_{th}, the ac component of v_{Th}, momentarily at zero ($t = 0$) and plot a load line for this instant on the diode characteristic. This particular load line is called the *dc load line*, and its intersection with the diode characteristic curve is called the *quiescent point* or *Q-point*. The solutions for i_D and v_D at the Q-point are respectively labeled I_{DQ} and V_{DQ} in Fig. 7-9(b).

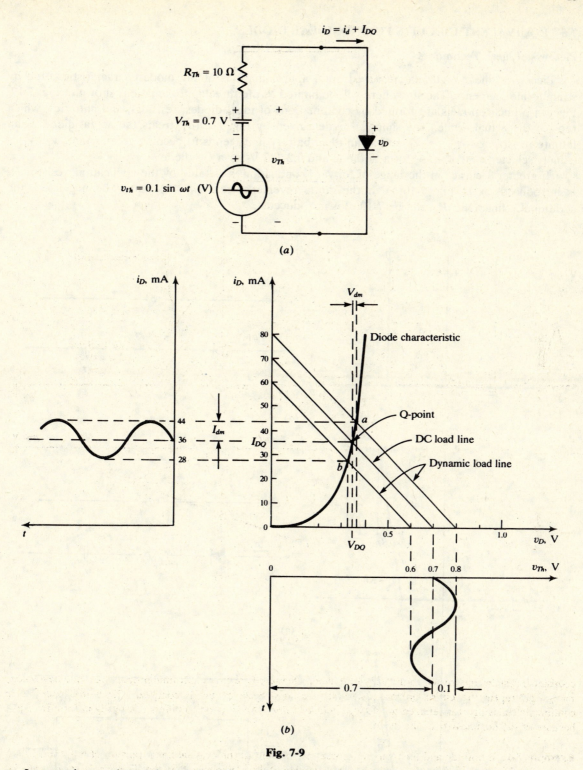

Fig. 7-9

In general, a number of dynamic load lines are needed to complete the analysis for i_D over a cycle of v_{th}. However, for the network under study, only dynamic load lines for the maximum and minimum values of v_{th} are required. The reason is that the diode characteristic curve is almost a straight line from a to b through Q in Fig. 7-9(b), so that negligible distortion of i_d, the ac component of i_D, will occur. Without distortion, i_d will be of the same form as v_{th}, i.e. sinusoidal, and can easily be sketched once the extremes of variation have been determined. The graphical solution for i_D is thus

$$i_D = I_{DQ} + i_d = I_{DQ} + I_{dm} \sin \omega t = 36 + 8 \sin \omega t \quad \text{(mA)}$$

7.5   EQUIVALENT CIRCUITS FOR NONIDEAL DIODES

Piecewise Linear Techniques

Piecewise linear methods are based upon approximations of the diode characteristic curve by straight-line segments. The study here will be limited to the three approximations shown in Fig. 7-10. Piecewise linear models are formed by combinations of ideal diodes, resistors, and batteries, which replace the actual diode. The simplest model, given by Fig. 7-10(a), treats the actual diode as an infinite resistance for $v_D < V_F$ and as an ideal battery if v_D tends to become greater than V_F. V_F is usually selected as 0.6–0.7 V for a Si diode and 0.2–0.3 V for a Ge diode.

If greater accuracy in the range of forward conduction is dictated by the application, resistor R_F is introduced, as in Fig. 7-10(b). If the diode reverse current ($i_D < 0$) cannot be neglected, the additional refinement, R_R, of Fig. 7-10(c) is introduced.

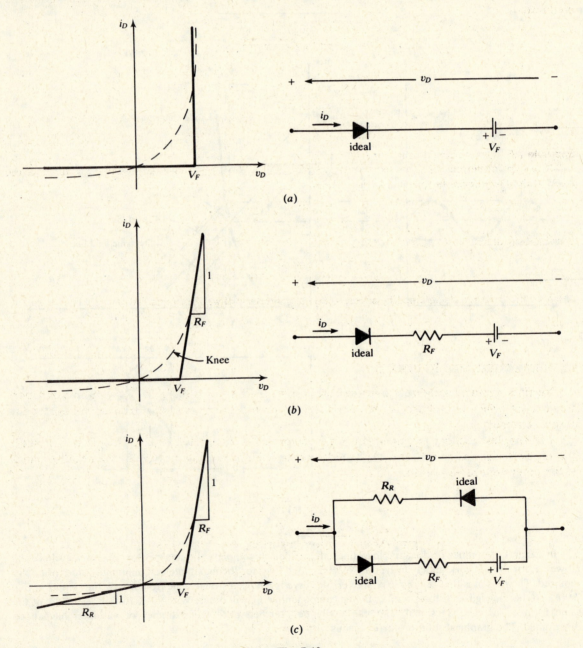

Fig. 7-10

Small-Signal Techniques

Figure 7-9 illustrates a diode circuit to which small-signal analysis may be successfully applied. Here, the amplitude of the ac signal v_{th} is so small that the curvature of the diode characteristic may be neglected over the entire range of operation, from b to a. Thus the diode voltage and current may each be written as the sum of a dc signal and an *undistorted* ac signal. Furthermore, the ratio of the diode ac voltage, v_d, to the diode ac current, i_d, will be constant:

$$\frac{v_d}{i_d} = \frac{2V_{dm}}{2I_{dm}} = \frac{v_D|_a - v_D|_b}{i_D|_a - i_D|_b} = \frac{\Delta v_D}{\Delta i_D}\bigg|_Q = \frac{dv_D}{di_D}\bigg|_Q \equiv r_d \qquad (7.4)$$

where r_d is known as the *dynamic resistance* of the diode. It follows that the ac components may be determined by analysis of an equivalent resistive circuit, Fig. 7-11. The dc or quiescent components must generally be determined by graphical methods, since, overall, the diode characteristic is nonlinear.

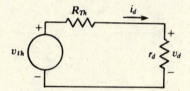

Fig. 7-11

Example 7.5 For the circuit of Fig. 7-9, determine i_D.

The Q-point current, I_{DQ}, has been determined as 36 mA (Example 7.4). The dynamic resistance of the diode at the Q-point can be evaluated graphically.

$$r_d = \frac{\Delta v_D}{\Delta i_D} = \frac{0.37 - 0.33}{0.044 - 0.028} = 2.5 \ \Omega$$

The small-signal circuit of Fig. 7-11 can be analyzed to find i_d:

$$i_d = \frac{v_{th}}{R_{Th} + r_d} = \frac{0.1 \sin \omega t}{10 + 2.5} = 0.008 \sin \omega t \quad \text{(A)}$$

The total diode current is obtained by superposition and checks well with Example 7.4.

$$i_D = I_{DQ} + i_d = 36 + 8 \sin \omega t \quad \text{(mA)}$$

7.6 RECTIFIER APPLICATIONS

Rectifier circuits capitalize upon the near-unilateral conduction of the diode to form a two-port network for which ac voltage is impressed across the input terminals to give a dc voltage at the output terminals.

The simplest rectifier circuit uses a single diode and is illustrated in Fig. 7-12. It is commonly called a *half-wave rectifier*, because the diode conducts either over the positive or the negative halves of the input voltage waveform.

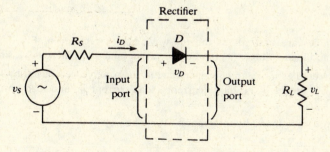

Fig. 7-12

Example 7.6 In Fig. 7-12, $v_S = V_m \sin \omega t$, and the diode is ideal. Calculate the average value of v_L.

Only one cycle of v_S need be considered. For the positive half-cycle, $i_D > 0$ and

$$v_L = \frac{R_L}{R_L + R_S}(V_m \sin \omega t) \equiv V_{Lm} \sin \omega t$$

For the negative half-cycle, the diode is reverse-biased, $i_D = 0$, and $v_L = 0$. Hence,

$$V_{Lavg} = \frac{1}{2\pi}\int_0^{2\pi} v_L(\omega t)\, d(\omega t) = \frac{1}{2\pi}\int_0^{\pi} V_{Lm}\sin \omega t\, d(\omega t) = \frac{V_{Lm}}{\pi}$$

Although the half-wave rectifier gives a dc output, current flows through R_L for only half the time. Furthermore, the average value of v_L is only $1/\pi = 0.318$ times the peak value. A *full-wave rectifier* can offer improvement when needed, and is possible with only two diodes and a center-tapped transformer, as illustrated in Fig. 7-13(a).

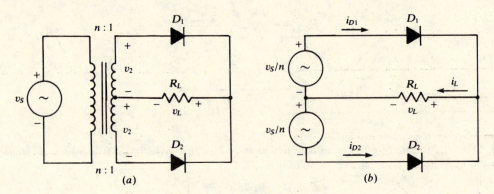

Fig. 7-13

Example 7.7 Calculate v_L for the full-wave rectifier circuit of Fig. 7-13(a), treating the transformer and diodes as ideal.

The two voltages in Fig. 7-13(a) labeled v_2 are identical in magnitude and phase. The ideal transformer and the voltage source v_S can be replaced by two identical voltage sources, as indicated in Fig. 7-13(b), without alteration of the electrical performance of the balance of the network. When v_S/n is positive, D_1 is forward-biased and conducts, but D_2 is reverse-biased and blocks. Conversely, when v_S/n is negative, D_2 conducts and D_1 blocks. In short,

$$i_{D1} = \begin{cases} \dfrac{v_S/n}{R_L} & (v_S/n \geq 0) \\ 0 & (v_S/n < 0 \end{cases} \qquad i_{D2} = \begin{cases} 0 & (v_S/n > 0) \\ -\dfrac{v_S/n}{R_L} & (v_S/n \leq 0) \end{cases}$$

By KCL,

$$i_L = i_{D1} + i_{D2} = \frac{|v_S/n|}{R_L}$$

and so $v_L = R_L i_L = |v_S/n|$.

Example 7.8 Four diodes are utilized for the full-wave bridge rectifier of Fig. 7-14(a). (a) Sketch diode currents, load current, and load voltage. (b) Find the average value of v_L if $v_S = V_m \sin \omega t$.

(a) If the diodes are treated as ideal, an equivalent circuit for positive v_S is given by Fig. 7-14(b); for negative v_S, by Fig. 7-14(c). Sketches of v_S, diode currents, load current i_L, and load voltage v_L are given in Fig. 7-14(d), where $v_L = |v_S|$.

(b) The average value of v_L is found by integration over a half-cycle of v_S:

$$V_{Lavg} = \frac{1}{\pi}\int_0^{\pi} v_L(\omega t)\, d(\omega t) = \frac{1}{\pi}\int_0^{\pi} V_m \sin \omega t\, d(\omega t) = \frac{2}{\pi}V_m \approx 0.636\, V_m$$

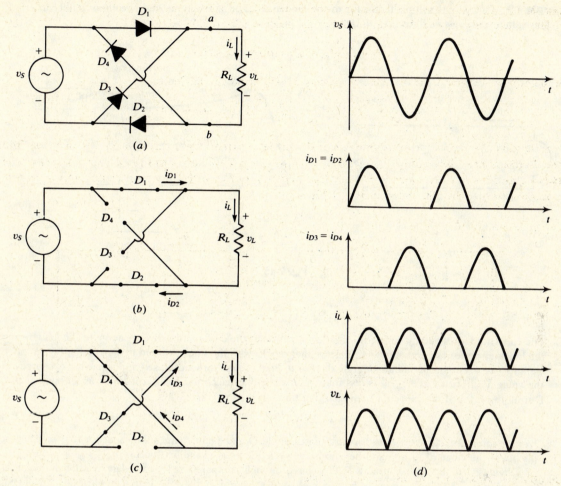

Fig. 7-14

When rectifiers are used as dc power supplies for electronic equipment, it is desirable that the average value of output voltage remain nearly constant as load varies. The degree of constancy is measured by

$$voltage\ regulation \equiv \frac{(\text{no-load }V_{L\text{avg}}) - (\text{full-load }V_{L\text{avg}})}{(\text{full-load }V_{L\text{avg}})} \qquad (7.5)$$

7.7 WAVEFORM FILTERING

The output of a rectifier alone does not usually suffice as a power supply due to its instantaneous variation. The situation is improved by placing between the rectifier and the load a *filter*, which acts to suppress the harmonics from the rectified waveform and to preserve the dc component. A measure of goodness for rectified waveforms is the *ripple factor*,

$$F_r \equiv \frac{\text{maximum variation in output voltage}}{\text{average value of output voltage}} = \frac{\Delta v_L}{V_{L\text{avg}}} \qquad (7.6)$$

A small value, $F_r \leq 0.05$, is usually attainable and practical.

Example 7.9 Calculate the ripple factor for the half-wave rectifier of Example 7.6 (*a*) without a filter inserted; (*b*) with a shunt capacitor as filter [Fig. 7-15(*a*)].

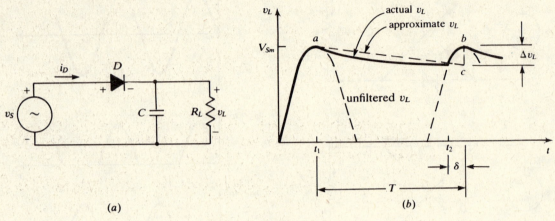

Fig. 7-15

(*a*)
$$F_r = \frac{\Delta v_L}{V_{Lavg}} = \frac{V_{Lm}}{V_{Lm}/\pi} = \pi \approx 3.14$$

(*b*) The capacitor stores energy during the conduction cycle of the diode and delivers energy to the load during the blocking cycle. The actual load voltage v_L that results with the filter inserted is sketched in Fig. 7-15(*b*), assuming $v_S = V_{Sm} \sin \omega t$ and an ideal diode. For $0 < t \le t_1$, D is forward-biased and capacitor C charges to the value V_{Sm}. For $t_1 < t \le t_2$, v_S is less than v_L, reverse-biasing D and causing it to act as an open circuit. During this interval the capacitor is discharging through the load R_L, giving

$$v_L = V_{Sm}e^{-(t-t_1)/R_L C} \qquad (t_1 < t \le t_2) \qquad (7.7)$$

Over the interval $t_2 < t \le t_2 + \delta$, v_S forward-biases diode D and again charges the capacitor to V_{Sm}. Then v_S falls below the value of v_L and another discharge cycle identical to the first occurs.

Obviously, if the time constant $R_L C$ is large enough compared to T to result in a decay like that indicated in Fig. 7-15(*b*), a major reduction in Δv_L and a major increase in V_{Lavg} will have been achieved, relative to the unfiltered rectifier. Introduction of two quite reasonable approximations leads to simple formulas for Δv_L and V_{Lavg}, and hence for F_r, that are sufficiently accurate for design and analysis work:

1. If Δv_L is to be small, then $\delta \to 0$ in Fig. 7-15(*b*) and $t_2 - t_1 \approx T$.

2. If Δv_L is small enough, then (7.7) can be represented over the interval $t_1 < t \le t_2$ by a straight line segment having a slope of magnitude $V_{Sm}/R_L C$.

The dashed line labeled "approximate v_L" in Fig. 7-15(*b*) is an implementation of these two approximations. From right triangle *abc*,

$$\frac{\Delta v_L}{T} = \frac{V_{Sm}}{R_L C} \qquad \text{or} \qquad \Delta v_L = \frac{V_{Sm}}{fR_L C}$$

where f is the frequency of v_S. Since, under this approximation,

$$V_{Lavg} = V_{Sm} - \frac{1}{2}\Delta v_L$$

and $R_L C/T = fR_L C$ is presumed large,

$$F_r = \frac{\Delta v_L}{V_{Lavg}} = \frac{2}{2fR_L C - 1} \approx \frac{1}{fR_L C} \qquad (7.8)$$

If a shunt capacitor filter is added to the full-wave rectifier of Fig. 7-14(*a*), then a similar approximate analysis (over a half-period of v_S) yields for the ripple voltage and ripple factor

$$\Delta v_L = \frac{V_{Sm}}{2fR_L C} \qquad F_r = \frac{2}{4fR_L C - 1} \approx \frac{1}{2fR_L C} \qquad (7.9)$$

It is seen from (7.8) and (7.9) that, for a given value of fR_LC, the full-wave rectifier output is twice as good as that of the half-wave rectifier.

7.8 CLIPPING AND CLAMPING OPERATIONS

Diode clipping circuits separate an input signal at a particular dc level and pass to the output, without distortion, the desired upper or lower portion of the original waveform. They are used to eliminate amplitude noise or to fabricate new waveforms from an existing signal.

Example 7.10 Figure 7-16(a) illustrates a *positive clipping* circuit, which removes any portion of the input signal v_i that is greater than V_b and passes as the output signal v_o any portion of v_i that is less than V_b. It is seen that v_D is negative for $v_i < V_b$, causing the ideal diode to act as an open circuit. With no path for current to flow through R, the value of v_i appears at the output terminals as v_o. However, if $v_i \geq V_b$, the diode conducts, acting as a short circuit and forcing $v_o = V_b$. Figure 7-16(b), the *transfer graph* or *transfer characteristic*, shows the relationship between the input, here taken as $v_i = 2V_b \sin \omega t$, and the output voltages.

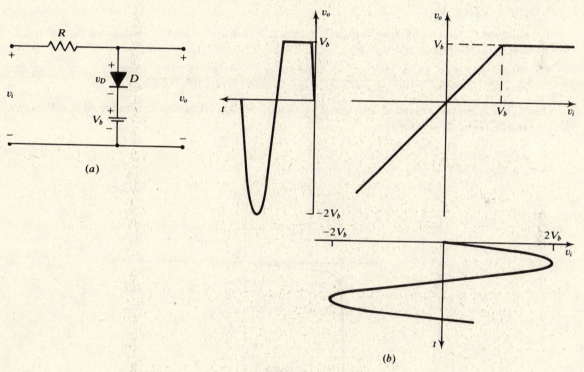

(a)

(b)

Fig. 7-16

Clamping is a process of setting the positive or the negative peaks of an input ac waveform at a specific dc level, regardless of any variation in those peaks. As a result, the output signal has a different average (dc) value from the input signal.

Example 7.11 An ideal clamping circuit is illustrated in Fig. 7-17(b) with a triangular ac waveform as v_i. Assuming the capacitor C to be initially uncharged, ideal diode D is forward-biased for $0 < t \leq T/4$, acting as a short circuit while the capacitor charges to $v_C = V_p$. At $t = T/4$, D open-circuits, breaking the only possible discharge path for the capacitor. Thus, the value $v_C = V_p$ is preserved; and since v_i can never exceed V_p, D remains reverse-biased for all $t > T/4$, giving $v_o = v_D = v_i - V_p$. The function v_o is sketched in Fig. 7-17(c); all positive peaks are clamped at zero, and the average value has been shifted from 0 to $-V_p$.

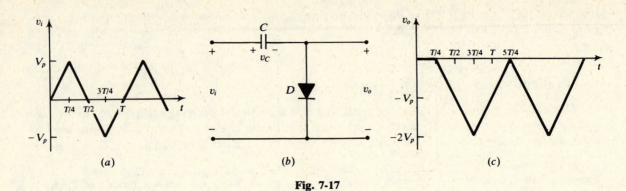

Fig. 7-17

7.9 ZENER DIODE

The *Zener diode*, symbolized in Fig. 7-18(a), finds primary usage as a voltage regulator or reference. The forward conduction characteristic of a Zener is the same as that of a rectifier diode; however, it is usually operated with reverse bias, where its characteristic is radically different. From Fig. 7-18(b), it is seen that:

1. Reverse voltage breakdown is rather sharp and is controlled by the manufacturing process to a reasonably predictable value.

2. When a Zener is in reverse breakdown, its voltage remains extremely close to the breakdown value, while current varies from rated (I_Z) to 10% of rated or less.

The Zener regulator should be designed so that $i_Z \geq 0.1 I_Z$ to assure constancy of v_Z; Problem 7.21 illustrates design technique.

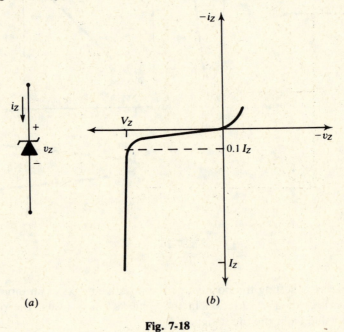

Fig. 7-18

7.10 VACUUM DIODES

The symbol of the *vacuum diode* is shown in Fig. 7-19(a). Conduction is by electrons that escape the *cathode* and travel through the internal vacuum space to be collected by the *anode*. The *filament* is an electric resistance heater used to elevate the cathode temperature to a level where thermionic emission occurs.

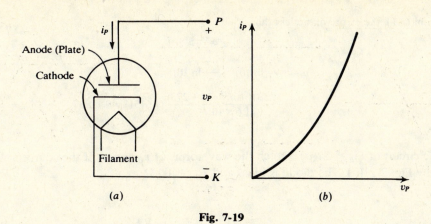

Fig. 7-19

A typical i_P-v_P characteristic curve (the *plate characteristic*) is depicted in Fig. 7-19(*b*). This characteristic is described by the *Childs-Langmuir three-halves-power law*:

$$i_P = \kappa v_P^{3/2} \tag{7.10}$$

where κ is the *perveance* (a constant that depends upon the mechanical design of the tube).

Like the semiconductor diode, the vacuum diode is a near-unilateral conductor. However, the forward voltage drop of the vacuum diode is typically several volts, rendering the ideal-diode approximation unjustifiable. Quite commonly, the vacuum diode is modeled in forward conduction by a resistance, $R_P = V_P/I_P$, obtained as the slope of a straight-line approximation of the plate characteristic (Fig. 7-20). A reverse-biased vacuum diode acts like an open circuit.

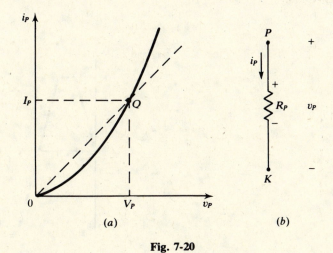

Fig. 7-20

Solved Problems

7.1   At a junction temperature of 25 °C, over what range of forward voltage drop (v_D) can (*7.1*) be approximated as

$$i_D \approx I_o e^{v_D/V_T}$$

with less than 1% error?

From (7.1), the error will be less than 1% if

$$e^{v_D/V_T} > 101$$

or
$$v_D > V_T \ln 101 = \frac{kT}{q} \ln 101$$

$$= \frac{(1.38 \times 10^{-23})(25 + 273)}{1.6 \times 10^{-19}} (4.6151) = 0.1186 \text{ V}$$

7.2 For the circuit of Fig. 7-21(a), sketch the waveforms of v_L and v_D if the source voltage v_S is as given by Fig. 7-21(b). The diode is ideal and $R_L = 100 \ \Omega$.

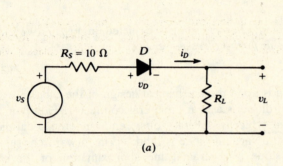

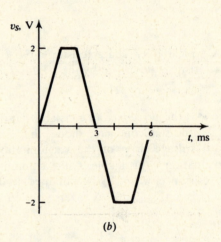

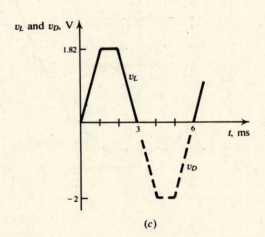

Fig. 7-21

If $v_S \geq 0$, D conducts, so that $v_D = 0$ and

$$v_L = \frac{R_L}{R_L + R_S} v_S = \frac{100}{100 + 10} v_S = 0.909 \, v_S$$

If $v_S < 0$, D blocks, so that

$$v_D = v_S \qquad \text{and} \qquad v_L = 0$$

The sketches of v_D and v_L are shown in Fig. 7-21(c).

7.3   For the circuit of Fig. 7-22, find i_{D1} and i_{D2} if D_1 and D_2 are ideal diodes.

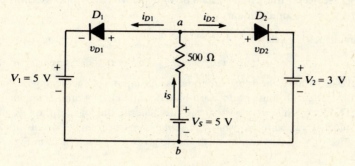

Fig. 7-22

Because of the polarities of D_1 and D_2, it is necessary that $i_S \geq 0$. Thus, $v_{ab} \leq V_S = V_1$. But $v_{D1} = v_{ab} - V_1$; therefore, $v_D \leq 0$ and so $i_{D1} \equiv 0$, regardless of conditions in the right-hand loop. It follows that $i_{D2} = i_S$. Using the analysis procedure of Section 7.3, assume D_2 forward-biased and replace it with a short circuit. By KVL,

$$i_{D2} = \frac{V_S - V_2}{500} = \frac{5-3}{500} = 4 \text{ mA}$$

Since $i_{D2} \geq 0$, D_2 is in fact forward-biased and the analysis is valid.

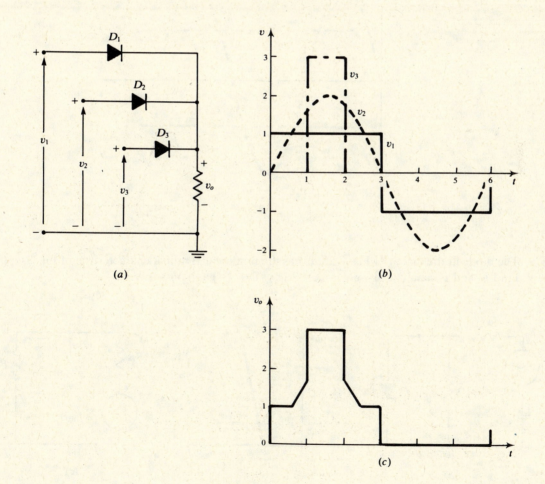

Fig. 7-23

7.4   The logic OR gate can be utilized to fabricate composite waveforms. Sketch the output (v_o) of the gate of Fig. 7-23(a) if the three signals of Fig. 7-23(b) are impressed on the input terminals. The diodes are to be assumed ideal.

For this circuit, KVL gives

$$v_1 - v_2 = v_{D1} - v_{D2} \qquad v_1 - v_3 = v_{D1} - v_{D3}$$

i.e., the diode voltages have the same ordering as the input voltages. Suppose that v_1 is positive and exceeds v_2 and v_3. Then D_1 must be forward-biased, with $v_{D1} = 0$ and, consequently, $v_{D2} < 0$, $v_{D3} < 0$. Hence D_2 and D_3 block, while v_1 is passed as v_o. In general, the logic of the OR gate is that the largest positive input signal is passed as v_o, while the remainder of the input signals are blocked. If all input signals are negative, $v_o = 0$. Application of this logic gives the sketch of v_o in Fig. 7-23(c).

7.5   Reverse the diodes in Fig. 7-23(a) and sketch the output voltage v_o, if $v_1 = 10 \sin \omega t$ (V), $v_2 = +5$ V, and $v_3 = -5$ V.

As in Problem 7.4, we determine that v_o equals the most negative of the three input signals. When all signals are positive or zero, $v_o = 0$. See Fig. 7-24.

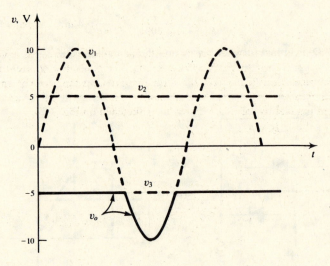

Fig. 7-24

7.6   The diode in the circuit of Fig. 7-25(a) has the nonlinear terminal characteristic of Fig. 7-25(b). Find i_D and v_D analytically, given $v_S = 0.1 \cos \omega t$ (V) and $V_b = 2$ V.

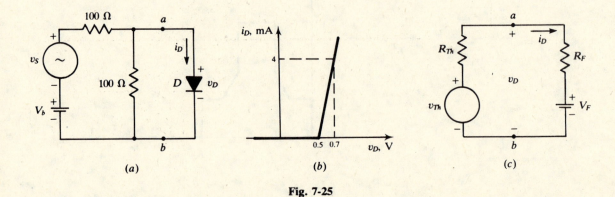

(a)   (b)   (c)

Fig. 7-25

The Thévenin's equivalent circuit looking to the left through terminals ab in Fig. 7-25(a) is found.

$$V_{Th} = \frac{100}{200}(2 + 0.1 \cos \omega t) = 1 + 0.05 \cos \omega t \quad (V) \qquad R_{Th} = \frac{(100)^2}{200} = 50 \ \Omega$$

The diode can be modeled as in Fig. 7-10(b), with $V_F = 0.5$ V and

$$R_F = \frac{0.7 - 0.5}{0.004} = 50 \ \Omega$$

The Thévenin's equivalent and the diode model are now combined in Fig. 7-25(c) to form the circuit for analytical solution. By Ohm's law,

$$i_D = \frac{V_{Th} - V_F}{R_{Th} + R_F} = \frac{(1 + 0.05 \cos \omega t) - 0.5}{50 + 50} = 5 + 0.5 \cos \omega t \quad (mA)$$

$$v_D = V_F + R_F i_D = 0.5 + 50(0.005 + 0.0005 \cos \omega t) = 0.75 + 0.025 \cos \omega t \quad (V)$$

7.7　　Solve Problem 7.6 graphically to find i_D.

The Thévenin's equivalent has already been determined in Problem 7.6. By (7.3), the dc load line is given by

$$i_D = \frac{V_{Th}}{R_{Th}} - \frac{v_D}{R_{Th}} = \frac{1}{50} - \frac{v_D}{50} = 20 - 20v_D \quad (mA) \tag{1}$$

In Fig. 7-26, (1) has been superimposed on the diode characteristic curve replotted from Fig. 7-25(b). As in Example 7.4, equivalent time scales for v_{Th} and i_D are laid out adjacent to the characteristic curve. Since the diode characteristic is linear about the Q-point over the range of operation, dynamic load lines need only be drawn corresponding to the maximum and minimum of v_{Th}. Once these two dynamic load lines are constructed parallel to the dc load line, i_D is sketched on Fig. 7-26.

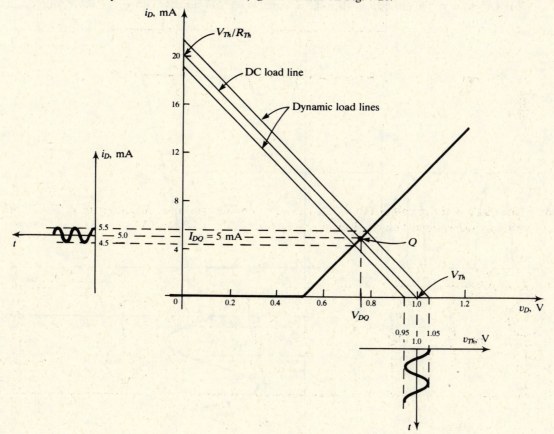

Fig. 7-26

7.8   Use small-signal equivalent-circuit methods (Section 7.5) to find i_D and v_D in Problem 7.6.

The Thévenin's equivalent circuit of Problem 7.6 is still valid. From the intersection of the dc load line with the diode characteristic curve in Fig. 7-26, it is noted that $I_{DQ} = 5$ mA and $V_{DQ} = 0.75$ V. Dynamic resistance is found in accordance with (7.4):

$$r_d = \frac{\Delta v_D}{\Delta i_D} = \frac{0.7 - 0.5}{0.004} = 50 \ \Omega$$

All values are now available for analysis using the small-signal circuit of Fig. 7-11. By Ohm's law,

$$i_d = \frac{v_{th}}{R_{Th} + r_d} = \frac{0.05 \cos \omega t}{50 + 50} = 0.5 \cos \omega t \quad \text{(mA)}$$

$$v_d = r_d i_d = 50(0.0005 \cos \omega t) = 0.025 \cos \omega t \quad \text{(V)}$$

$$i_D = I_{DQ} + i_d = 5 + 0.5 \cos \omega t \quad \text{(mA)}$$

$$v_D = V_{DQ} + v_d = 0.75 + 0.025 \cos \omega t \quad \text{(V)}$$

7.9   A voltage source, $v_S = 0.4 + 0.2 \sin \omega t$ (V), is placed directly across a diode characterized by Fig. 7-25(*b*). The source has no internal impedance and is of proper polarity to forward-bias the diode. (*a*) Sketch the resulting diode current i_D. (*b*) Determine the value of quiescent current I_{DQ}.

(*a*)   With zero resistance between the ideal voltage source and the diode, the dc load line has an infinite slope and $v_D = v_S$. A scaled plot of v_S has been laid out adjacent to the v_D-axis of the diode characteristic in Fig. 7-27. i_D is found by a point-by-point projection of v_S onto the diode characteristic, followed by reflection in the i_D-axis. Notice that i_D is extremely distorted, bearing little resemblance to v_S.

(*b*)   Quiescent conditions obtain when the ac signal is zero. In this case, when $v_S = 0.4$ V, $i_D = I_{DQ} = 0$.

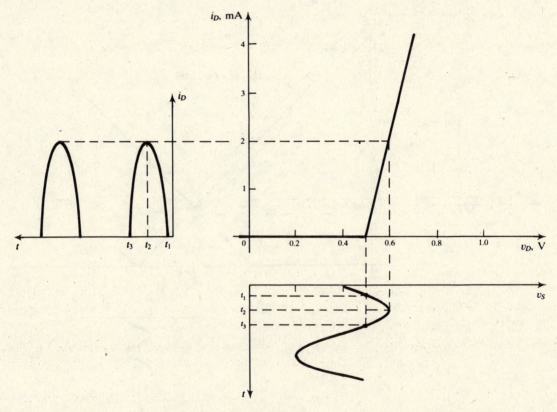

Fig. 7-27

7.10 For the circuit of Fig. 7-28, v_S is a 10-V square wave of period 4 ms, $R = 100\ \Omega$, and $C = 20\ \mu F$. Sketch v_C for the first two cycles of v_S if the capacitor is initially uncharged and the diode is ideal.

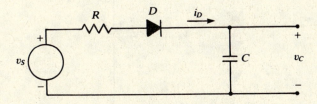

Fig. 7-28

For $0 \le t < 2$ ms, $v_C(t) = v_S(1 - e^{-t/RC}) = 10(1 - e^{-500t})$ (V) (see Example 4.2).
For $2 \le t < 4$ ms, D blocks, and the capacitor voltage remains at

$$v_C(2\ \text{ms}) = 10[1 - e^{-500(0.002)}] = 6.32\ \text{V}$$

For $4 \le t < 6$ ms,

$$v_C(t) = v_S - (v_S - 6.32)e^{-(t-0.004)/RC}$$
$$= 10 - (10 - 6.32)e^{-500(t-0.004)}\quad \text{(V)}$$

For $6 \le t < 8$ ms, D again blocks, and the capacitor voltage remains at

$$v_C(6\ \text{ms}) = 10 - (10 - 6.32)e^{-500(0.002)} = 8.654\ \text{V}$$

The waveforms of v_S and v_C are sketched in Fig. 7-29.

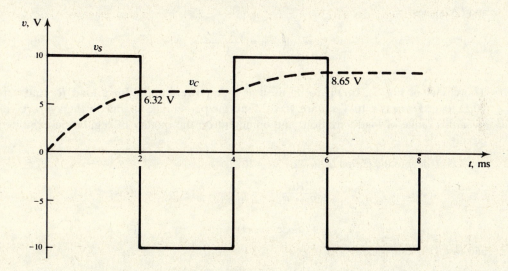

Fig. 7-29

7.11 The circuit of Fig. 7-30(a) is an "inexpensive" voltage regulator. If all diodes are identical, having the characteristic of Fig. 7-25(b), find the regulation of v_o when V_b increases from its nominal value, 4 V, to the value 6 V. Take $R = 2\ k\Omega$.

It has already been determined in Problem 7.6 that each diode can be modeled by a series battery, $V_F = 0.5$ V, and resistor, $R_F = 50\ \Omega$. Combining the diode strings between points a and b and points b and c gives the circuit of Fig. 7-30(b), where

$$V_{F1} = 2V_F = 1\ \text{V} \qquad V_{F2} = 4V_F = 2\ \text{V} \qquad R_{F1} = 2R_F = 100\ \Omega \qquad R_{F2} = 4R_F = 200\ \Omega$$

By KVL,

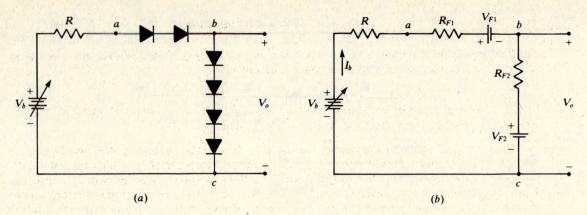

Fig. 7-30

$$I_b = \frac{V_b - V_{F1} - V_{F2}}{R + R_{F1} + R_{F2}}$$

whence

$$V_o = V_{F2} + I_b R_{F2} = V_{F2} + \frac{(V_b - V_{F1} - V_{F2})R_{F2}}{R + R_{F1} + R_{F2}}$$

For $V_{b1} = 4$ V and $V_{b2} = 6$ V,

$$V_{o1} = 2 + \frac{(4 - 1 - 2)(200)}{2000 + 100 + 200} = 2.09 \text{ V} \qquad V_{o2} = 2 + \frac{(6 - 1 - 2)(200)}{2000 + 100 + 200} = 2.26 \text{ V}$$

and (7.5) gives

$$\% \text{ regulation} = \frac{V_{o2} - V_{o1}}{V_{o1}} \times 100\% = 8.1\%$$

7.12   The circuit of Fig. 7-21(a) is to be used as a dc power supply for a load R_L that varies from 10 Ω to 1 kΩ; v_S is a 10-V square wave. Find the percentage change in the average value of v_L over this range of load variation, and comment on the quality of regulation exhibited by this circuit.

Let T denote the period of v_S. For $R_L = 10$ Ω,

$$v_L = \begin{cases} \dfrac{R_L}{R_L + R_S} v_S = \dfrac{10}{10 + 10}(10) = 5 \text{ V} & (0 \le t < T/2) \\[2mm] 0 \ (\textit{diode blocks}) & (T/2 \le t < T) \end{cases}$$

and so

$$V_{L\text{avg}} = \frac{(5)(T/2) + (0)T/2}{T} = 2.5 \text{ V}$$

For $R_L = 1$ kΩ,

$$v_L = \begin{cases} \dfrac{R_L}{R_L + R_S} v_S = \dfrac{1000}{1010}(10) = 9.9 \text{ V} & (0 \le t < T/2) \\[2mm] 0 \ (\textit{diode blocks}) & (T/2 \le t < T) \end{cases}$$

and

$$V_{L\text{avg}} = \frac{(9.9)(T/2) + 0}{T} = 4.95 \text{ V}$$

By (7.5), using $R_L = 10$ Ω as full-load,

$$\% \text{ regulation} = \frac{4.95 - 2.5}{2.5} \times 100\% = 98\%$$

This large value of regulation is prohibitive for most applications. Either another circuit or a filter network would be necessary to make this power supply useful.

7.13 Find the average value of the load current i_L in Fig. 7-14, if $v_S = 10 \sin t$ (V) and $R_L = 2 \text{ k}\Omega$. Assume the diodes are modeled as in Fig. 7-10(a), with $V_F = 0.7$ V.

The equivalent circuit for the positive half-cycles of v_S would be like Fig. 7-14(b), except that the short circuits of D_1 and D_2 would each be replaced by a 0.7-V battery opposing v_S. Similarly, replace D_3 and D_4 in Fig. 7-14(c) for the negative half-cycles of v_S. With the two 0.7-V batteries in the circuit, the appropriate diodes will not conduct until $v_S > 2V_F = 1.4$ V. As a result, the waveform of v_L has regions of constant zero value, as illustrated in Fig. 7-31. Time t_1 for onset of diode conduction can be calculated as follows:

$$10 \sin t_1 = 1.4 \qquad \text{or} \qquad t_1 = \arcsin 0.14 \approx 0.14 \text{ s}$$

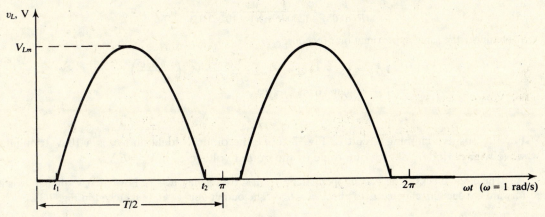

Fig. 7-31

By symmetry, $t_2 = \pi - t_1 = 3.00$ s. The average value of v_L is found by integrating the load voltage over $[t_1, t_2]$ and dividing by $T/2 = \pi$.

$$V_{L\text{avg}} = \frac{1}{\pi} \int_{t_1}^{t_2} (-1.4 + 10 \sin t) \, dt = \frac{1}{\pi} [-1.4t - 10 \cos t]_{0.14}^{3.00} = 5.029 \text{ V}$$

Now Ohm's law gives

$$I_{L\text{avg}} = \frac{V_{L\text{avg}}}{R_L} = \frac{5.029}{2 \times 10^3} = 2.514 \text{ mA}$$

7.14 The circuit of Fig. 7-32 adds a dc level to v_L, although v_S is a pure ac signal. If v_S is a 10-V square wave of period T, $R_L = R_1 = 10 \ \Omega$, and the diode is ideal, find the average value of v_L. Is this circuit a good one to use for adding a dc level to a signal?

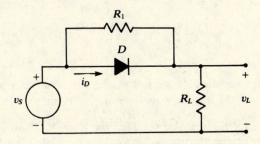

Fig. 7-32

For $v_L > 0$, D is forward-biased and $v_L = v_S = 10$ V; for $v_L < 0$, D is reverse-biased and

$$v_L = \frac{R_L}{R_L + R_1} v_S = \frac{10}{10 + 10}(-10) = -5 \text{ V}$$

Thus

$$V_{Lavg} = \frac{(10)(T/2) + (-5)(T/2)}{T} = 2.5 \text{ V}$$

For many input signals, the resulting output signal would be unsymmetric about the average value; see Problem 7.37.

7.15 Size the filter capacitor in the rectifier circuit of Fig. 7-15 so that the ripple voltage is approximately 5% of the average value of output voltage. The diode is ideal, $R_L = 1$ kΩ, and $v_S = 90 \sin 2000t$ (V). Calculate the average value of v_L for this filter.

With $F_r = 0.05$, (7.8) gives

$$C \approx \frac{1}{fR_L(0.05)} = \frac{1}{(2000/2\pi)(1 \times 10^3)(0.05)} = 62.83 \ \mu\text{F}$$

Then, under the same approximation procedure,

$$V_{Lavg} = V_{Sm} - \tfrac{1}{2}\Delta v_L = V_{Sm} - \frac{V_{Sm}}{2fR_LC} \approx V_{Sm}\left(1 - \frac{0.05}{2}\right)$$
$$= (90)(0.975) = 87.75 \text{ V}$$

7.16 For the positive clipping circuit of Fig. 7-16(a), the diode is ideal and v_i is a 10-V triangular wave with period T. Sketch one cycle of the output voltage v_o if $V_b = 6$ V.

The diode blocks or acts as an open circuit for $v_i < 6$ V, giving $v_o = v_i$. For $v_i \geq 6$ V, the diode is in forward conduction, clipping v_i to effect $v_o = 6$ V. The output voltage v_o is sketched in Fig. 7-33.

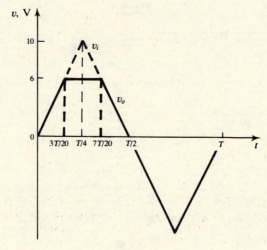

Fig. 7-33

7.17 Draw a transfer characteristic relating v_o to v_i for the positive clipper network of Problem 7.16. Also, sketch a cycle of the output waveform if $v_i = 10 \sin \omega t$ (V).

The diode blocks for $v_i < 6$ V and conducts for $v_i \geq 6$ V. Thus, $v_o = v_i$ for $v_i < 6$ V and $v_o = 6$ V for $v_i \geq 6$ V. The transfer characteristic is displayed in Fig. 7-34(a). For the input signal given, the output is a sine wave with the positive peak clipped at 6 V, as shown in Fig. 7-34(b).

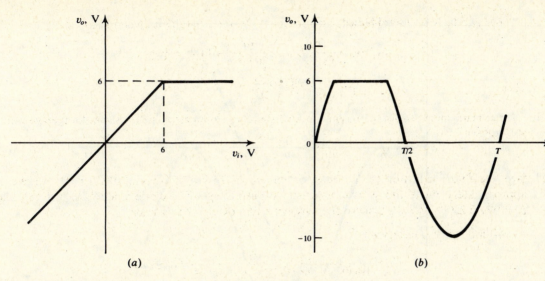

Fig. 7-34

7.18   Reverse the diode in Fig. 7-16(a) to create a negative clipping network. (a) If $V_b = 6$ V, draw
the network transfer characteristic. (b) Sketch one cycle of the output waveform if $v_S = 10 \sin \omega t$ (V).

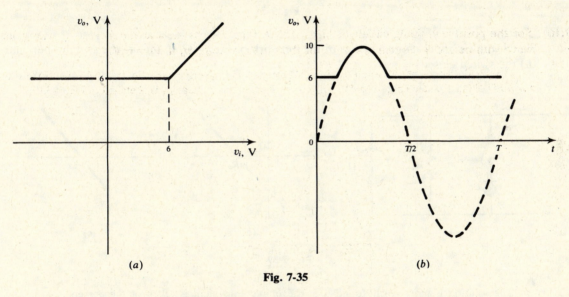

Fig. 7-35

(a)   The diode conducts for $v_i \leq 6$ V and blocks for $v_i > 6$ V. Consequently, $v_o = v_i$ for $v_i > 6$ V and
$v_o = 6$ V for $v_i \leq 6$ V. The transfer characteristic is drawn in Fig. 7-35(a).

(b)   With negative clipping, the output is made up of the positive peaks of $10 \sin \omega t$ above 6 V and held
at 6 V otherwise. Figure 7-35(b) displays the output waveform.

7.19   A signal, $v_i = 10 \sin \omega t$ (V), is applied to the negative clamping circuit of Fig. 7-17(b). Treating
the diode as ideal, sketch the output waveform for $1\frac{1}{2}$ cycles of v_i. The capacitor is initially
uncharged.

For $0 \leq t \leq T/4$, the diode is forward-biased, giving $v_o = 0$ as the capacitor charges to $v_C = +10$ V.
For $t > T/4$, $v_o \leq 0$, and thus the diode remains in the blocking mode, resulting in

$$v_o = -v_C + v_i = -10 + v_i = -10(1 - \sin \omega t)   \text{(V)}$$

The waveform of v_o is sketched in Fig. 7-36.

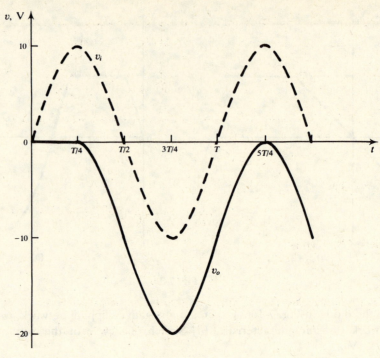

Fig. 7-36

7.20 Sketch the *i-v* input characteristics of the network of Fig. 7-37(*a*) for (*a*) the switch open and (*b*) the switch closed.

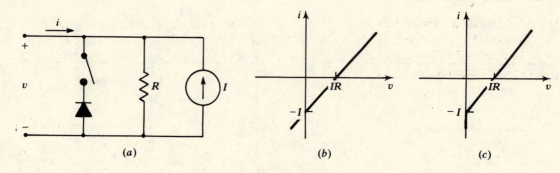

(*a*)   (*b*)   (*c*)

Fig. 7-37

The solution is more easily found if a Thévenin's equivalent is taken of the current source and resistor to give $V_{Th} = IR$ and $R_{Th} = R$.

(*a*) KVL gives $v = iR_{Th} + V_{Th} = iR + IR$, which is the equation of a straight line intersecting the *i*-axis at $-I$ and the *v*-axis at $+IR$. The slope of the line is $1/R$. The characteristic is sketched in Fig. 7-37(*b*).

(*b*) The diode is reverse-biased or acts as an open circuit for $v > 0$. It follows that the *i-v* characteristic is identical to that with the switch open if $v > 0$. On the other hand, if $v \leq 0$, the diode is forward-biased, acting as a short circuit. Consequently, v can never reach negative values, and current i can increase negatively without limit. The corresponding *i-v* plot is sketched in Fig. 7-37(*c*).

7.21 The Zener diode in the voltage regulator circuit of Fig. 7-38 has a constant reverse breakdown voltage, $V_Z = 8.2$ V, for 75 mA $\leq i_Z \leq 1$ A. If $R_L = 9\ \Omega$, size R_S so that v_L regulates to 8.2 V while V_b varies $\pm 10\%$ from its nominal value, 12 V.

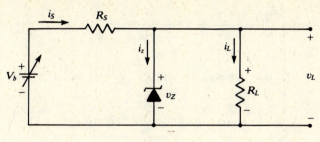

Fig. 7-38

By Ohm's law,

$$i_L = \frac{v_Z}{R_L} = \frac{V_Z}{R_L} = \frac{8.2}{9} = 0.911 \text{ A}$$

Application of KVL gives

$$R_S = \frac{V_b - V_Z}{i_Z + i_L} \qquad (1)$$

Size R_S by (1) for maximum Zener current (I_Z) at the greatest value of V_b.

$$R_S = \frac{(1.1)(12) - 8.2}{1 + 0.911} = 2.62 \text{ }\Omega$$

Check to see if $i_Z \geq 75$ mA at lowest value of V_b:

$$i_Z = \frac{V_b - v_Z}{R_S} - i_L = \frac{(0.9)(12) - 8.2}{2.62} - 0.911 = 81.3 \text{ mA}$$

Since $i_Z > 75$ mA, $v_Z = V_Z = 8.2$ V and regulation is preserved.

7.22 The vacuum diodes in the rectifier circuit of Fig. 7-39(a) are identical and can be modeled by $R_F = 400$ Ω in the forward direction and by an infinite resistance in the reverse direction. If $C = 0$, $R_L = 5$ kΩ, $V_{Lavg} = 110$ V, and $v_S = 120\sqrt{2}\sin 120\pi t$ (V), calculate the turns ratio of the ideal transformer.

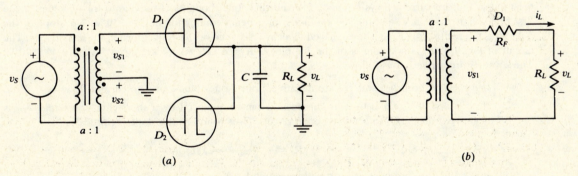

(a)   (b)

Fig. 7-39

By the symmetry of the problem, it is necessary to analyze only one leg of the rectifier circuit, as indicated in Fig. 7-39(b). Since the output is a rectified sine wave, for which (Example 7.8) $V_{Lavg} = 0.636 V_{Lm}$, the following expression for load voltage can be written:

$$v_L(t) = \frac{110}{0.636}\sin 120\pi t = 172.96 \sin 120\pi t \quad \text{(V)} \qquad [0 \leq t < (1/120) \text{ s}]$$

By Ohm's law,

$$i_L(t) = \frac{v_L(t)}{R_L} = \frac{172.96}{5000} \sin 120\pi t = 0.03459 \sin 120\pi t \quad \text{(A)} \qquad [0 \le t < (1/120) \text{ s}]$$

Also,

$$v_{D1}(t) = R_F i_L(t) = (400)(0.03459 \sin 120\pi t) = 13.84 \sin 120\pi t \quad \text{(V)} \qquad [0 \le t < (1/120) \text{ s}]$$

Now,

$$v_{S1}(t) = v_{D1}(t) + v_L(t) = 186.8 \sin 120\pi t \quad \text{(V)} \qquad [0 \le t < (1/120) \text{ s}]$$

and the turns ratio is given by

$$a = \frac{v_S}{v_{S1}} = \frac{120\sqrt{2}}{186.8} = 0.908$$

7.23 Synthesize a function generator circuit using ideal diodes, resistors, and batteries that will yield the i-v characteristic illustrated in Fig. 7-40(a).

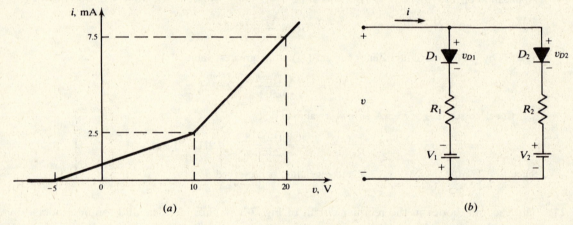

(a)   (b)

Fig. 7-40

Since the i-v characteristic has two breakpoints, two diodes are required. Both diodes must be oriented so that no current flows for $v < -5$ V. Further, one diode must move into forward bias at the first breakpoint, $v = -5$ V, and the second diode must begin conduction at $v = +10$ V. The slope of the i-v plot is the reciprocal of the Thévenin's equivalent resistance of the active portion of the network. The circuit of Fig. 7-40(b) is a synthesis of the i-v plot if $R_1 = 6$ kΩ, $R_2 = 3$ kΩ, $V_1 = 5$ V, and $V_2 = 10$ V. These values are arrived at as follows:

1. If $v < -5$ V, both v_{D1} and v_{D2} are negative, both diodes block, and no current flows.

2. If $-5 \le v < 10$ V, D_1 is forward-biased and can be replaced by a short circuit, whereas v_{D2} is negative, causing D_2 to act as an open circuit. R_1 is found as the slope reciprocal:

$$R_1 = \frac{10 - (-5)}{0.0025} = 6 \text{ k}\Omega$$

3. If $v \ge 10$ V, both diodes are forward-biased and

$$R_{Th} = \frac{R_1 R_2}{R_1 + R_2} = \frac{\Delta v}{\Delta i} = \frac{20 - 10}{(7.5 - 2.5) \times 10^{-3}} = 2 \text{ k}\Omega$$

Solving for R_2 gives

$$R_2 = \frac{R_1 R_{Th}}{R_1 - R_{Th}} = \frac{(6 \times 10^3)(2 \times 10^3)}{4 \times 10^3} = 3 \text{ k}\Omega$$

Supplementary Problems

7.24 A Si diode has a saturation current $I_o = 10$ nA at $T = 300$ K. (a) Find the forward current (i_D) if the forward drop (v_D) is 0.5 V. (b) If this diode is rated for a maximum current of 5 A, what is the junction temperature at rated current if the forward drop is 0.7 V?   *Ans.*   (a) 2.47 A; (b) 405.4 K

7.25 The diode of Problem 7.24 is operating in a circuit where it has dynamic resistance $r_d = 100$ Ω. What must be the quiescent conditions?   *Ans.* $V_{DQ} = 0.263$ V, $I_{DQ} = 0.259$ mA

7.26 The diode of Problem 7.24 has a forward current $i_D = 2 + 0.004 \sin \omega t$ (mA). Find the total voltage across the diode, $v_D = V_{DQ} + v_d$.   *Ans.* $v_D = 339.5 + 0.0207 \sin \omega t$ (mV)

7.27 Find the power dissipated in the load resistor, $R_L = 100$ Ω, of the circuit of Fig. 7-21(a) if the diode is ideal and $v_S = 10 \sin \omega t$ (V).   *Ans.*   206.6 mW

7.28 The logic AND gate of Fig. 7-41(a) has trains of input pulses arriving at the gate inputs as indicated by Fig. 7-41(b). Signal v_2 is erratic, dropping below nominal logic level on occasion. Determine v_o.   *Ans.*   10 V   ($1 \le t \le 2$ ms), 5 V   ($4 \le t \le 5$ ms), zero otherwise.

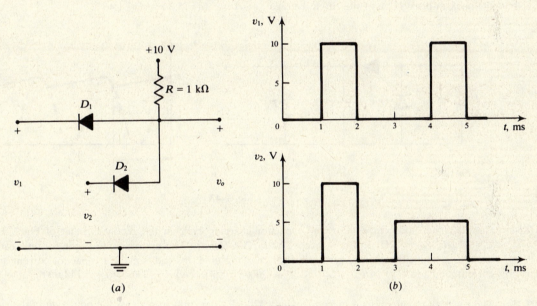

(a)   (b)

Fig. 7-41

7.29 The logic AND gate of Fig. 7-41(a) is to be used to generate a crude pulse train by letting $v_1 = 10 \sin \omega t$ (V) and $v_2 = 5$ V. Determine (a) the amplitude and (b) the period of the pulse train appearing as v_o.   *Ans.*   (a) 5 V; (b) $2\pi/\omega$

7.30 For the circuit of Fig. 7-28, v_S is a 10-V square wave of 4 ms period. The diode is nonideal, with the characteristic of Fig. 7-25(b). If the capacitor is initially uncharged, determine v_C for the first cycle of v_S.
$$\textit{Ans.} \quad v_C = \begin{cases} 9.5(1 - e^{-333.3t}) & (V) \quad (0 \le t < 2 \text{ ms}) \\ 4.62 \text{ V} & (2 \le t < 4 \text{ ms}) \end{cases}$$

7.31 Solve Problem 7.16 if the diode is nonideal, having the characteristic displayed in Fig. 7-42.
    Ans. v_o as in Fig. 7-33, except in ($3T/20$, $7T/20$). v_o increases linearly from 6 V at $3T/20$ to 7.33 V at $T/4$, and then decreases linearly to 6 V at $7T/20$.

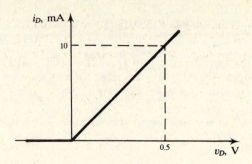

Fig. 7-42

. 7.32 The forward voltage across the diode of Problem 7.24 is $v_D = 0.3 + 0.060 \cos \omega t$ (V). Find the ac component of diode current (i_d).   *Ans.* $2.52 \cos \omega t$ (mA)

7.33 The circuit of Fig. 7-43(a) is a voltage doubler circuit, sometimes used as a low-level power supply when the load R_L is reasonably constant. It is called a "doubler" because the steady-state peak value of v_L is twice the peak value of the sinusoidal source voltage. Figure 7-43(b) is a sketch of the steady-state output voltage for $v_S = 10 \cos \omega t$ (V). Assume ideal diodes, $\omega = 377$ rad/s, $C_1 = 200$ μF, $C_2 = 10$ μF, and $R_L = 1$ kΩ. (a) Solve by trial and error the transcendental equation for the decay time t_d. (b) Calculate the peak-to-peak value of the ripple voltage.   *Ans.* (a) 15.2 ms; (b) 1.46 V

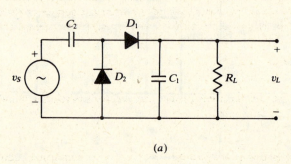

(a)   (b)

Fig. 7-43

7.34 Calculate the diode current for one capacitor charging cycle in the rectifier circuit of Fig. 7-15(a), if $C = 47$ μF, $R_L = 1$ kΩ, and $v_S = 90 \cos 2000t$ (V). (*Hint:* The approximate ripple formula cannot be used, as it implicitly assumes zero capacitor charging time. Instead, solve for capacitor current and load current, and add.)   *Ans.* $i_D = -8.49 \sin (2000t - 0.6°)$ (A)   ($2.966 \le t < 3.142$ ms)

7.35 Make a peak rectifier circuit by connecting a 470-μF filter capacitor across the points ab in the circuit of Fig. 7-14(a). If $R_L = 2$ kΩ and $v_S = 100 \sin \omega t$ (V), (a) find the magnitude (peak-to-peak) of the ripple voltage, and (b) calculate the average value of v_L.   *Ans.* (a) 3.34 V; (b) 98.33 V

7.36 For the circuit of Fig. 7-39(a), $v_S = 120\sqrt{2} \sin 120\pi t$ (V), $R_L = 10$ kΩ, and $a = 2$. The capacitor is sized for 2% ripple. The identical diodes can be modeled by 300 Ω forward resistance and infinite reverse resistance. Find (a) the value of C and (b) the average value of v_L.
Ans. (a) 41.67 μF; (b) 81.55 V

7.37 For the circuit of Fig. 7-32, $R_1 = R_L = 10$ Ω. If the diode is ideal and $v_S = 10 \sin \omega t$ (V), find the average value of load voltage v_L.   *Ans.* 3.18 V

7.38 Work Problem 7.19 if the diode of Fig. 7-17(b) is reversed, all else remaining unchanged. (The circuit is now a positive clamping circuit.)

$$Ans. \quad v_o = \begin{cases} 10 \sin \omega t \quad (V) & (0 \le t < T/2) \\ 0 & (T/2 \le t < 3T/4) \\ 10(1 - \sin \omega t) \quad (V) & (t \ge 3T/4) \end{cases}$$

7.39 In a practical application, the output of the clamping circuit of Fig. 7-17(b) must be connected to a load (R_L) to give the circuit of Fig. 7-44. If $C = 4$ μF, $R_L = 5$ kΩ, the diode has the characteristic of Fig. 7-42, and v_S is a 10-V square wave of period 2 ms. Find expressions for (a) v_L and (b) v_C, over the first cycle of v_S.

Ans.   (a) $v_L = \begin{cases} 10\,e^{-5050t} & \text{(V)} \qquad (0 \le t < 1 \text{ ms}) \\ 19.94\,e^{-50(t-0.001)} & \text{(V)} \qquad (1 \le t < 2 \text{ ms}) \end{cases}$

   (b) $v_C = |10 - v_L|$ (V)

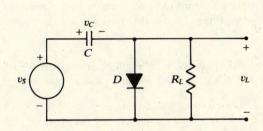

Fig. 7-44

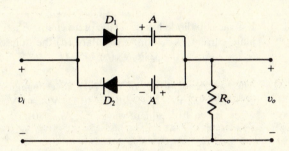

Fig. 7-45

7.40 A *level-discriminator circuit* (Fig. 7-45) has an output of zero, regardless of the polarity of the input signal, until the input reaches a threshold. Above the threshold value, the output duplicates the input. Such a circuit can sometimes be used to eliminate the effects of low-level noise at the expense of slight distortion. Relate v_o to v_i.

Ans. $v_o = \begin{cases} v_i(1 - A/|v_i|) & (|v_i| > A) \\ 0 & (|v_i| \le A) \end{cases}$

7.41 If the diode of Fig. 7-37(a) is reversed and all else remains the same, write equations relating v and i for (a) switch open, and (b) switch closed.
Ans.   (a) $v = R(i + I)$; (b) $v = R(i + I)$ for $i < I$, $v = 0$ for $i \ge I$

7.42 The Zener diode in the voltage regulator circuit of Fig. 7-38 has $v_Z = V_Z = 18.6$ V at a minimum i_Z of 15 mA. If $V_b = 24 \pm 3$ V and R_L varies from 250 Ω to 2 kΩ, (a) find the maximum value of R_S to maintain regulation, and (b) specify the minimum power rating of the Zener diode.
Ans.   (a) 26.8 Ω; (b) 5.65 W

7.43 The regulator circuit of Fig. 7-38 is modified by replacing the Zener diode by two Zener diodes in series to obtain a regulated voltage of 20 V. The characteristics of the two Zeners are:

   Zener 1 $V_Z = 9.2$ V   ($15 \le i_Z \le 300$ mA)   **Zener 2** $V_Z = 10.8$ V   ($12 \le i_Z \le 240$ mA)

(a) If i_L varies from 10 mA to 90 mA and V_b varies from 22 V to 26 V, size R_S so that regulation is preserved. (b) Can either Zener exceed rated current?
Ans.   (a) 19.6 Ω. (b) For $V_b = 26$ V, $i_{Z1} = i_{Z2} = 296$ mA, which exceeds the rating of Zener 2.

7.44 The vacuum diode of Fig. 7-46 has the plate characteristic $i_P = 0.08 v_P^{3/2}$ (mA). If $E_{PP} = 50$ V and $R = 2.5$ kΩ, calculate the value of plate current.   Ans.   10 mA

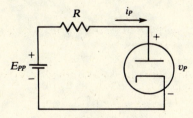

Fig. 7-46

7.45   On a common set of axes sketch the forward characteristics of a vacuum diode described by $i_P = 1.00\,v_P^{3/2}$ (mA) and a semiconductor diode described by $i_D = 10^{-5}e^{40v_D}$ (mA), for values of forward current from zero to 100 mA. What major difference between the devices is apparent from your sketch?
      Ans. $(i, v_P, v_D) = (0, 0, 0), (20, 7.4, 0.362), (60, 15.3, 0.39), (100, 21.5, 0.403)$

7.46   Analogous to (*7.4*), the dynamic resistance of a vacuum diode is defined by

$$r_p \equiv \frac{dv_P}{di_P}\bigg|_Q = \frac{V_P}{I_P}$$

If a vacuum diode has the static plate characteristic $i_P = 0.08\,v_P^{3/2}$ (mA) and is operating with a quiescent plate current $I_P = 15$ mA, calculate (*a*) the quiescent plate voltage (V_P) and (*b*) the dynamic resistance (r_p).   *Ans.*   (*a*) 32.76 V; (*b*) 1456 Ω

7.47   The two Zener diodes of Fig. 7-47 have negligible forward drops and both regulate at constant V_Z for $50 \le i_Z \le 500$ mA. If $R_1 = R_L = 10\ \Omega$, $V_{Z1} = 8$ V, and $V_{Z2} = 5$ V, find the average value of load voltage when v_i is a 10-V square wave.   *Ans.*   0.75 V

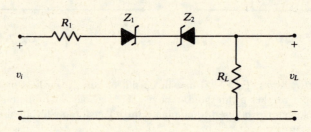

Fig. 7-47

Chapter 8

Bipolar Junction Transistors

8.1 CONSTRUCTION AND SYMBOLS OF THE BJT

The *bipolar junction transistor* (BJT) is a three-element device (*emitter, base, collector*) made up of alternating layers of n and p semiconductor materials joined metallurgically as illustrated in Fig. 8-1.

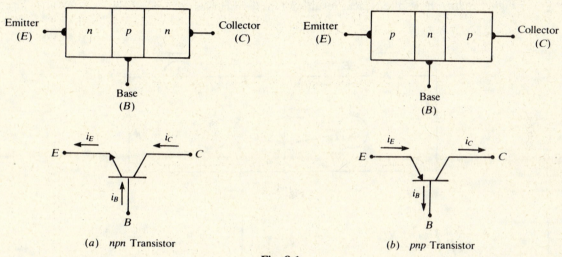

(a) *npn* Transistor (b) *pnp* Transistor

Fig. 8-1

The transistor can be *pnp-type* (principal conduction by positive holes) or *npn-type* (principal conduction by negative electrons). The physical arrangement of each type is displayed in Fig. 8-1, along with the schematic symbols indicating positive directions for currents. In reference to terminal voltages, the double-subscript notation is utilized; e.g., v_{BE} symbolizes the increase in potential from emitter terminal E to base terminal B. For reasons later apparent, terminal currents and voltages commonly consist of superimposed dc and ac components (usually sinusoidal signals). Table 8-1 establishes notation for terminal voltages and currents.

Table 8-1

Type of Value	Symbol		Examples
	Variable	Subscript	
total instantaneous	lowercase	uppercase	i_B, v_{BE}
dc	uppercase	uppercase	I_B, V_{BE}
quiescent-point	uppercase	uppercase plus Q	I_{BQ}, V_{BEQ}
ac instantaneous	lowercase	lowercase	i_b, v_{be}
rms	uppercase	lowercase	I_b, V_{be}
maximum (sinusoid)	uppercase	lowercase plus m	I_{bm}, V_{bem}

8.2 COMMON-BASE TERMINAL CHARACTERISTICS

The *common-base* (CB) connection refers to a two-port transistor arrangement whereby the base shares a common point with the input and output terminals. The independent input variables are emitter current (i_E) and base-to-emitter voltage (v_{EB}). Corresponding independent output variables are collector current (i_C) and base-to-collector voltage (v_{CB}). Practical CB transistor analysis is based on two experimentally determined sets of curves:

1. *Input* or *transfer characteristics* relate i_E and v_{EB} (port input variables), with v_{CB} (port output variable) held constant. The method of laboratory measurement is indicated in Fig. 8-2(*a*), and the typical form of the resulting family of curves is depicted in Fig. 8-2(*b*).

2. *Output* or *collector characteristics* give i_C as a function of v_{CB} (port output variables) for constant values of i_E (port input variable), using the test principles illustrated by Fig. 8-2(*a*). Figure 8-2(*c*) shows the typical resulting family of curves.

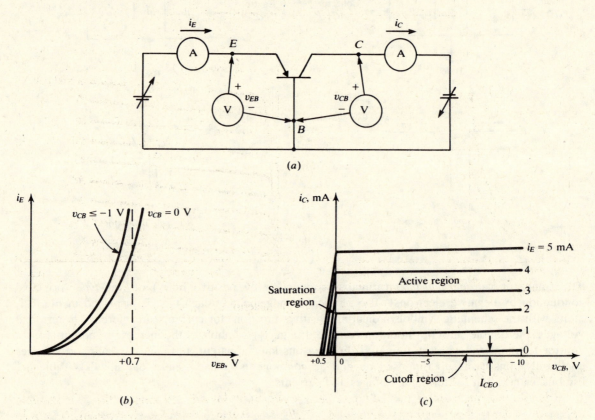

Fig. 8-2. Common-Base Characteristics (*pnp*, Si Device)

8.3 COMMON-EMITTER TERMINAL CHARACTERISTICS

A two-port transistor arrangement (widely used because of its large current amplification) where the emitter shares a common point with the input and output terminals is the *common-emitter* (CE) connection. Independent port input variables are base current (i_B) and emitter-to-base voltage (v_{BE}), and independent port output variables are collector current (i_C) and emitter-to-collector voltage (v_{CE}). Like CB analysis, CE analysis is based on:

1. *Input* or *transfer characteristics* that relate port input variables (i_B and v_{BE}), with v_{CE} held constant. Figure 8-3(*a*) shows the measurement setup and Fig. 8-3(*b*) the resulting input characteristics.

2. *Output* or *collector characteristics* that show functional relationship between port output variables (i_C and v_{CE}) for constant i_B, utilizing the test principles suggested by Fig. 8-3(a). Typical collector characteristics are displayed in Fig. 8-3(c).

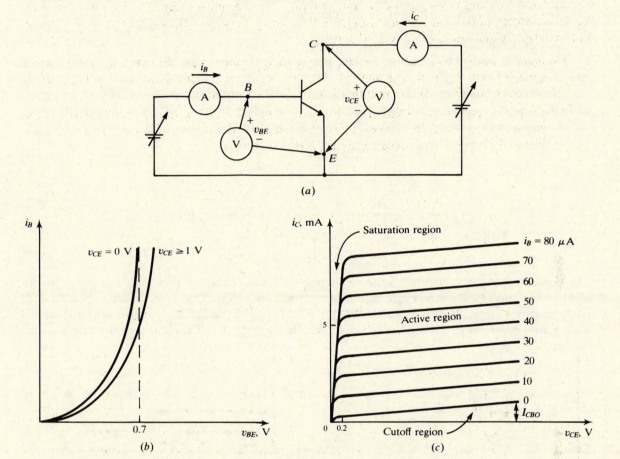

Fig. 8-3.   Common-Emitter Characteristics (*npn*, Si Device)

8.4   CURRENT FLOW AND AMPLIFICATION

The two *pn* junctions of the BJT can be independently biased, resulting in the four possible *operating modes* of the transistor summarized in Table 8-2. A rise in potential when passing from *n* to *p* material is the condition for forward junction bias; conversely, if the *n* region is at higher potential than the *p* region, the junction is reverse-biased.

Table 8-2

Emitter-Base Bias	Collector-Base Bias	Operating Mode
forward	forward	saturation
reverse	reverse	cutoff
reverse	forward	inverse
forward	reverse	linear or active

Saturation denotes a region of operation ($|v_{CE}| \approx 0.2$ V and $|v_{BC}| \approx 0.5$ V for Si devices) where maximum collector current flows and the transistor acts much like a closed switch from collector to emitter terminals.

Cutoff denotes a region of operation near the voltage axis of the collector characteristics where the transistor acts much like an open switch. Only leakage current (similar to I_o of the diode) flows in this mode of operation; thus, $i_C = I_{CEO} \approx 0$ for CB connection, and $i_C = I_{CBO} \approx 0$ for CE connection. Figures 8-2(c) and 8-3(c) indicate leakage currents.

The *inverse* mode is a little-used, inefficient active mode with the emitter and collector interchanged.

The *active* or *linear* mode describes transistor operation in the region to the right of saturation and above cutoff in Figs. 8-2(c) and 8-3(c); here, near-linear relationships exist between terminal currents. Constants of proportionality are defined for dc currents:

$$\alpha (\equiv h_{FB}) \equiv \frac{I_C - I_{CBO}}{I_E} \qquad (8.1)$$

$$\beta (\equiv h_{FE}) \equiv \frac{\alpha}{1 - \alpha} \equiv \frac{I_C - I_{CEO}}{I_B} \qquad (8.2)$$

where the thermally generated leakage currents are related by

$$I_{CEO} = (\beta + 1)I_{CBO} \qquad (8.3)$$

The constant $\alpha < 1$ is a measure of the proportion of majority carriers (holes for *pnp*, electrons for *npn*) injected into the base region from the emitter that are received by the collector. Equation (8.2) is the dc current amplification characteristic of the BJT; neglecting leakage current, the base current is increased or amplified β times to become the collector current. KCL applied to the transistor for dc conditions gives

$$I_E = I_C + I_B \qquad (8.4)$$

which, in conjunction with (8.1) through (8.3), completely describes the dc current relationships of the BJT in the active mode.

Example 8.1 A BJT has $\alpha = 0.99$, $i_B = I_B = 25$ μA, and $I_{CBO} = 200$ nA. Find (a) the dc collector current, and (b) the dc emitter current. (c) Find the percent error in emitter current when leakage current is neglected.

(a) With $\alpha = 0.99$, (8.2) gives

$$\beta = \frac{\alpha}{1 - \alpha} = 99$$

Using (8.3) in (8.2) then gives

$$I_C = \beta I_B + (\beta + 1)I_{CBO} = 99(25 \times 10^{-6}) + (99 + 1)(200 \times 10^{-9}) = 2.495 \text{ mA}$$

(b) The dc emitter current follows from (8.1):

$$I_E = \frac{I_C - I_{CBO}}{\alpha} = \frac{2.495 \times 10^{-3} - 200 \times 10^{-9}}{0.99} = 2.518 \text{ mA}$$

(c) Neglecting the leakage current, $I_B = (25 \times 10^{-6}) - (200 \times 10^{-9}) = 24.8$ μA.

Thus $\qquad I_C = \beta I_B = 99(24.8 \times 10^{-6}) = 2.455 \text{ mA} \qquad I_E = \frac{I_C}{\alpha} = \frac{2.455}{0.99} = 2.48 \text{ mA}$

giving an emitter current error of

$$\frac{2.518 - 2.48}{2.518} \times 100\% = 1.51\%$$

Constants of proportionality, or *short-circuit, forward-gain constants*, relating ac or small signals are defined analogous to (8.1) and (8.2):

$$h_{fb} \equiv \frac{i_c}{i_e} \tag{8.5}$$

$$h_{fe} \equiv \frac{i_c}{i_b} \tag{8.6}$$

In general, $h_{fe} \neq h_{FE}$ and $h_{fb} \neq h_{FB}$; however, unless transistor operation occurs near the boundary of the active region, numerical equality is, for simplicity, assumed.

8.5   CONSTANT-EMITTER-CURRENT BIAS

Bias (specific dc terminal voltages and currents of a transistor to set a desired point of active-mode operation, called the *quiescent point* or *Q-point*) can be categorized as *β-dependent* (see Problems 8.9 and 8.12) or *β-independent*. The latter type is most desirable and widely used.

The universal bias arrangement of Fig. 8-4(a) requires only one dc power supply (V_{CC}) to establish active-mode operation of the transistor. The Thévenin equivalent looking to the left through *ab* leads to the circuit of Fig. 8-4(b), where

$$R_B = \frac{R_1 R_2}{R_1 + R_2} \qquad V_{BB} = \frac{R_1}{R_1 + R_2} V_{CC} \tag{8.7}$$

Neglect leakage current [$I_{EQ} = (\beta + 1)I_{BQ}$] and assume that emitter-to-base voltage V_{BEQ} is constant (≈ 0.7 V and ≈ 0.3 V for Si and Ge, respectively). Then KVL around the emitter loop of Fig. 8-4(b) yields

$$V_{BB} = \frac{I_{EQ}}{\beta + 1} R_B + V_{BEQ} + I_{EQ}R_E \tag{8.8}$$

which can be represented by the emitter-loop equivalent bias circuit of Fig. 8-4(c). Solving (8.8) for I_{EQ}, and noting that

$$I_{EQ} = \frac{I_{CQ}}{\alpha} \approx I_{CQ}$$

yields

$$I_{CQ} \approx I_{EQ} = \frac{V_{BB} - V_{BEQ}}{\dfrac{R_B}{\beta + 1} + R_E} \tag{8.9}$$

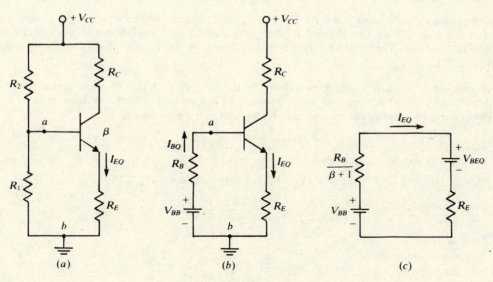

(a)   (b)   (c)

Fig. 8-4

If components and worst-case β-value are such that

$$\frac{R_B}{\beta + 1} \approx \frac{R_B}{\beta} \ll R_E \qquad (8.10)$$

then I_{EQ} (and thus I_{CQ}) is nearly constant, regardless of changes in β; the amplifier then has β-independent bias.

8.6   DC LOAD LINE AND COLLECTOR BIAS

From Fig. 8-3(c), it is apparent that the family of collector characteristics may be expressed by a mathematical relationship $i_C = f(v_{CE}, i_B)$ in the independent variable v_{CE} and the parameter i_B. We suppose that the collector circuit can be biased to place the Q-point anywhere in the active region. A typical setup is shown in Fig. 8-5(a), from which

$$I_{CQ} = -\frac{V_{CEQ}}{R_{dc}} + \frac{V_{CC}}{R_{dc}}$$

Thus, if the equation of the *dc load line*,

$$i_C = -\frac{v_{CE}}{R_{dc}} + \frac{V_{CC}}{R_{dc}} \qquad (8.11)$$

and the specification

$$i_B = I_{BQ} \qquad (8.12)$$

are combined with the equation of the collector characteristics, the resulting system can be solved (graphically) for the collector quiescent quantities I_{CQ} and V_{CEQ}.

Example 8.2   The signal source switch of Fig. 8-5(a) is closed and the transistor base current becomes

$$i_B = I_{BQ} + i_b = 40 + 20 \sin \omega t \quad (\mu A)$$

Collector characteristics of the transistor are displayed in Fig. 8-5(b). If $V_{CC} = 12$ V and $R_{dc} = 1$ kΩ, graphically determine (a) I_{CQ} and V_{CEQ}, (b) i_c and v_{ce}, (c) $h_{FE}(=\beta)$ at the Q-point, and (d) h_{fe}.

(a)   The dc load line is constructed on Fig. 8-5(b) with ordinate intercept $V_{CC}/R_{dc} = 12$ mA and abscissa intercept $V_{CC} = 12$ V. The Q-point is the intersection of the load line with the characteristic curve $i_B = I_{BQ} = 40\ \mu A$. The collector quiescent quantities are read from the axes as $I_{CQ} = 4.9$ mA and $V_{CEQ} = 7.2$ V.

(b)   A time scale is constructed perpendicular to the load line at the Q-point, and a scaled sketch of $i_b = 20 \sin \omega t$ (μA) is displayed. As i_b swings $\pm 20\ \mu A$ along the load line from point a to b, the ac components of collector current and voltage are seen to have the values

$$i_c = 2.25 \sin \omega t \quad (\text{mA}) \qquad v_{ce} = -2.37 \sin \omega t \quad (\text{V})$$

The negative sign for v_{ce} signifies a 180° phase shift.

(c)   From (8.2) with $I_{CEO} = 0$ [the $i_B = 0$ curve is noted to be coincidental with the v_{CE}-axis in Fig. 8-5(b)],

$$h_{FE} = \frac{I_{CQ}}{I_{BQ}} = \frac{4.9 \times 10^{-3}}{40 \times 10^{-6}} = 122.5$$

(d)   From (8.6),

$$h_{fe} = \frac{i_c}{i_b} = \frac{I_{cm}}{I_{bm}} = \frac{2.25 \times 10^{-3}}{20 \times 10^{-6}} = 112.5$$

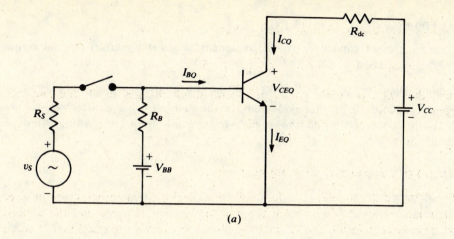

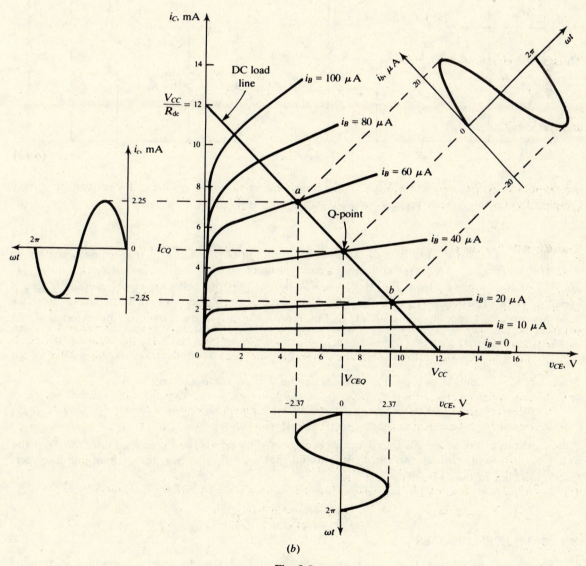

Fig. 8-5

8.7. CAPACITORS AND AC LOAD LINES

Two common uses of capacitors (sized to appear as short circuits to signal frequencies) are illustrated by the circuit of Fig. 8-6(a):

1. Coupling capacitors (C_C) confine dc quantities to the transistor and its bias circuitry.
2. Bypass capacitors (C_E) effectively remove the gain-reducing emitter resistor insofar as ac signals are concerned, while allowing R_E to play its role in establishing β-independent bias (Section 8.5).

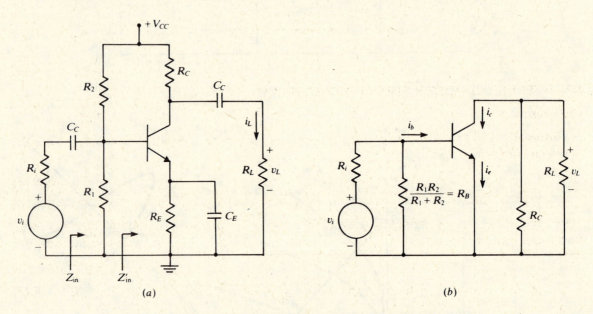

(a)   (b)

Fig. 8-6. Capacitor-Coupled and Bypassed Amplifier

Shorting of capacitors in Fig. 8-6(a) allows presentation of the circuit as it appears to ac signals, Fig. 8-6(b). From Fig. 8-6(a), the collector circuit resistance seen by the dc bias current I_{CQ} ($\approx I_{EQ}$) is $R_{dc} = R_C + R_E$. However, from Fig. 8-6(b) it is apparent that the collector signal current i_c sees a collector circuit resistance $R_{ac} = R_C R_L/(R_C + R_L)$. Since $R_{ac} \neq R_{dc}$ in general, the concept of an *ac load line* arises. By application of KVL to Fig. 8-6(b), the v-i characteristic of the external signal circuitry is given by

$$v_{ce} = i_c R_{ac} \qquad (8.13)$$

Since $i_c = i_C - I_{CQ}$ and $v_{ce} = v_{CE} - V_{CEQ}$, (8.13) can be written analogous to (8.11) as

$$i_C = -\frac{v_{CE}}{R_{ac}} + \left(\frac{V_{CEQ}}{R_{ac}} + I_{CQ}\right) \qquad (8.14)$$

All excursions of the ac signals i_c and v_{ce} are represented by points on the ac load line, (8.14). If the value $i_C = I_{CQ}$ is substituted into (8.14), it is found that $v_{CE} = V_{CEQ}$; thus, the ac load line intersects the dc load line at the Q-point.

8.8 AMPLIFIER CLASSES

From previous Sections it is clear that amplifiers can be biased for operation at any point along the dc load line. Amplifiers are classified according to the percentage of the signal cycle over which the amplifier operates in the linear or active region; see Table 8-3.

Table 8-3

Class	Percent of Active-Region Signal Excursion
A	100
AB	between 50 and 100
B	50
C	less than 50

8.9 HYBRID-PARAMETER EQUIVALENT CIRCUITS

CE Transistor

From Fig. 8-3(b) and (c), it is seen that if i_C and v_{BE} are taken as dependent variables of the CE transistor connection, then

$$v_{BE} = f_1(i_B, v_{CE}) \tag{8.15}$$

$$i_C = f_2(i_B, v_{CE}) \tag{8.16}$$

If the total emitter-to-base voltage v_{BE} experiences *small* excursions (ac signals) about the Q-point, then $\Delta v_{BE} = v_{be}$, $\Delta i_C = i_c$, etc. Therefore, applying the chain rule to (8.15) and (8.16), respectively, we have

$$v_{be} = \Delta v_{BE} \approx dv_{BE} = \frac{\partial v_{BE}}{\partial i_B}\bigg|_Q i_b + \frac{\partial v_{BE}}{\partial v_{CE}}\bigg|_Q v_{ce} \tag{8.17}$$

$$i_c = \Delta i_C \approx di_C = \frac{\partial i_C}{\partial i_B}\bigg|_Q i_b + \frac{\partial i_C}{\partial v_{CE}}\bigg|_Q v_{ce} \tag{8.18}$$

The four partial derivatives, evaluated at the Q-point, that occur in (8.17) and (8.18) are called *hybrid parameters*, denoted as follows:

$$\textbf{\textit{input resistance}} \quad h_{ie} \equiv \frac{\partial v_{BE}}{\partial i_B}\bigg|_Q \approx \frac{\Delta v_{BE}}{\Delta i_B}\bigg|_Q \tag{8.19}$$

$$\textbf{\textit{reverse voltage ratio}} \quad h_{re} \equiv \frac{\partial v_{BE}}{\partial v_{CE}}\bigg|_Q \approx \frac{\Delta v_{BE}}{\Delta v_{CE}}\bigg|_Q \tag{8.20}$$

$$\textbf{\textit{forward current gain}} \quad h_{fe} \equiv \frac{\partial i_C}{\partial i_B}\bigg|_Q \approx \frac{\Delta i_C}{\Delta i_B}\bigg|_Q \tag{8.21}$$

$$\textbf{\textit{output admittance}} \quad h_{oe} \equiv \frac{\partial i_C}{\partial v_{CE}}\bigg|_Q \approx \frac{\Delta i_C}{\Delta v_{CE}}\bigg|_Q \tag{8.22}$$

The equivalent circuit for (8.17) and (8.18) is shown in Fig. 8-7(a). The circuit is valid for use with signals whose excursion about the Q-point is sufficiently small so that the h-parameters may be treated as constants.

CB Transistor

If v_{EB} and i_C are taken as the dependent variables for the CB transistor characteristics of Fig. 8-2(b) and (c), then equations can be found relating small excursions about the Q-point, in a manner similar to the CE case. The results are

$$v_{eb} = h_{ib}i_e + h_{rb}v_{cb} \tag{8.23}$$

$$i_c = h_{fb}i_e - h_{ob}v_{cb} \tag{8.24}$$

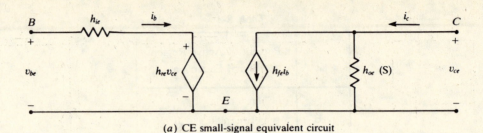

(a) CE small-signal equivalent circuit

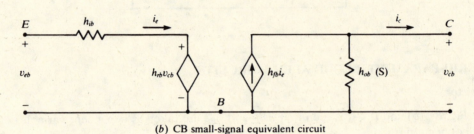

(b) CB small-signal equivalent circuit

Fig. 8-7

The partial-derivative definitions of the CB h-parameters are:

$$\textit{input resistance} \quad h_{ib} \equiv \left.\frac{\partial v_{EB}}{\partial i_E}\right|_Q \approx \left.\frac{\Delta v_{EB}}{\Delta i_E}\right|_Q \tag{8.25}$$

$$\textit{reverse voltage ratio} \quad h_{rb} \equiv \left.\frac{\partial v_{EB}}{\partial v_{CB}}\right|_Q \approx \left.\frac{\Delta v_{EB}}{\Delta v_{CB}}\right|_Q \tag{8.26}$$

$$\textit{forward current gain} \quad h_{fb} \equiv \left.\frac{\partial i_C}{\partial i_E}\right|_Q \approx \left.\frac{\Delta i_C}{\Delta i_E}\right|_Q \tag{8.27}$$

$$\textit{output admittance} \quad h_{ob} \equiv -\left.\frac{\partial i_C}{\partial v_{CB}}\right|_Q \approx -\left.\frac{\Delta i_C}{\Delta v_{CB}}\right|_Q \tag{8.28}$$

A small-signal, h-parameter equivalent circuit for the CB transistor satisfying (8.23) and (8.24) is given in Fig. 8-7(b).

CC Amplifier

The *common-collector* (CC) or *emitter-follower* (EF) amplifier, with the universal bias circuitry of Fig. 8-8(a), can be modeled for small-signal ac analysis by replacing the CE transistor by its h-parameter model, Fig. 8-7(a). Assuming, for simplicity, that $h_{re} = h_{oe} = 0$, we obtain the equivalent circuit of Fig. 8-8(b).

An even simpler model can be obtained by finding a Thévenin equivalent looking to the right through aa in Fig. 8-8(b). Application of KVL around the outer loop gives

$$v = i_b h_{ie} + i_e R_E = i_b h_{ie} + (h_{fe} + 1) i_b R_E \tag{8.29}$$

The Thévenin impedance is the driving-point impedance,

$$R_{Th} = \frac{v}{i_b} = h_{ie} + (h_{fe} + 1) R_E \tag{8.30}$$

The Thévenin voltage is zero upon opening terminals at aa; thus, the equivalent circuit consists only of R_{Th}. This is shown, in a base-current frame of reference, by Fig. 8-8(c).

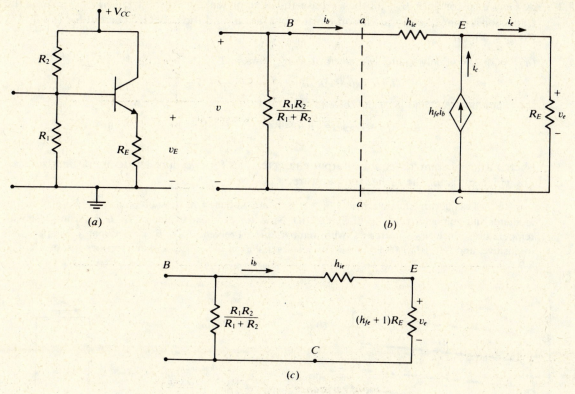

Fig. 8-8. CC Amplifier

8.10   POWER AND EFFICIENCY OF AMPLIFIERS

The *efficiency* of an amplifier is the percentage ratio of the average value of signal output power (P_O) to the power supplied by the bias supplies (P_S):

$$\eta \equiv \frac{P_O}{P_S} \times 100\% \tag{8.31}$$

An amplifier has a zero efficiency if no signal is present, and efficiency is maximum at maximum signal excursion.

Average power dissipated in the collector of a transistor (P_C) is the basis of device thermal rating; it is most easily found as

$$P_C = P_S - (I^2R \text{ losses}) - P_O \tag{8.32}$$

Solved Problems

8.1   Leakage current approximately doubles for every 10 °C increase (6 °C increase) in the temperature of a Ge transistor (Si transistor). If a Si transistor has $I_{CBO} = 500$ nA at 25 °C, find its leakage current at 90 °C.

$$I_{CBO} = (500 \times 10^{-9})2^{(90-25)/6} = (500 \times 10^{-9})(117.4) = 58.68 \; \mu A$$

8.2 A Ge transistor with $\beta = 100$ has a base-to-collector leakage current (I_{CBO}) of 5μA. If the transistor is connected for common-emitter operation, find the collector current for (a) $I_B = 0$, (b) $I_B = 40 \mu$A.

(a) With $I_B = 0$, only emitter-to-collector leakage flows, and by (8.3),

$$I_{CEO} = (\beta + 1)I_{CBO} = (100 + 1)(5 \times 10^{-6}) = 505 \ \mu\text{A}$$

(b) Substitute (8.3) into (8.2) and solve for I_C:

$$I_C = \beta I_B + (\beta + 1)I_{CBO} = (100)(40 \times 10^{-6}) + (101)(5 \times 10^{-6}) = 4.505 \text{ mA}$$

8.3 Sketch a set of common-emitter output characteristics for two different temperatures, indicating which set is for the higher temperature.

The CE collector characteristics, Fig. 8-3(c), are obtained as sets of points (I_C, V_{CE}) from the ammeter and voltmeter readings of Fig. 8-3(a). Now, for each fixed value of I_B, I_C must increase with temperature, since I_{CBO} increases with temperature (Problem 8.1), β is reasonably temperature-insensitive, and $I_C = \beta I_B + (\beta + 1)I_{CBO}$. The resultant shift in the collector characteristics is shown in Fig. 8-9.

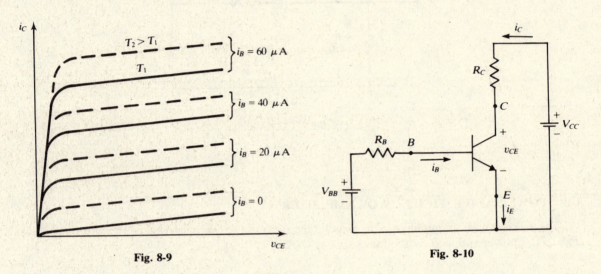

Fig. 8-9   Fig. 8-10

8.4 The transistor of Fig. 8-10 has $\alpha = 0.98$ and a base current of 30μA. Find (a) β, (b) I_{CQ}, and (c) I_{EQ}. Assume negligible leakage current.

(a)

$$\beta = \frac{\alpha}{1 - \alpha} = \frac{0.98}{1 - 0.98} = 49$$

(b) From (8.2), with $I_{CEO} = 0$, $I_{CQ} = \beta I_{BQ} = (49)(30 \times 10^{-6}) = 1.47$ mA.

(c) From (8.1), with $I_{CBO} = 0$,

$$I_{EQ} = \frac{I_{CQ}}{\alpha} = \frac{1.47}{0.98} = 1.50 \text{ mA}$$

8.5 The transistor circuit of Fig. 8-10 is to be operated with a base current of 40μA and $V_{BB} = 6$ V. The Si transistor ($V_{BEQ} = 0.7$ V) has negligible leakage current. Find the value of R_B.

By KVL around the base-emitter loop,

$$V_{BB} = I_{BQ}R_B + V_{BEQ} \qquad \text{or} \qquad R_B = \frac{V_{BB} - V_{BEQ}}{I_{BQ}} = \frac{6 - 0.7}{40 \times 10^{-6}} = 132.5 \text{ k}\Omega$$

8.6   For the circuit of Fig. 8-10, $\beta = 100$, $I_{BQ} = 20$ μA, $V_{CC} = 15$ V, and $R_C = 3$ kΩ. If $I_{CBO} = 0$, find
(*a*) I_{EQ} and (*b*) V_{CEQ}. (*c*) If R_C is changed to 6 kΩ and all else remains the same, find V_{CEQ}.

(*a*)
$$\alpha = \frac{\beta}{\beta+1} = \frac{100}{101} = 0.9901$$

Using (*8.1*) and (*8.2*) with $I_{CBO} = I_{CEO} = 0$,

$$I_{CQ} = \beta I_{BQ} = (100)(20 \times 10^{-6}) = 2 \text{ mA}$$
$$I_{EQ} = \frac{I_{CQ}}{\alpha} = \frac{2 \times 10^{-3}}{0.9901} = 2.02 \text{ mA}$$

(*b*)   From application of KVL around the collector circuit,

$$V_{CEQ} = V_{CC} - I_{CQ}R_C = 15 - (2)(3) = 9 \text{ V}$$

(*c*)   If I_{BQ} is unchanged, then I_{CQ} is unchanged. The solution proceeds as in (*b*).

$$V_{CEQ} = V_{CC} - I_{CQ}R_C = 15 - (2)(6) = 3 \text{ V}$$

8.7   The transistor of Fig. 8-11 is a Si device with a base current of 40 μA and $I_{CBO} = 0$. If $V_{BB} = 6$ V,
$R_E = 1$ kΩ, and $\beta = 80$, find (*a*) I_{EQ} and (*b*) R_B. (*c*) If $V_{CC} = 15$ V and $R_C = 3$ kΩ, find V_{CEQ}.

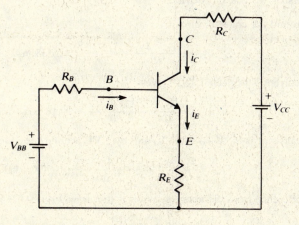

Fig. 8-11

(*a*)
$$\alpha = \frac{\beta}{\beta+1} = \frac{80}{81} = 0.9876$$

Combining (*8.1*) and (*8.2*) with $I_{CBO} = I_{CEO} = 0$ gives

$$I_{EQ} = \frac{I_{BQ}}{1-\alpha} = \frac{40 \times 10^{-6}}{1 - 0.9876} = 3.226 \text{ mA}$$

(*b*)   Applying KVL around the base-emitter loop gives

$$V_{BB} = I_{BQ}R_B + V_{BEQ} + I_{EQ}R_E$$

or
$$R_B = \frac{V_{BB} - V_{BEQ} - I_{EQ}R_E}{I_{BQ}} = \frac{6 - 0.7 - (3.226)(1)}{40 \times 10^{-6}} = 51.85 \text{ k}\Omega$$

(*c*)   From (*8.2*), with $I_{CEO} = 0$,

$$I_{CQ} = \beta I_{BQ} = (80)(40 \times 10^{-6}) = 3.2 \text{ mA}$$

By KVL around collector circuit,

$$V_{CEQ} = V_{CC} - I_{EQ}R_E - I_{CQ}R_C = 15 - (3.24)(1) - (3.2)(3) = 2.16 \text{ V}$$

8.8   The CE collector characteristics of Fig. 8-5(*b*) apply to the transistor of Fig. 8-10. If $I_{BQ} = 30\ \mu A$, $V_{CEQ} = 8$ V, and $V_{CC} = 14$ V, find graphically (*a*) I_{CQ}, (*b*) R_C, (*c*) I_{EQ}, and (*d*) β if leakage current is negligible.

(*a*)   The Q-point is the intersection of $i_B = I_{BQ} = 30\ \mu A$ and $v_{CE} = V_{CEQ} = 8$ V. The dc load line must pass through the Q-point and intersect the v_{CE}-axis at $V_{CC} = 14$ V. Construct the dc load line on Fig. 8-5(*b*) and read $I_{CQ} = 2.7$ mA as the i_C-coordinate of the Q-point.

(*b*)   The i_C-intercept of the dc load line is $V_{CC}/R_{dc} = V_{CC}/R_C$, which, from Fig. 8-5(*b*), has the value 6.25 mA; thus,

$$R_C = \frac{V_{CC}}{6.25 \times 10^{-3}} = \frac{14}{6.25 \times 10^{-3}} = 2.24\ k\Omega$$

(*c*)   By (*8.4*), $I_{EQ} = I_{CQ} + I_{BQ} = (2.7 \times 10^{-3}) + (30 \times 10^{-6}) = 2.73$ mA.

(*d*)   With $I_{CEO} = 0$, (*8.2*) yields

$$\beta = \frac{I_{CQ}}{I_{BQ}} = \frac{2.7 \times 10^{-3}}{30 \times 10^{-6}} = 90$$

8.9   The Si transistor of Fig. 8-12 is biased for constant base current (β-dependent method). If $\beta = 80$, $V_{CEQ} = 8$ V, $R_C = 3\ k\Omega$, and $V_{CC} = 15$ V, find (*a*) I_{CQ} and (*b*) the value of R_B. (*c*) Find R_B if the transistor were a Ge device.

(*a*)   By KVL around the collector circuit,

$$I_{CQ} = \frac{V_{CC} - V_{CEQ}}{R_C} = \frac{15 - 8}{3 \times 10^3} = 2.333\ mA$$

(*b*)   If leakage current is neglected, (*8.2*) gives

$$I_{BQ} = \frac{I_{CQ}}{\beta} = \frac{2.333 \times 10^{-3}}{80} = 29.16\ \mu A$$

Since the transistor is Si, $V_{BEQ} = 0.7$ V, and by KVL around the outer loop,

$$R_B = \frac{V_{CC} - V_{BEQ}}{I_{BQ}} = \frac{15 - 0.7}{29.16 \times 10^{-6}} = 490.4\ k\Omega$$

(*c*)   The only difference from the above solution is that now $V_{BEQ} = 0.3$ V; thus,

$$R_B = \frac{15 - 0.3}{29.16 \times 10^{-6}} = 504.1\ k\Omega$$

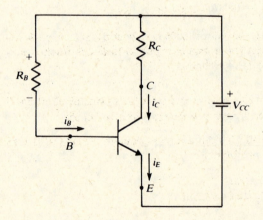

Fig. 8-12

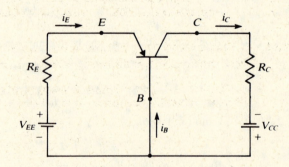

Fig. 8-13

8.10 The Si transistor of Fig. 8-13 has $\alpha = 0.99$ and $I_{CEO} = 0$. Also, $V_{EE} = 4$ V and $V_{CC} = 12$ V. (a) If $I_{EQ} = 1.1$ mA, find R_E. (b) If $V_{CEQ} = -7$ V, find R_C.

(a) By KVL around the emitter-base loop,

$$R_E = \frac{V_{EE} + V_{BEQ}}{I_{EQ}} = \frac{4 + (-0.7)}{1.1 \times 10^{-3}} = 3 \ \Omega$$

(b) By KVL around the transistor terminals,

$$V_{CBQ} = V_{CEQ} - V_{BEQ} = -7 - (-0.7) = -6.3 \ \text{V}$$

With negligible leakage current, (8.1) gives

$$I_{CQ} = \alpha I_{EQ} = (0.99)(1.1 \times 10^{-3}) = 1.089 \ \text{mA}$$

By KVL around the base-collector loop,

$$R_C = \frac{V_{CC} + V_{CBQ}}{I_{CQ}} = \frac{12 - 6.3}{1.089 \times 10^{-3}} = 5.234 \ \text{k}\Omega$$

8.11 For the circuit of Fig. 8-4(a), $R_C = 300 \ \Omega$, $R_E = 200 \ \Omega$, $R_1 = 2 \ \text{k}\Omega$, $R_2 = 15 \ \text{k}\Omega$, $V_{CC} = 15$ V, and $\beta = 110$ for the Si transistor. Assume that $I_{CQ} \approx I_{EQ}$ and $V_{CEsat} \approx 0$. Find the maximum symmetrical swing in collector current (a) if an ac base current is injected; (b) if V_{CC} is changed to 10 V but all else remains the same.

(a) Using (8.7) and (8.9),

$$R_B = \frac{(2 \times 10^3)(15 \times 10^3)}{17 \times 10^3} = 1.765 \ \text{k}\Omega$$

$$V_{BB} = \frac{R_1}{R_1 + R_2} V_{CC} = \frac{2 \times 10^3}{17 \times 10^3} (15) = 1.765 \ \text{V}$$

$$I_{CQ} \approx I_{EQ} = \frac{V_{BB} - V_{BEQ}}{\dfrac{R_B}{\beta + 1} + R_E} = \frac{1.765 - 0.7}{(1765/111) + 200} = 4.93 \ \text{mA}$$

By KVL around the collector circuit, and using $I_{CQ} \approx I_{EQ}$,

$$V_{CEQ} = V_{CC} - I_{CQ}(R_C + R_E) = 15 - (4.93 \times 10^{-3})(200 + 300) = 12.535 \ \text{V}$$

Since $V_{CEQ} > V_{CC}/2 = 7.5$ V, cutoff occurs before saturation, and i_C can swing ± 4.93 mA about I_{CQ} and remain in the active region.

(b)

$$V_{BB} = \frac{R_1}{R_1 + R_2} V_{CC} = \frac{2 \times 10^3}{17 \times 10^3} (10) = 1.1765 \ \text{V}$$

$$I_{CQ} \approx I_{EQ} = \frac{V_{BB} - V_{BEQ}}{\dfrac{R_B}{\beta + 1} + R_E} = \frac{1.1765 - 0.7}{(1765/111) + 200} = 2.206 \ \text{mA}$$

$$V_{CEQ} = V_{CC} - I_{CQ}(R_C + R_E) = 10 - (2.206 \times 10^{-3})(0.5) = 8.79 \ \text{V}$$

Since $V_{CEQ} > V_{CC}/2 = 5$ V, cutoff occurs first, and i_C can swing ± 2.206 mA about I_{CQ} and remain in the active region of operation. Here, the 33.3% reduction in power supply voltage has resulted in a reduction of over 50% in symmetrical collector current swing.

8.12 If a Si transistor were removed from the circuit of Fig. 8-4(a) and a Ge transistor of identical β were substituted, would the Q-point move in the direction of saturation or of cutoff?

Since R_1, R_2, and V_{CC} are unchanged, R_B and V_{BB} would remain unchanged. However, due to different emitter-to-base forward drops for Si (0.7 V) and Ge (0.3 V),

$$I_{CQ} \approx \frac{V_{BB} - V_{BEQ}}{\dfrac{R_B}{\beta + 1} + R_E}$$

would increase upon insertion of the Ge transistor. Thus, the Q-point would move in the direction of saturation.

8.13   The circuit of Fig. 8-14 uses current feedback bias. The Si transistor has $I_{CEO} \approx 0$, $V_{CEsat} \approx 0$, and $h_{FE} = 100$. If $R_C = 2\,k\Omega$ and $V_{CC} = 12\,V$, size R_F for maximum symmetrical swing ($V_{CEQ} = V_{CC}/2$).

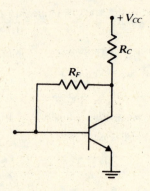

Fig. 8-14

Application of KVL to the collector circuit gives

$$(I_{BQ} + I_{CQ})R_C = V_{CC} - V_{CEQ} \tag{1}$$

But $I_{CQ} = h_{FE}I_{BQ}$, which when substituted in (*1*) leads to

$$I_{BQ} = \frac{V_{CC} - V_{CEQ}}{(h_{FE} + 1)R_C} = \frac{12 - 6}{(100 + 1)(2 \times 10^3)} = 29.7\ \mu A$$

By KVL around the transistor terminals,

$$R_F = \frac{V_{CEQ} - V_{BEQ}}{I_{BQ}} = \frac{6 - 0.7}{29.7 \times 10^{-6}} = 178.5\ k\Omega$$

8.14   Collector characteristics for the Ge transistor of Fig. 8-13 are given in Fig. 8-15. If $V_{EE} = 2\,V$, $V_{CC} = 12\,V$, and $R_C = 2\,k\Omega$, size R_E so that $V_{CEQ} = -6.4\,V$.

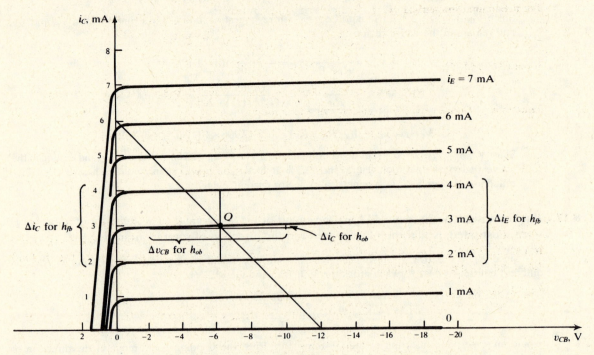

Fig. 8-15

Construct a dc load line on Fig. 8-15, having v_{CB}-intercept $-V_{CC} = -12$ V and i_C-intercept $V_{CC}/R_C = 6$ mA. The abscissa of the Q-point is given by KVL around the transistor terminals:

$$V_{CBQ} = V_{CEQ} - V_{BEQ} = -6.4 - (-0.3) = -6.1 \text{ V}$$

With the Q-point defined, read $I_{EQ} = 3$ mA from the graph. Application of KVL around the emitter-base loop leads to

$$R_E = \frac{V_{EE} + V_{BEQ}}{I_{EQ}} = \frac{2 + (-0.3)}{3 \times 10^{-3}} = 566.7 \ \Omega$$

8.15 For the circuit of Fig. 8-11, $I_{BQ} = 30 \ \mu\text{A}$, $R_E = 1 \text{ k}\Omega$, $V_{CC} = 15$ V, and $\beta = 80$. Find the minimum value of R_C to maintain the transistor quiescent point at saturation, if $V_{CE\text{sat}} = 0.2$ V, β is constant, and leakage current is negligible.

$$\alpha = \frac{\beta}{\beta + 1} = \frac{80}{81} = 0.9876$$

Use of (8.1) and (8.2), with negligible leakage current, yields

$$I_{CQ} = \beta I_{BQ} = (80)(30 \times 10^{-6}) = 2.4 \text{ mA} \qquad I_{EQ} = \frac{I_{CQ}}{\alpha} = \frac{2.4 \times 10^{-3}}{0.9876} = 2.43 \text{ mA}$$

KVL around the collector circuit leads to the minimum value of R_C to assure saturation.

$$R_C = \frac{V_{CC} - V_{CE\text{sat}} - I_{EQ}R_E}{I_{CQ}} = \frac{15 - 0.2 - (2.43)(1)}{2.4 \times 10^{-3}} = 5.154 \text{ k}\Omega$$

8.16 For the circuit of Fig. 8-4(a), $R_E = 300 \ \Omega$, $R_C = 500 \ \Omega$, $V_{CC} = 15$ V, and $\beta = 100$. This Si transistor is β-independent biased. Size R_1 and R_2 for maximum symmetrical swing, if $V_{CE\text{sat}} \approx 0$.

Using the standard factor of 10 to establish the margin of inequality in (8.10),

$$R_B = \frac{\beta R_E}{10} = \frac{(100)(300)}{10} = 3 \text{ k}\Omega \tag{1}$$

For maximum symmetrical swing, the quiescent collector current is

$$I_{CQ} = \frac{1}{2}\left(\frac{V_{CC}}{R_E + R_C}\right) = \frac{15}{2(300 + 500)} = 9.375 \text{ mA}$$

From (8.9) and (1),

$$V_{BB} \approx V_{BEQ} + I_{CQ}(1.1 R_E) = 0.7 + (9.375 \times 10^{-3})(330) = 3.794 \text{ V}$$

Equations (8.7) may now be solved simultaneously to give

$$R_1 = \frac{R_B}{1 - (V_{BB}/V_{CC})} = \frac{3 \times 10^3}{1 - (3.794/15)} = 4.02 \text{ k}\Omega$$

$$R_2 = R_B \frac{V_{CC}}{V_{BB}} = (3 \times 10^3)\left(\frac{15}{3.794}\right) = 11.86 \text{ k}\Omega$$

8.17 For the circuit of Fig. 8-6(a) with a Si transistor, $R_E = 200 \ \Omega$, $R_2 = 10R_1 = 10 \text{ k}\Omega$, $R_L = R_C = 2 \text{ k}\Omega$, $\beta = 100$, and $V_{CC} = 15$ V. Assume that C_C and C_E approach ∞, and that $V_{CE\text{sat}} \approx 0$, $i_C = 0$ at cutoff. Find (a) I_{CQ}, (b) V_{CEQ}, (c) the slope of the ac load line, (d) the slope of the dc load line, and (e) the peak value of undistorted i_L.

(a) Using (8.7) and (8.9),

$$R_B = \frac{(1 \times 10^3)(10 \times 10^3)}{11 \times 10^3} = 909 \ \Omega \qquad V_{BB} = \frac{1 \times 10^3}{11 \times 10^3}(15) = 1.364 \text{ V}$$

$$I_{CQ} \approx \frac{V_{BB} - V_{BEQ}}{[R_B/(\beta + 1)] + R_E} = \frac{1.364 - 0.7}{(909/101) + 200} = 3.177 \text{ mA}$$

(b)   KVL around the collector circuit, with $I_{CQ} \approx I_{EQ}$, gives

$$V_{CEQ} = V_{CC} - I_{CQ}(R_E + R_C) = 15 - (3.177 \times 10^{-3})(2.2 \times 10^3) = 8.01 \text{ V}$$

(c)

$$\text{slope} = \frac{1}{R_{ac}} = \frac{1}{R_C} + \frac{1}{R_L} = 2\left(\frac{1}{2 \times 10^3}\right) = 1 \text{ mS}$$

(d)

$$\text{slope} = \frac{1}{R_{dc}} = \frac{1}{R_C + R_E} = \frac{1}{2.2 \times 10^3} = 0.454 \text{ mS}$$

(e)   The ac load line intersects the v_{CE}-axis at

$$v_{CE\max} = V_{CEQ} + I_{CQ}R_{ac} = 8.01 + (3.177 \times 10^{-3})(1 \times 10^3) = 11.187 \text{ V}$$

Since $v_{CE\max} < 2V_{CEQ}$, cutoff occurs first. With the large capacitors appearing as ac shorts,

$$i_L = \frac{v_L}{R_L} = \frac{v_{ce}}{R_L}$$

or, expressed in terms of peak values,

$$I_{Lm} = \frac{V_{cem}}{R_L} = \frac{v_{CE\max} - V_{CEQ}}{R_L} = \frac{11.187 - 8.01}{2 \times 10^3} = 1.588 \text{ mA}$$

8.18   The amplifier of Fig. 8-16 uses a Si transistor for which $V_{BEQ} = 0.7$ V. Assuming that the collector circuit presents no limitation, classify the amplifier according to bias (class A, AB, B, or C), if (a) $V_B = 1.0$ V, $v_S = 0.25 \cos \omega t$ (V); (b) $V_B = 1.0$ V, $v_S = 0.5 \cos \omega t$ (V); (c) $V_B = 0.5$ V, $v_S = 0.6 \cos \omega t$ (V); (d) $V_B = 0.7$ V, $v_S = 0.5 \cos \omega t$ (V).

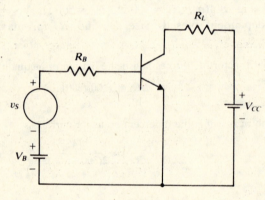

Fig. 8-16

So long as $v_S + V_B > 0.7$ V, the emitter-base junction is forward-biased; thus classification becomes a matter of determining the portion of the period of v_S over which the above inequality holds.

(a) $v_S + V_B \geq 0.75$ V; thus the transistor is always in the active region, and the amplifier is class A.

(b) $0.5 \leq v_S + V_B \leq 1.5$ V; thus the transistor is cut off for a portion of the negative excursion of v_S. Since cutoff is for less than 180°, operation is class AB.

(c) $-0.1 \leq v_S + V_B \leq 1.1$ V, which gives conduction for less than 180° of v_S, or class C operation.

(d) $v_S + V_B \geq 0.7$ V over exactly 180° of v_S, or class B operation.

8.19   Answer the following questions relating to a CE-connected transistor. (a) How are input characteristics (i_B versus v_{BE}) affected if negligible feedback of v_{CE} is present? (b) What is a possible effect of a small or borderline emitter-base junction bias? (c) If the transistor has an infinite output impedance, what is the nature of the output characteristics? (d) With reference to Fig. 8-5(b), does the current gain of the transistor increase or decrease as the mode of operation approaches saturation from the active region?

(a)   The family of input characteristics degenerates to a single curve—a frequently-made approximation.

(b)   If I_{BQ} is so small that operation occurs near the knee of an input characteristic curve, distortion results.

(c)   Slopes of the output characteristic curves are zero in the active region.

(d) Δi_C decreases for a constant Δi_B; hence, current gain decreases.

8.20   Use a small-signal, h-parameter equivalent circuit to analyze the amplifier of Fig. 8-6(a), given $R_C = R_L = 800\ \Omega$, $R_i = 0$, $R_1 = 1.2\ k\Omega$, $R_2 = 2.7\ k\Omega$, $h_{re} \approx 0$, $h_{oe} = 100\ \mu S$, $h_{fe} = 90$, and $h_{ie} = 200\ \Omega$. Calculate (a) the voltage gain (A_v), and (b) the current gain (A_i).

The small-signal circuit is shown in Fig. 8-17, where $R_B = R_1 R_2/(R_1 + R_2) = 831\ \Omega$.

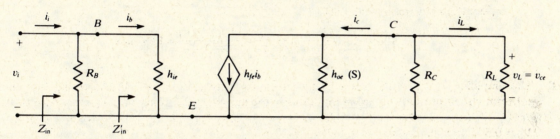

Fig. 8-17

(a)   Applying the current-division rule, (2.11), to the collector circuit,

$$-i_L = \frac{R_C(1/h_{oe})}{R_C(1/h_{oe}) + R_L(1/h_{oe}) + R_L R_C}(h_{fe}i_b)$$

The voltage gain is then

$$A_v \equiv \frac{v_L}{v_i} = \frac{R_L i_L}{h_{ie} i_b} = -\frac{h_{fe} R_L R_C}{h_{ie}(R_C + R_L + h_{oe} R_L R_C)}$$

$$= -\frac{(90)(800)^2}{200[1600 + (100 \times 10^{-6})(800)^2]} = -173.08$$

(b)   By current division,

$$i_b = \frac{R_B}{R_B + h_{ie}}i_i$$

and so

$$A_i \equiv \frac{i_L}{i_i} = \frac{R_B}{R_B + h_{ie}}\frac{i_L}{i_b} = \frac{R_B h_{ie}}{R_L(R_B + h_{ie})}A_v$$

$$= \frac{(831)(200)(-173.08)}{(800)(1031)} = -34.87$$

8.21   If the emitter-base junction of a transistor is modeled as a forward-biased diode, express h_{ie} in terms of emitter current.

Using transistor notation in (7.1) gives

$$i_B = I_{CBO}(e^{v_{BE}/V_T} - 1) \tag{1}$$

Then, by (8.19),

$$\frac{1}{h_{ie}} = \frac{\partial i_B}{\partial v_{BE}}\bigg|_Q = \frac{1}{V_T}I_{CBO}e^{V_{BEQ}/V_T} \tag{2}$$

But, by (1) and Problem 7.1,

$$I_{BQ} = I_{CBO}(e^{V_{BEQ}/V_T} - 1) \approx I_{CBO}e^{V_{BEQ}/V_T} \tag{3}$$

and also,

$$I_{BQ} = \frac{I_{EQ}}{\beta + 1} \tag{4}$$

Equations (2), (3), and (4) imply

$$h_{ie} = \frac{V_T(\beta + 1)}{I_{EQ}}$$

8.22 For the CB amplifier of Problem 8.14, determine graphically (a) h_{fb} and (b) h_{ob}.

The Q-point, established in Problem 8.14, is indicated Fig. 8-15.

(a) By (8.27),

$$h_{fb} \approx \frac{\Delta i_C}{\Delta i_E}\bigg|_{V_{CBQ}=-6.1\,V} = \frac{(3.97 - 2.0) \times 10^{-3}}{(4 - 2) \times 10^{-3}} = 0.985$$

(b) By (8.28),

$$h_{ob} \approx -\frac{\Delta i_C}{\Delta v_{CB}}\bigg|_{I_{EQ}=3\,mA} = -\frac{(3.05 - 2.95) \times 10^{-3}}{-10 - (-2)} = 12.5\ \mu S$$

8.23 Find the input impedance, Z_{in}, of the circuit of Fig. 8-6(a) in terms of the h-parameters, all of which are nonzero.

The small-signal circuit of Fig. 8-17, with $R_B = R_1 R_2/(R_1 + R_2)$, is applicable if a dependent source $h_{re}v_{ce}$ is added, as in Fig. 8-7(a). The admittance of the collection circuit is given by

$$G = h_{oe} + \frac{1}{R_L} + \frac{1}{R_C}$$

and, by Ohm's law,

$$v_{ce} = \frac{-h_{fe}i_b}{G} \tag{1}$$

By KVL applied to the input circuit,

$$i_b = \frac{v_i - h_{re}v_{ce}}{h_{ie}} \tag{2}$$

Eliminate v_{ce} between (1) and (2) to obtain

$$Z'_{in} \equiv \frac{v_i}{i_b} = h_{ie} - \frac{h_{re}h_{fe}}{G} \tag{3}$$

Then,

$$Z_{in} = \frac{R_B Z'_{in}}{R_B + Z'_{in}} = \frac{R_B[h_{ie} - (h_{re}h_{fe}/G)]}{R_B + h_{ie} - (h_{re}h_{fe}/G)} \tag{4}$$

8.24 In the two-stage amplifier of Fig. 8-18, both transistors are identical, having $h_{ie} = 1500\ \Omega$, $h_{fe} = 40$, $h_{re} \approx 0$, and $h_{oe} = 30\ \mu S$. Also, $R_i = 1\ k\Omega$, $R_{C2} = 20\ k\Omega$, $R_{C1} = 10\ k\Omega$, and

$$R_{B1} \equiv \frac{R_{11}R_{12}}{R_{11} + R_{12}} = 5\ k\Omega \qquad R_{B2} = \frac{R_{21}R_{22}}{R_{21} + R_{22}} = 5\ k\Omega$$

Find (a) final-stage voltage gain, $A_{v2} \equiv v_o/v_{o1}$; (b) final-stage input impedance, Z_{in2}; (c) initial-stage voltage gain, $A_{v1} \equiv v_{o1}/v_{in}$; (d) amplifier input impedance, Z_{in1}; (e) amplifier voltage gain, $A_v \equiv v_o/v_i$.

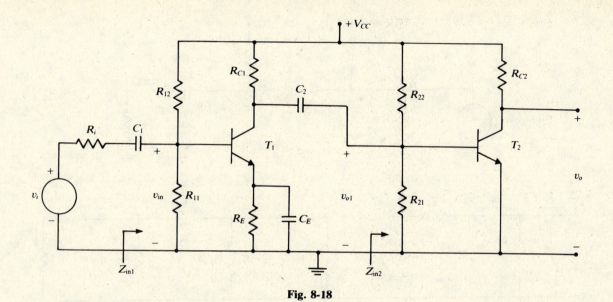

Fig. 8-18

(a) Final-stage voltage gain is given by the result of Problem 8.20(a) if the parallel combination of R_L and R_C is replaced by R_{C2}.

$$A_{v2} = -\frac{h_{fe}R_{C2}}{h_{ie}(1 + h_{oe}R_{C2})} = -\frac{(40)(20 \times 10^3)}{(1500)[1 + (30 \times 10^{-6})(20 \times 10^3)]} = -333.3$$

(b) From (4) of Problem 8.23, with $h_{re} \approx 0$,

$$Z_{in2} = \frac{R_{B2}h_{ie}}{R_{B2} + h_{ie}} = \frac{(5 \times 10^3)(1500)}{(5 \times 10^3) + 1500} = 1.154 \text{ k}\Omega$$

(c) Initial-stage voltage gain is given by the result of Problem 8.20(a) if R_C and R_L are respectively replaced by R_{C1} and Z_{in2}.

$$A_{v1} = -\frac{h_{fe}Z_{in2}R_{C1}}{h_{ie}(R_{C1} + Z_{in2} + h_{oe}Z_{in2}R_{C1})} = -\frac{(40)(1154)(10^4)}{(1500)(10^4 + 1154 + 346.2)} = -26.8$$

(d) As in (b),

$$Z_{in1} = \frac{R_{B1}h_{ie}}{R_{B1} + h_{ie}} = 1.154 \text{ k}\Omega$$

(e) By voltage division,

$$\frac{v_{in}}{v_i} = \frac{Z_{in1}}{Z_{in1} + R_i} = \frac{1154}{1154 + 1000} = 0.5357$$

$$A_v \equiv \frac{v_o}{v_i} = \frac{v_{in}}{v_i}A_{v1}A_{v2} = (0.5357)(-26.8)(-333.3) = 4786$$

8.25 In terms of the CB h-parameters for the amplifier of Fig. 8-19(a), find (a) the input impedance, Z_{in}; (b) the voltage gain, A_v; (c) the current gain, A_i.

(a) The h-parameter equivalent circuit is given in Fig. 8-19(b). By Ohm's law,

$$v_{cb} = \frac{h_{fb}i_e}{h_{ob} + (1/R_C) + (1/R_L)} \equiv \frac{h_{fb}i_e}{G} \tag{1}$$

Application of KVL at the input gives

$$v_S = h_{rb}v_{cb} + h_{ib}i_e \tag{2}$$

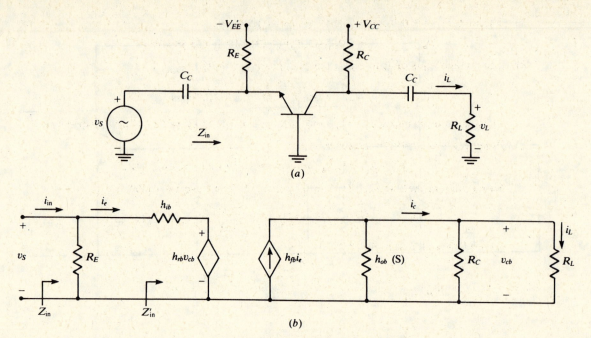

Fig. 8-19

Substitute (1) into (2) and solve for $Z'_{in} \equiv v_S/i_e$. Finally, Z_{in} is found as the parallel combination of Z'_{in} and R_E.

$$Z_{in} = \frac{R_E(h_{ib}G + h_{rb}h_{fb})}{R_E G + h_{ib}G + h_{rb}h_{fb}} \qquad (3)$$

(b) By elimination of i_e between (1) and (2),

$$A_v = \frac{v_{cb}}{v_S} = \frac{h_{fb}}{h_{ib}G + h_{rb}h_{fb}}$$

(c) By (1),

$$i_L = \frac{v_{cb}}{R_L} = \frac{h_{fb}i_e}{R_L G} \qquad (4)$$

By KCL at the emitter node,

$$i_e = i_{in} - \frac{v_S}{R_E} = i_{in} - \frac{i_{in}Z_{in}}{R_E} = i_{in}\left(1 - \frac{Z_{in}}{R_E}\right) \qquad (5)$$

Eliminate i_e between (4) and (5) to obtain

$$A_i = \frac{i_L}{i_{in}} = \frac{h_{fb}}{R_L G}\left(1 - \frac{Z_{in}}{R_E}\right)$$

8.26 Add a large, emitter bypass capacitor to the circuit of Fig. 8-4(a), and let $V_{CC} = 24$ V, $R_C = 5$ kΩ, and $R_E = 200$ Ω. Assume that R_1 and R_2 are selected for maximum symmetrical swing but have negligible power dissipation. Also, let $V_{CE\,sat} \approx 0$ and $I_{CBO} = 0$. (a) Find the greatest possible amplitude of undistorted collector current. Calculate, for 0, 50%, and 100% of this maximum undistorted collector current, (b) P_S, (c) P_O, (d) P_C, and (e) efficiency. The output signal is taken from the common to the collector terminal.

(a)   maximum amplitude $= I_{CO} = \dfrac{1}{2}\left(\dfrac{V_{CC}}{R_C + R_E}\right) = \dfrac{1}{2}\left(\dfrac{24}{5000 + 200}\right) = 2.308$ mA

Thus calculations are to be made for $I_{cm} = 0$, 1.154, 2.308 mA.

(b) The supply power is independent of I_{cm} and is given by

$$P_S = V_{CC}I_{CO} = (24)(2.308 \times 10^{-3}) = 55.38 \text{ mW}$$

(c) The average value of ac or signal power delivered to R_C is

$$P_O = I_c^2 R_C = \frac{I_{cm}^2 R_C}{2} = 2500 I_{cm}^2 \quad \text{(W)}$$

which when evaluated at $I_{cm} = 0, 1.154, 2.308$ mA gives $0, 3.33$, and 13.32 mW, respectively.

(d) From (8.32), the collector power is

$$P_C = P_S - I_{CQ}^2(R_C + R_E) - P_O = (27.7 \times 10^{-3}) - P_O \quad \text{(W)}$$

Substitution of the three values from (c) gives 27.7, 24.37, and 14.38 mW.

(e)
$$\eta = \frac{P_O}{P_S} \times 100\% = 0, \frac{333\%}{55.38}, \frac{1332\%}{55.38} = 0, 6.01\%, 24.04\%$$

All results are plotted in Fig. 8-20.

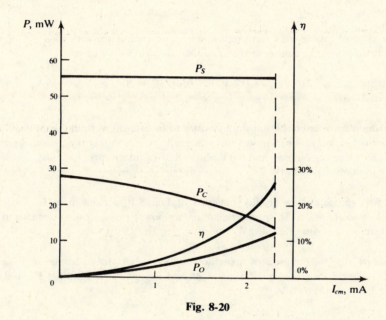

Fig. 8-20

Supplementary Problems

8.27 If the leakage currents of a transistor are $I_{CBO} = 5$ μA and $I_{CEO} = 0.4$ mA, and if $I_B = 30$ μA, determine the value of I_C.   *Ans.* 2.77 mA

8.28 A transistor has a 10-V dc source across the collector and emitter terminals with proper polarity to reverse-bias the collector-base junction; the base remains open-circuit. A microammeter in the collector circuit reads 800 μA. (a) If $\alpha = 0.98$ for this transistor, what is the value of leakage current I_{CBO}? (b) If the transistor is a Si device and $I_{CBO} = 3$ μA at 25 °C, what was its temperature when the data were taken?   *Ans.* (a) 16 μA; (b) 39.5 °C

8.29   Collector-to-base leakage current can be modeled by a current source as in Fig. 8-21, it being understood that transistor action relates currents I'_C, I'_B, and I_E ($I'_C = \alpha I_E$ and $I'_C = \beta I'_B$). Prove that

$$(a) \quad I_C = \beta I_B + (\beta + 1)I_{CBO} \qquad (b) \quad I_B = \frac{I_E}{\beta + 1} - I_{CBO} \qquad (c) \quad I_E = \frac{\beta + 1}{\beta}(I_C - I_{CBO})$$

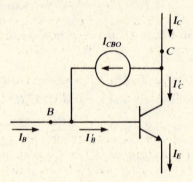

Fig. 8-21

8.30   If the transistor of Problem 8.4 were replaced by a new transistor with an α 1% greater, what would be the percentage change in emitter current?   *Ans.*   96.07% increase

8.31   For the circuit of Fig. 8-10, $V_{CEsat} = 0.2$ V, $\alpha = 0.99$, $I_{BQ} = 20$ μA, $V_{CC} = 15$ V, and $R_C = 15$ kΩ. What is the value of V_{CEQ}?   *Ans.* $V_{CEQ} = V_{CEsat} = 0.2$ V

8.32   Many switching applications permit the transistor to be utilized without a heat sink, since $P_C \approx 0$ in cutoff and P_C is small in saturation. Support this statement by calculating the collector power dissipated in (*a*) Problem 8.6 (active-region bias) and (*b*) Problem 8.31 (saturation-region bias).
Ans.   (*a*) 18 mW; (*b*) 0.39 mW

8.33   The collector characteristics of the transistor of Fig. 8-10 are given in Fig. 8-5(*b*). If $I_{BQ} = 40$ μA, $V_{CC} = 15$ V, and $R_C = 2.2$ kΩ, specify the minimum power rating of the transistor to assure no danger of thermal damage.   *Ans.*   22.54 mW

8.34   The circuit of Fig. 8-22 illustrates a method to bias a CB transistor using a single dc source. The transistor is a Si device ($V_{BEQ} = 0.7$ V), $\beta = 99$, and $I_{BQ} = 30$ μA. Find (*a*) R_2, and (*b*) V_{CEQ}.
Ans.   (*a*) 3.36 kΩ; (*b*) 6.06 V

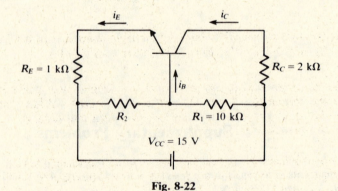

Fig. 8-22

8.35   Work Problem 8.11(*a*) if $R_2 = 5$ kΩ and all else remains unchanged.
Ans. ± 13.16 mA about $I_{CQ} = 16.84$ mA

8.36   Because of a poor solder joint, the resistor R_1 of Problem 8.11(*a*) becomes open-circuit. Calculate the percentage change in I_{CQ} that will be observed.   *Ans.*   +508.5%

8.37   The circuit of Problem 8.11(a) has β-independent bias ($R_E \geq 10 R_B/\beta$). Find the allowable range of β if I_{CQ} can change at most $\pm 2\%$ from its value for $\beta = 110$.   *Ans.* $86.4 \leq \beta \leq 149.7$

8.38   Show that the ac load line intercepts the axes of the collector characteristics at $I_{CQ} + (V_{CEQ}/R_{ac})$ and $V_{CEQ} + I_{CQ}R_{ac}$.

8.39   For the circuit of Fig. 8-16, $v_S = 0.25 \cos \omega t$ (V), $R_B = 30$ kΩ, $V_B = 1$ V, and $V_{CC} = 12$ V. The transistor is a Si device with negligible base-to-emitter resistance. Assume that $V_{CEsat} \approx 0$ and $I_{CBO} = 0$. Find the range of R_L for class A operation.   *Ans.* $R_L \leq 6.545$ kΩ

8.40   A CE transistor amplifier is operating in the active region, with $V_{CC} = 12$ V and $R_{dc} = 2$ kΩ. If the collector characteristics are given by Fig. 8-5(c) and the quiescent base current is 30 μA, determine (a) h_{fe} and (b) h_{oe}.   *Ans.*   (a) 190; (b) 83.33 μS

8.41   For the circuit of Fig. 8-23, $h_{re} = 10^{-4}$, $h_{ie} = 200$ Ω, $h_{fe} = 100$, and $h_{oe} = 100$ μS. (a) Find the power gain as $A_p = |A_i A_v|$, the product of the current and voltage gains. (b) Determine the numerical value of R_L that maximizes the power gain.   *Ans.*   (a) $h_{fe}^2/|(h_{oe}R_L + 1)(h_{oe}h_{ie} - h_{re}h_{fe} + h_{ie}R_L^{-1})|$; ($b$) 14.14 k$\Omega$

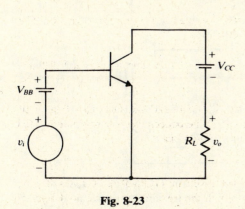

Fig. 8-23

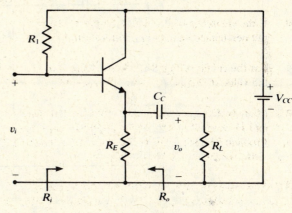

Fig. 8-24

8.42   If a large, emitter bypass capacitor is added to the circuit of Fig. 8-4(a) and if the power dissipated in R_1 and R_2 is negligible, show that the power dissipated in the transistor collector is

$$P_C = V_{CEQ}I_{CQ} - \frac{I_{cm}^2 R_C}{2}$$

8.43   On the assumption that $h_{re} = h_{oe} = 0$, show that the input impedance of the CB h-parameter equivalent circuit is given by $h_{ib} = h_{ie}/(h_{fe} + 1)$. (*Hint:*   Use (8.23), with $v_{cb} = 0$.)

8.44   The constant-base-current-biased, emitter-follower amplifier of Fig. 8-24 utilizes a Si transistor with negligible leakage current and $\beta = 59$. Also, $V_{CC} = 15$ V, $V_O = 3$ V (dc component of v_o), and $R_E = 1.5$ kΩ. Calculate (a) R_1; (b) output resistance, R_o; (c) input resistance, R_i.
Ans.   (a) 339 kΩ; (b) 1.185 kΩ; (c) 50.98 kΩ

8.45   For the circuit of Problem 8.11(a), find the quiescent power dissipated in (a) R_1, (b) R_2, (c) R_C, (d) R_E, (e) the collector junction. (f) How much power is delivered by the dc supply?
Ans.   (a) 1.42 mW; (b) 11.8 mW; (c) 7.16 mW; (d) 4.86 mW; (e) 61.24 mW; (f) 86.55 mW

8.46   The Si transistor in the circuit of Fig. 8-12 has negligible leakage current and $\beta = 90$. Let $R_B = 500$ kΩ, $R_C = 3$ kΩ, and $V_{CC} = 15$ V. If the transistor is rated at 50 mW (collector power) at 25 °C, and by the manufacturer's specification must be derated by 1.2 mW/°C for temperatures above 25 °C, what is the maximum ambient temperature in which the circuit can be safely operated? Assume temperature-independent resistance values.   *Ans.*   51 °C

8.47   The *stability factor*,

$$S_I \equiv \frac{\partial I_C}{\partial I_{CBO}}$$

of an amplifier circuit is a measure of the sensitivity of its Q-point to changes in temperature. Using the results of Problem 8.29, find the stability factor for the circuit of Fig. 8-21 (a common arrangement).

Ans. $S_I = \dfrac{R_E + R_B}{R_E + (1 - \alpha)R_B}$

8.48   For the Darlington-pair emitter-follower of Fig. 8-25, $h_{re1} = h_{re2} = h_{oe1} = h_{oe2} = 0$. In terms of the (nonzero) h-parameters, find expressions for (a) Z_{in}; (b) voltage gain, $A_v \equiv v_E/v_S$; (c) current gain, $A_i \equiv i_{e2}/i_{in}$; (d) Z_{in}; and (e) Z_o (if the signal source has internal resistance R_S).

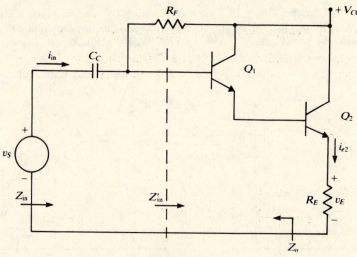

Fig. 8-25

Ans.   (a) $Z'_{in} = h_{ie1} + (h_{fe1} + 1)[h_{ie2} + (h_{fe2} + 1)R_E]$

(b) $A_v = \dfrac{(h_{fe1} + 1)(h_{fe2} + 1)R_E}{Z'_{in}}$

(c) $A_i = \dfrac{(h_{fe1} + 1)(h_{fe2} + 1)R_F}{R_F + Z'_{in}}$

(d) $Z_{in} = \dfrac{R_F Z'_{in}}{R_F + Z'_{in}}$

(e) $Z_o = h_{ie2} + \dfrac{[R_S R_F/(R_S + R_F)] + h_{ie1}}{(h_{fe1} + 1)(h_{fe2} + 1)}$

Chapter 9

Field-Effect Transistors

The operation of the *field-effect transistor* (FET) can be explained in terms of only majority-carrier (one-polarity) charge flow; the transistor is therefore called *unipolar*. Two kinds of field-effect devices are widely used, the *junction field-effect transistor* (JFET) and the *metal-oxide-semiconductor field-effect transistor* (MOSFET).

9.1 CONSTRUCTION AND SYMBOLS OF THE JFET

Physical arrangements and symbols of the JFET are shown in Fig. 9-1. Conduction is by passage of charge carriers from *source* (S) to *drain* (D) through the path (*channel*) between the *gate* elements.

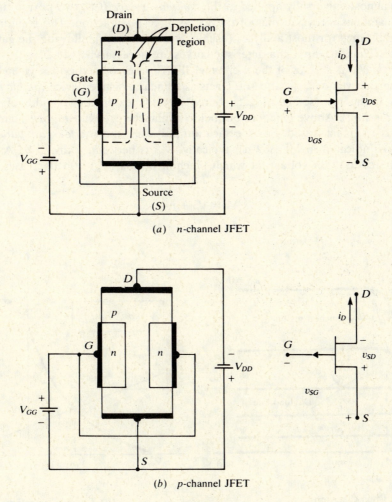

(a)   *n*-channel JFET

(b)   *p*-channel JFET

Fig. 9-1

137

The transistor can be *n-channel* (conduction by electrons) or *p-channel* (conduction by holes); any discussion of *n*-channel applies to *p*-channel devices if *complementary* (opposite in sign) voltages and currents are used. Analogies between the JFET and the BJT are drawn in Table 9-1. Current and voltage symbology for FETs parallels Table 8-1.

Table 9-1

JFET	BJT
source (S)	emitter (E)
drain (D)	collector (C)
gate (G)	base (B)
drain supply (V_{DD})	collector supply (V_{CC})
gate supply (V_{GG})	base supply (V_{BB})
drain current (i_D)	collector current (i_C)

9.2 TERMINAL CHARACTERISTICS OF THE JFET

The JFET is almost universally applied in the *common-source* (CS) two-port arrangement of Fig. 9-1, where v_{GS} maintains a reverse bias of the gate-source *pn* junction. The resulting gate leakage current is negligibly small for most analyses (usually less than $1\ \mu$A), allowing the gate to be treated as an open circuit. Thus, no input characteristic curves are necessary.

Output or *drain characteristics* of the *n*-channel JFET in CS connection are given by Fig. 9-2(a) if $v_{GS} \le 0$. For a constant value of v_{GS}, the JFET acts as a linear resistive device (*ohmic region*) until the *depletion region* of the reverse-biased gate-source junction extends the width of the channel (a condition called *pinchoff*). Above pinchoff, but below avalanche breakdown, drain current i_D remains nearly constant as v_{DS} is increased. For specification purposes, the *shorted-gate parameters* I_{DSS} and V_{p0} are defined as indicated in Fig. 9-2(a); typically, V_{p0} is between 4 and 5 V. As gate potential decreases, the source-to-drain voltage at which pinchoff occurs (V_p) also decreases, approximately obeying the equation

$$V_p = V_{p0} + v_{GS} \qquad (9.1)$$

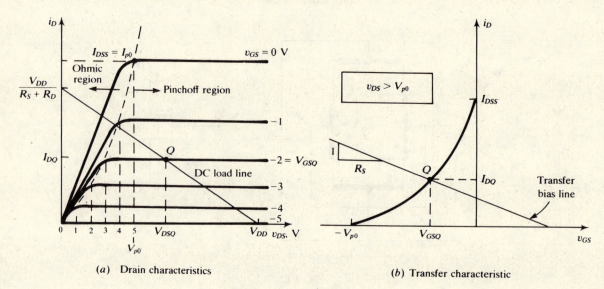

(*a*) Drain characteristics

(*b*) Transfer characteristic

Fig. 9-2. CS *n*-Channel JFET

Under reasonable approximations, the drain current shows a square-law dependence on source-to-gate voltage for constant values of v_{DS} in the pinchoff region:

$$i_D = I_{DSS}\left(1 + \frac{v_{GS}}{V_{p0}}\right)^2 \tag{9.2}$$

This accounts for the unequal vertical spacing of the characteristic curves in Fig. 9-2(a). Figure 9-2(b) is the graph of (9.2), known as the *transfer characteristic*, utilized in bias determination. The transfer characteristic is also determined by the intersections of the drain characteristics with a fixed vertical line, v_{DS} = constant. To the extent that the drain characteristics actually are horizontal in the pinchoff region, one and the same transfer characteristic will be found for all $v_{DS} > V_{p0}$. (See Fig. 9-4 for a slightly nonideal case.)

9.3 BIAS LINE AND LOAD LINE

The commonly used, *voltage divider* bias arrangement of Fig. 9-3(a) can be reduced to the equivalent in Fig. 9-3(b), where the Thévenin parameters are given by

$$R_G = \frac{R_1 R_2}{R_1 + R_2} \qquad V_{GG} = \frac{R_1}{R_1 + R_2}\,V_{DD} \tag{9.3}$$

With $i_G = 0$, application of KVL around the gate-source loop of Fig. 9-3(b) yields the equation of the *transfer bias line*,

$$i_D = \frac{V_{GG}}{R_S} - \frac{v_{GS}}{R_S} \tag{9.4}$$

which can be solved simultaneously with (9.2) or plotted as indicated on Fig. 9-2(b) to yield I_{DQ} and V_{GSQ}, two of the necessary three quiescent variables.

KVL around the drain-source loop of Fig. 9-3(b) leads to the equation of the dc load line,

$$i_D = \frac{V_{DD}}{R_S + R_D} - \frac{v_{DS}}{R_S + R_D} \tag{9.5}$$

which when plotted on the drain characteristics of Fig. 9-2(a) yields the remaining quiescent value, V_{DSQ}. Alternatively, with I_{DQ} already determined,

$$V_{DSQ} = V_{DD} - (R_S + R_D)I_{DQ}$$

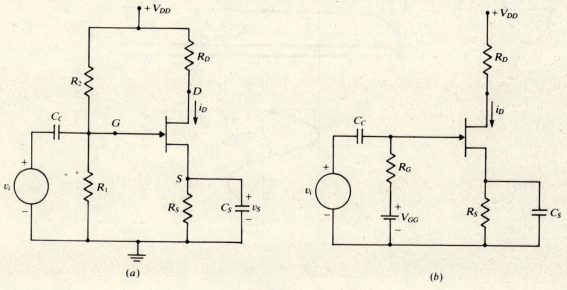

(a)   (b)

Fig. 9-3

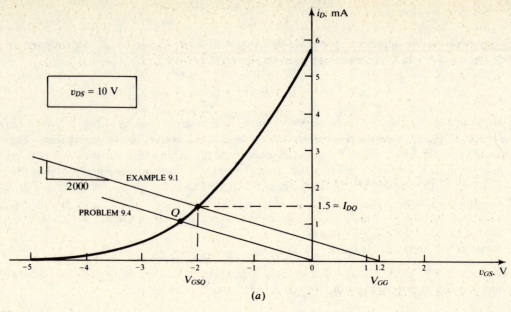

$v_{DS} = 10$ V

EXAMPLE 9.1

PROBLEM 9.4

$1.5 = I_{DQ}$

V_{GSQ}

V_{GG}

(a)

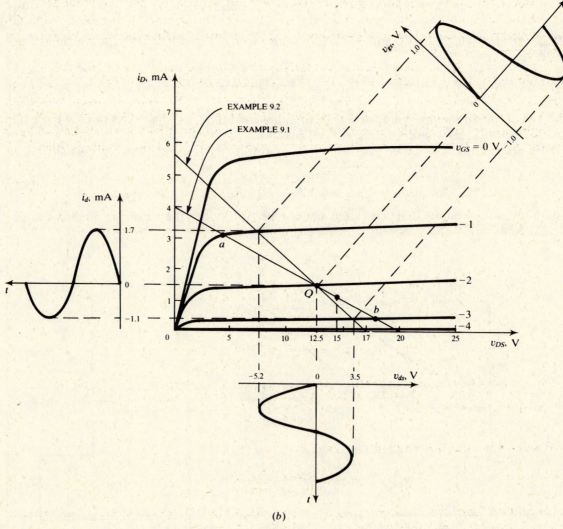

(b)

Fig. 9-4

Example 9.1 For the amplifier of Fig. 9-3(a), $V_{DD} = 20$ V, $R_1 = 1$ MΩ, $R_2 = 15.7$ MΩ, $R_D = 3$ kΩ, and $R_S = 2$ kΩ. If the JFET characteristics are given by Fig. 9-4, find (a) I_{DQ}, (b) V_{GSQ}, and (c) V_{DSQ}.

(a) By (9.3),

$$V_{GG} = \frac{R_1}{R_1 + R_2} V_{DD} = \frac{1 \times 10^6}{16.7 \times 10^6} (20) = 1.2 \text{ V}$$

On Fig. 9-4(a), construct the transfer bias line, (9.4); it intersects the transfer characteristic at the Q-point, giving $I_{DQ} = 1.5$ mA.

(b) The Q-point of Fig. 9-4(a) also gives $V_{GSQ} = -2$ V.

(c) Construct the dc load line on the drain characteristics, with v_{DS}-intercept $V_{DD} = 20$ V and i_D-intercept $V_{DD}/(R_S + R_D) = 4$ mA. The Q-point was established at $I_{DQ} = 1.5$ mA in (a) [and at $V_{GSQ} = -2$ V in (b)]; its abscissa is $V_{DSQ} = 12.5$ V. Analytically,

$$V_{DSQ} = V_{DD} - (R_S + R_D)I_{DQ} = 20 - (5 \times 10^3)(1.5 \times 10^{-3}) = 12.5 \text{ V}$$

Bias to Establish Bounds on Q-Point Location

Analogous to β-variations in the BJT, the shorted-gate parameters, I_{DSS} and V_{p0}, of the FET may vary widely within devices of the same classification. As will now be shown, it is possible to set the gate-source bias so that, in face of this variation, the Q-point, and hence the quiescent drain current, is confined to fixed limits.

The extremes of parameter variation are usually specified by the manufacturer, and (9.2) may be used to establish upper and lower (*worst-case*) transfer characteristics, as displayed in Fig. 9-5. The upper and lower quiescent points, Q_{max} and Q_{min}, are determined by their ordinates, I_{DQmax} and I_{DQmin}; we assign I_{DQmax} and I_{DQmin} on the basis of the maximum allowable variation of I_{DQ} along a dc load line superimposed on the family of nominal drain characteristics. This dc load line is established by so choosing $R_D + R_S$ that v_{DS} remains within a desired region of the nominal drain characteristics.

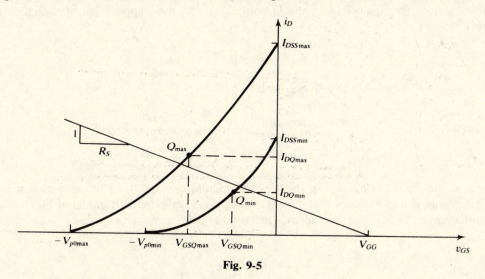

Fig. 9-5

If now a value of R_S is selected such that

$$R_S \geq \frac{|V_{GSQmax} - V_{GSQmin}|}{I_{DQmax} - I_{DQmin}} \tag{9.6}$$

and the transfer bias line with slope $-1/R_S$ and v_{GS}-intercept $V_{GG} \geq 0$ is drawn as in Fig. 9-5, then the nominal Q-point is forced to lie beneath Q_{max} and above Q_{min}, so that

$$I_{DQmin} \leq I_{DQ} \leq I_{DQmax}$$

as desired.

With R_S, R_D, and V_{GG} already assigned, R_G is chosen large enough to give a satisfactory input impedance, and then R_1 and R_2 are determined from equations (9.3). Generally, R_S will be comparable in magnitude to R_D; in order to obtain desirable ac gains, a bypass capacitor must be used with R_S and an ac load line introduced, with analysis techniques similar to those of Section 8.7.

9.4 GRAPHICAL ANALYSIS OF JFET AMPLIFIERS

Graphical analysis is favored for large-ac-signal conditions in the JFET since the square-law relation between v_{GS} and i_D leads to signal distortion.

Example 9.2 For the amplifier of Example 9.1, let $v_i = \sin t$ ($\omega = 1$ rad/s) and $C_S \to \infty$. Graphically determine v_{ds} and i_d.

Since C_S appears as a short to ac signals, an ac load line must be added to Fig. 9-4(b), passing through the Q-point and intersecting the v_{DS}-axis at

$$V_{DSQ} + I_{DQ}R_{ac} = 12.5 + (1.5)(3) = 17 \text{ V}$$

Construct an auxiliary time axis through Q perpendicular to the ac load line. Additional auxiliary axes have been constructed in Fig. 9-4(b) to display the excursions of i_d and v_{ds} as $v_{gs} = v_i$ swings ± 1 V along the ac load line. Note the distortion in both signals introduced by the square-law behavior of the JFET characteristics.

9.5 SMALL-SIGNAL EQUIVALENT CIRCUITS FOR THE JFET

From Fig. 9-2(a), it is seen that if i_D is taken as the dependent variable of the CS-connected JFET, then

$$i_D = f(v_{GS}, v_{DS}) \qquad (9.7)$$

For small excursions (ac signals) about the Q-point, $\Delta i_D = i_d$; thus, application of the chain rule to (9.7) leads to

$$i_d = \Delta i_D \approx di_D = g_m v_{gs} + \frac{1}{r_{ds}} v_{ds} \qquad (9.8)$$

where the partial derivatives are denoted as follows:

$$\textit{\textbf{transconductance}} \quad g_m \equiv \frac{\partial i_D}{\partial v_{GS}}\bigg|_Q \approx \frac{\Delta i_D}{\Delta v_{GS}}\bigg|_Q \qquad (9.9)$$

$$\textit{\textbf{source-drain resistance}} \quad r_{ds} \equiv \frac{\partial v_{DS}}{\partial i_D}\bigg|_Q \approx \frac{\Delta v_{DS}}{\Delta i_D}\bigg|_Q \qquad (9.10)$$

So long as the JFET is operated in the pinchoff region, $i_G = i_g = 0$, or the gate acts as an open circuit. This, along with (9.8), leads to the current-source equivalent circuit model of Fig. 9-6(a). The voltage-source model of Fig. 9-6(b) is derived in Problem 9.5.

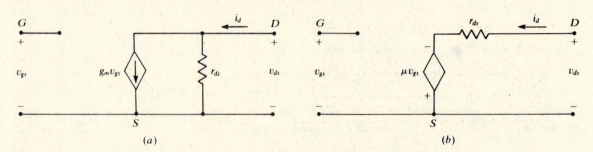

Fig. 9-6. Small-Signal Models of the CS JFET

9.6   CONSTRUCTION AND SYMBOLS OF THE MOSFET

The *n*-channel MOSFET (Fig. 9-7) has only a single *p* region (called the *substrate*), one side of which acts as a conducting channel. A metallic gate is separated by an insulating metal oxide (usually SiO_2), whence the name *insulated-gate FET* (IGFET) for the device. The *p*-channel MOSFET formed by interchanging *p* and *n* semiconductor materials is described by complementary voltages and currents.

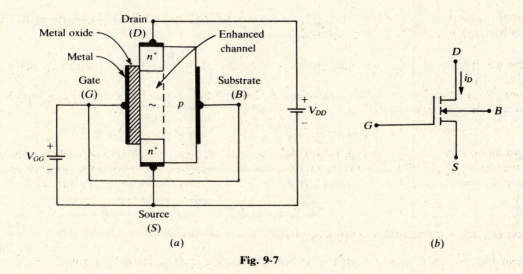

Fig. 9-7

9.7   TERMINAL CHARACTERISTICS OF A MOSFET

In an *n*-channel MOSFET, the gate (positive plate), metal oxide film (dielectric), and substrate (negative plate) form a capacitor, the electric field of which controls channel resistance. When the positive potential of the gate reaches *threshold voltage* V_T (typically 2 to 4 V), sufficient free electrons are attracted to the region immediately beside the metal oxide film (this is called *enhancement-mode* operation) to induce a conducting channel of low resistivity. If the source-to-drain voltage is increased, the enhanced channel is depleted of free charge carriers in the area near the drain, and pinchoff occurs as in the JFET. Typical drain and transfer characteristics are displayed in Fig. 9-8,

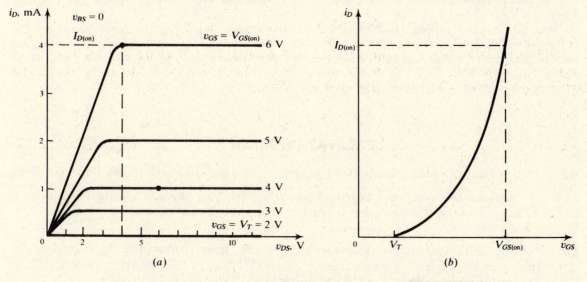

Fig. 9-8.   Characteristics of *n*-Channel Enhancement-Mode MOSFET

wherein $V_T = 2$ V is used for illustration. Commonly, the manufacturer specifies V_T and a value of pinchoff current, $I_{D(on)}$; the corresponding value of source-to-gate voltage is $V_{GS(on)}$.

The enhancement-mode MOSFET, operating in the pinchoff region, is described by (9.1) and (9.2) if V_{p0} and I_{DSS} are respectively replaced by $-V_T$ and $I_{D(on)}$, and if the substrate is shorted to the source, as in Fig. 9-9(a).

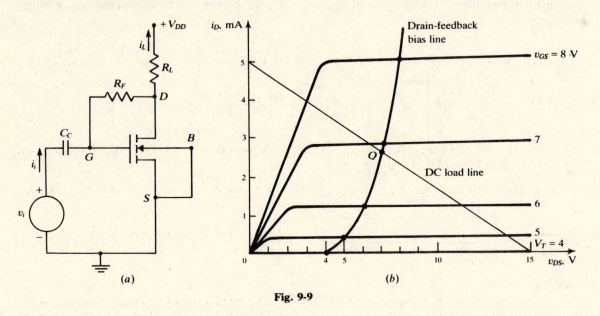

(a)   (b)

Fig. 9-9

9.8   BIAS OF ENHANCEMENT-MODE MOSFET

The voltage divider bias (Fig. 9-3) is readily applicable to the enhancement-mode MOSFET; however, since V_{GSQ} and V_{DSQ} are of the same polarity, the method of *drain-feedback bias*, illustrated in Fig. 9-9(a), can be utilized to compensate partially for variations in MOSFET characteristics.

Example 9.3   For the amplifier of Fig. 9-9(a), $V_{DD} = 15$ V, $R_L = 3$ kΩ, and $R_F = 50$ MΩ. If the MOSFET drain characteristics are given by Fig. 9-9(b), determine the values of the quiescent quantities.

The dc load line is constructed on Fig. 9-9(b) with v_{DS}-intercept $V_{DD} = 15$ V and i_D-intercept $V_{DD}/R_L = 5$ mA.

With gate current negligible, no voltage appears across R_F, and so $V_{GS} = V_{DS}$. The *drain-feedback bias line* of Fig. 9-9(b) is the locus of all points for which $V_{GS} = V_{DS}$. Since the Q-point must lie on both the dc load line and the drain-feedback bias line, their intersection is the Q-point. From Fig. 9-9(b), $I_{DQ} \approx 2.65$ mA and $V_{DSQ} = V_{GSQ} \approx 6.90$ V.

Solved Problems

9.1   If $C_S = 0$ and all else is unchanged in Example 9.1, find the extremes between which v_S swings.

Voltage v_{gs} will swing along the dc load line of Fig. 9-4(b) from point a to b, giving as extremes of i_D 3.1 mA and 0.4 mA. The corresponding extremes of $v_S = i_D R_S$ are 6.2 V and 0.8 V.

9.2   For the JFET amplifier of Example 9.1, use the drain characteristic to determine the small-signal equivalent circuit constants (a) g_m and (b) r_{ds}. (c) Alternatively, evaluate g_m from the transfer characteristic.

(a) Let v_{gs} change by ± 1 V about the Q-point of Fig. 9-4(b); then, by (9.9),

$$g_m \approx \left.\frac{\Delta i_D}{\Delta v_{GS}}\right|_Q = \frac{(3.3 - 0.3) \times 10^{-3}}{2} = 1.5 \text{ mS}$$

(b) At the Q-point of Fig. 9-4(b), while v_{DS} changes from 5 V to 20 V, i_D changes from 1.4 mA to 1.6 mA; by (9.10),

$$r_{ds} \approx \left.\frac{\Delta v_{DS}}{\Delta i_D}\right|_Q = \frac{20 - 5}{(1.6 - 1.4) \times 10^{-3}} = 75 \text{ k}\Omega$$

(c) At the Q-point of Fig. 9-4(a), while i_D changes from 1 mA to 2 mA, v_{GS} changes from -2.4 V to -1.75 V; by (9.9),

$$g_m \approx \left.\frac{\Delta i_D}{\Delta v_{GS}}\right|_Q = \frac{(2 - 1) \times 10^{-3}}{-1.75 - (-2.4)} = 1.54 \text{ mS}$$

9.3 For the MOSFET amplifier of Example 9.3, let $V_{GSQ} = 6.90$ V. Calculate I_{DQ} from the analog of (9.2) developed in Section 9.7.

From the drain characteristics of Fig. 9-9(b), it is seen that $V_T = 4$ V and that $I_{D(\text{on})} = 5$ mA at $V_{GS(\text{on})} = 8$ V. Thus,

$$I_{DQ} = I_{D(\text{on})}\left(1 - \frac{V_{GSQ}}{V_T}\right)^2 = (5 \times 10^{-3})\left(1 - \frac{6.90}{4}\right)^2 = 2.63 \text{ mA}$$

(compare Example 9.3).

9.4 By a method called *self-bias*, the Q-point of a JFET amplifier may be established using only a single resistor from gate to ground [Fig. 9-3(b), with $V_{GG} = 0$]. If $R_D = 3$ kΩ, $R_S = 2$ kΩ, $R_G = 5$ MΩ, $V_{DD} = 20$ V, and the JFET characteristics are given by Fig. 9-4, find (a) I_{DQ}, (b) V_{GSQ}, and (c) V_{DSQ}.

(a) On Fig. 9-4(a) construct a transfer bias line having v_{GS}-intercept $V_{GG} = 0$ and slope $-1/R_S = -0.5$ mS; the ordinate of its intersection with the transfer characteristic is $I_{DQ} = 1.15$ mA.

(b) The abscissa of the Q-point of Fig. 9-4(a) is $V_{GSQ} = -2.3$ V.

(c) The dc load line from Example 9.1, already constructed on Fig. 9-4(b), is applicable. The Q-point was established at $I_{DQ} = 1.15$ mA in (a); the corresponding abscissa is $V_{DSQ} \approx 14.2$ V.

9.5 From the current-source small-signal model of Fig. 9-6(a) derive the voltage-source model of Fig. 9-6(b).

Find the Thévenin equivalent looking to the left through the output terminals of Fig. 9-6(a). If all independent sources are deactivated, $v_{gs} = 0$; thus, $g_m v_{gs} = 0$, or the dependent source also is deactivated (open circuit for current source), and the Thévenin resistance is $R_{Th} = r_{ds}$. The open-circuit voltage appearing at the output terminals is $v_{Th} = v_{ds} = -g_m v_{gs} r_{ds} = -\mu v_{gs}$ where a new equivalent-circuit constant has been defined as

$$\textit{\textbf{amplification factor}} \quad \mu \equiv g_m r_{ds}$$

Proper series arrangement of v_{Th} and R_{Th} leads to Fig. 9-6(b).

9.6 Replace the JFET of Fig. 9-3 with an *n*-channel enhancement-mode MOSFET characterized by Fig. 9-8. Let $V_{DD} = 8$ V, $V_{GSQ} = 4$ V, $I_{DQ} = 1$ mA, $R_1 = 5$ MΩ, and $R_2 = 3$ MΩ. Find (a) V_{DSQ}, (b) V_{GG}, (c) R_S, and (d) R_D.

(a) The Q-point can be located on the drain characteristics of Fig. 9-8(a) at $v_{GS} = V_{GSQ} = 4$ V and $i_D = I_{DQ} = 1$ mA; the abscissa of the Q-point is $V_{DSQ} = 6$ V.

(b) By (9.3), $V_{GG} = R_1 V_{DD}/(R_1 + R_2) = 5$ V.

(c)   Application of KVL around the gate-source loop of Fig. 9-3(b) with $i_G = 0$ leads to

$$R_S = \frac{V_{GG} - V_{GSQ}}{I_{DQ}} = \frac{5 - 4}{1 \times 10^{-3}} = 1 \text{ k}\Omega$$

(d)   Using KVL around the drain-source loop of Fig. 9-3(b) and solving for R_D yields

$$R_D = \frac{V_{DD} - V_{DSQ} - I_{DQ}R_S}{I_{DQ}} = \frac{8 - 6 - (1 \times 10^{-3})(1 \times 10^3)}{1 \times 10^{-3}} = 1 \text{ k}\Omega$$

9.7   For the drain-feedback-biased amplifier of Fig. 9-9(a), $R_F = 5$ MΩ, $R_L = 14$ kΩ, $r_{ds} = 40$ kΩ, and $g_m = 1$ mS. Find (a) A_v ($= v_{ds}/v_i$), (b) Z_{in}, (c) Z_o looking back through the drain-source terminals, and (d) A_i ($= i_i/i_L$).

(a)   The voltage-source small-signal equivalent circuit is given in Fig. 9-10. With v_{ds} as a node voltage,

$$\frac{v_i - v_{ds}}{R_F} = \frac{v_{ds}}{R_L} + \frac{v_{ds} + \mu v_i}{r_{ds}} \tag{1}$$

Solve (1) for the voltage gain ratio and use $\mu = g_m r_{ds}$ to get

$$A_v = \frac{v_{ds}}{v_i} = \frac{R_L r_{ds}(1 - R_F g_m)}{R_F r_{ds} + R_L r_{ds} + R_L R_F}$$

$$= \frac{(14 \times 10^3)(40 \times 10^3)[1 - (5 \times 10^6)(1 \times 10^{-3})]}{(5 \times 10^6)(40 \times 10^3) + (14 \times 10^3)(40 \times 10^3) + (14 \times 10^3)(5 \times 10^6)} = -10.35$$

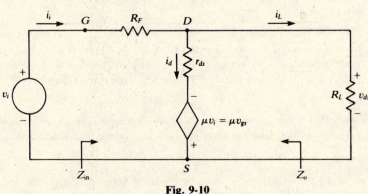

Fig. 9-10

(b)   KVL around the outer loop of Fig. 9-10 gives $v_i = i_i R_F + v_{ds} = i_i R_F + A_v v_i$

whence

$$Z_{\text{in}} = \frac{v_i}{i_i} = \frac{R_F}{1 - A_v} = \frac{5 \times 10^6}{1 - (-10.35)} = 440 \text{ k}\Omega$$

(c)   The driving-point impedance, Z_o, is found after deactivating the independent source, v_i. With $v_i = 0$, $\mu v_{gs} = \mu v_i = 0$, and

$$Z_o = \frac{r_{ds} R_F}{r_{ds} + R_F} = \frac{(40 \times 10^3)(5000 \times 10^3)}{5040 \times 10^3} = 39.68 \text{ k}\Omega$$

(d) $$A_i = \frac{i_L}{i_i} = \frac{v_{ds}/R_L}{v_i/Z_{\text{in}}} = \frac{A_v Z_{\text{in}}}{R_L} = \frac{(-10.35)(440 \times 10^3)}{14 \times 10^3} = -325.3$$

9.8   For the JFET amplifier of Fig. 9-11, $g_m = 2$ mS, $r_{ds} = 30$ kΩ, $R_S = 3$ kΩ, $R_D = R_L = 2$ kΩ, $R_1 = 200$ kΩ, $R_2 = 800$ kΩ, and $r_i = 5$ kΩ. If C_C and C_S are large and the amplifier is biased in the pinchoff region, find (a) Z_{in}, (b) A_v ($= v_L/v_i$), and (c) A_i ($= i_L/i_i$).

(a)   The current-source small-signal equivalent circuit is drawn in Fig. 9-12. Since the gate draws negligible current,

$$Z_{\text{in}} = R_G = \frac{R_1 R_2}{R_1 + R_2} = \frac{(200 \times 10^3)(800 \times 10^3)}{1000 \times 10^3} = 160 \text{ k}\Omega$$

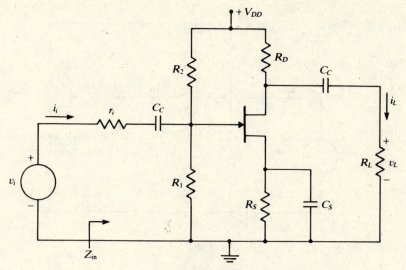

Fig. 9-11

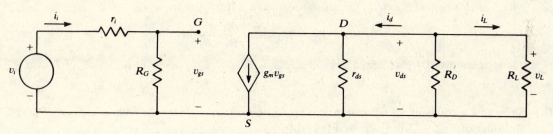

Fig. 9-12

(b)   By voltage division at the input loop,

$$v_{gs} = \frac{R_G}{R_G + r_i} v_i = \frac{160 \times 10^3}{165 \times 10^3} v_i = 0.97\, v_i \qquad (1)$$

The dependent current source drives into R_{ep}, where

$$\frac{1}{R_{ep}} = \frac{1}{r_{ds}} + \frac{1}{R_D} + \frac{1}{R_L} = \frac{1}{30 \times 10^3} + \frac{1}{2 \times 10^3} + \frac{1}{2 \times 10^3} = \frac{1}{967.74}\ \text{S}$$

and so

$$v_L = -g_m v_{gs} R_{ep} \qquad (2)$$

Eliminate v_{gs} between (1) and (2) to find

$$A_v = \frac{v_L}{v_i} = -g_m \left(\frac{R_G}{R_G + r_i} \right) R_{ep} = -(2 \times 10^{-3})(0.97)(967.74) = -1.88$$

(c) $$A_i = \frac{i_L}{i_i} = \frac{v_L/R_L}{v_i/(R_G + r_i)} = \frac{A_v(R_G + r_i)}{R_L} = \frac{(-1.88)(165 \times 10^3)}{2 \times 10^3} = -155.1$$

9.9   The JFET amplifier of Fig. 9-13 introduces a method of bias (*self-bias*) that allows extremely high input impedance even if low values of gate-source bias voltage are required. Find the Thévenin equivalent voltage and resistance looking to the left through *ab*.

With *ab* open, there is no voltage drop across R_3, and the voltage appearing at the open-circuited terminals is determined by the R_1-R_2 voltage divider, or

$$v_{Th} = V_{GG} = \frac{R_1}{R_1 + R_2} V_{DD}$$

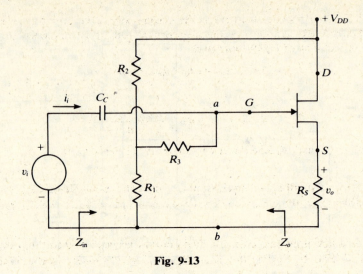

Fig. 9-13

With V_{DD} deactivated (shorted), the resistance looking left through ab is

$$R_{Th} = R_G = R_3 + \frac{R_1 R_2}{R_1 + R_2}$$

It is apparent that if R_3 is selected large, then $R_G = Z_{in}$ is large regardless of the values of R_1 and R_2.

9.10   Show that a small-signal equivalent circuit for the source-follower or common-drain (CD) FET amplifier of Fig. 9-13 is given by Fig. 9-14(b).

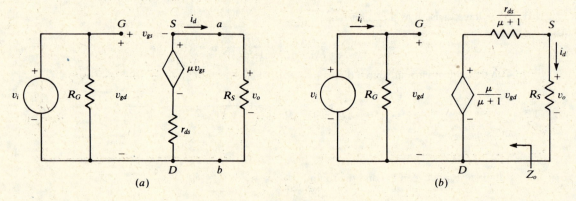

Fig. 9-14

The voltage-source model of Fig. 9-6(b) is inserted in the ac equivalent of Fig. 9-13 and the result redrawn to give the circuit of Fig. 9-14(a), where R_G is determined as in Problem 9.9. Voltage v_{gd}, which is more easily determined than v_{gs}, has been labeled. With terminals ab opened in Fig. 9-14(a), KVL around the S-G-D loop yields

$$v_{gs} = \frac{v_{gd}}{\mu + 1} \tag{1}$$

Using (1), the Thévenin voltage at open-circuit terminals ab is

$$v_{Th} = \mu v_{gs} = \frac{\mu}{\mu + 1} v_{gd} \tag{2}$$

The Thévenin impedance is found as the driving-point impedance to the left through ab (with v_i deactivated or shorted), as seen by a source v_{ab} driving current i_a into terminal a. Since $v_{gs} = -v_{ab}$, KVL around the output loop of Fig. 9-14(a) gives

$$v_{ab} = \mu v_{gs} + i_a r_{ds} = -\mu v_{ab} + i_a r_{ds} \tag{3}$$

From (3),

$$R_{Th} = \frac{v_{ab}}{i_a} = \frac{r_{ds}}{\mu + 1} \tag{4}$$

The expressions (2) and (4) lead directly to the circuit of Fig. 9-14(b).

9.11 Figure 9-15(a) is a small-signal equivalent circuit (voltage-source model) of a *common-gate* (CG) JFET amplifier. Use the circuit to verify two *rules of impedance and voltage reflection* for FET amplifiers:

(*a*) Voltages and impedances in the drain circuit are reflected to the source circuit divided by $\mu + 1$. [Verify by finding the Thévenin equivalent looking to the right through aa' in Fig. 9-15(a) and showing that Fig. 9-15(b) results.]

(*b*) Voltages and impedances in the source circuit are reflected to the drain circuit multiplied by $\mu + 1$. [Verify by finding the Thévenin equivalent looking to the left through bb' of Fig. 9-15(a) and showing that Fig. 9-15(c) results.]

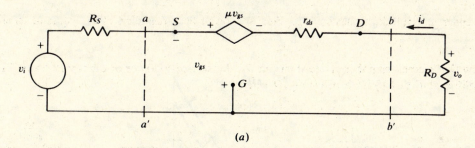

(a)

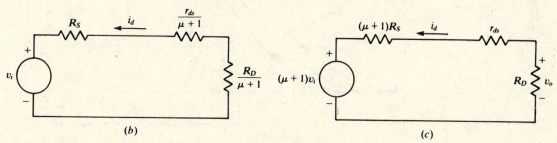

(b)   (c)

Fig. 9-15

(*a*) With aa' open, $i_d = 0$; hence, $v_{gs} = 0$ and $v_{Th} = 0$. After connecting a driving-point source ($v_{aa'}$) to terminals aa', which drives current i_a into terminal a, KVL gives

$$v_{aa'} = \mu v_{gs} + i_a(r_{ds} + R_D) \tag{1}$$

But $v_{gs} = -v_{aa'}$, which can be substituted into (1) to give

$$R_{Th} = \frac{v_{aa'}}{i_a} = \frac{r_{ds}}{\mu + 1} + \frac{R_D}{\mu + 1} \tag{2}$$

With $v_{Th} = 0$, insertion of R_{Th} in place of the network to the right of aa' in Fig. 9-15(a) leads directly to Fig. 9-15(b).

(*b*) Applying KVL to the left in Fig. 9-15(a) with bb' open, and noting that $v_i = -v_{gs}$, yields

$$v_{Th} = v_i - \mu v_{gs} = (\mu + 1)v_i \tag{3}$$

Deactivate (short) v_i, connect a driving-point source ($v_{bb'}$) to terminals bb' that drives current i_b into terminal b, note that $v_{gs} = -i_b R_S$, and apply KVL around the outer loop of Fig. 9-15(a) to find

$$v_{bb'} = i_b(r_{ds} + R_S) - \mu v_{gs} = i_b[r_{ds} + (\mu + 1)R_S] \tag{4}$$

The Thévenin impedance follows from (4) as

$$R_{Th} = \frac{v_{bb'}}{i_b} = r_{ds} + (\mu + 1)R_S \tag{5}$$

When the Thévenin source of (3) and impedance of (5) are used to replace the balance of the network to the left of bb', the circuit of Fig. 9-15(c) results.

9.12 The manufacturer's specification sheet for a certain kind of n-channel JFET has nominal and worst-case shorted-gate parameters as follows:

	I_{DSS}, mA	V_{p0}, V
Maximum	7	4.2
Nominal	6	3.6
Minimum	5	3.0

Sketch the nominal and worst-case transfer characteristics that can be expected from a large sample of the device.

Values can be calculated for the nominal, maximum, and minimum transfer characteristics using (9.2) over the range $-V_{p0} \leq v_{GS} \leq 0$. The results are plotted in Fig. 9-16.

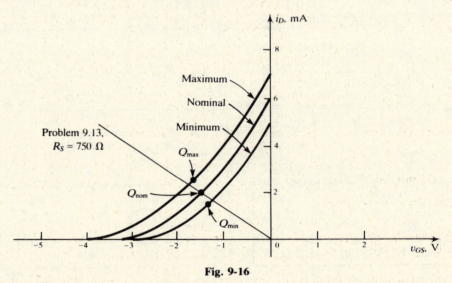

Fig. 9-16

9.13 A self-bias JFET amplifier (Fig. 9-13) is to be designed with $V_{DSQ} = 15$ V and $V_{DD} = 24$ V, using a device described by Problem 9.12. In order to control gain variation, the quiescent drain current must satisfy $I_{DQ} = 2 \pm 0.4$ mA regardless of the particular JFET utilized from a batch. Determine appropriate values of R_S and R_D.

Quiescent points are established on the transfer characteristics of Fig. 9-16: Q_{max} at $I_{DQ} = 2.4$ mA, Q_{nom} at $I_{DQ} = 2.0$ mA, and Q_{min} at $I_{DQ} = 1.6$ mA. A transfer bias line is constructed passing through the origin (i.e., choose $V_{GG} = 0$) and Q_{nom}. Since its slope is $-1/R_S$, the source resistor value is determined as

$$R_S = \frac{0 - (-3)}{(4 - 0)} = 750\ \Omega$$

The drain resistor value is found by applying KVL around the drain-source loop and solving for R_D:

$$R_D = \frac{V_{DD} - V_{DSQ} - I_{DQ}R_S}{I_{DQ}} = \frac{24 - 15 - (0.002)(750)}{0.002} = 3.75\ \text{k}\Omega$$

Since Q_{max} lies above the transfer bias line and Q_{min} lies below it, the condition on I_{DQ} is satisfied.

Supplementary Problems

9.14 For the JFET amplifier of Example 9.1, R_1 is changed to 2 MΩ to increase the input impedance. If R_D, R_S, and V_{DD} are unchanged, what value of R_2 is needed to maintain the original Q-point?
Ans. 31.33 MΩ

9.15 Find the voltage across R_S in Example 9.1. *Ans.* 3 V

9.16 Find the input impedance as seen by the source v_i of Example 9.1 if C_C is large. *Ans.* 940 kΩ

9.17 The method of *source bias*, illustrated by Fig. 9-17, can be employed for both JFETs and MOSFETs. In the case of a JFET with characteristics given by Fig. 9-4, and with $R_D = 1$ kΩ, $R_S = 4$ kΩ, and $R_G = 10$ MΩ, determine V_{DD} and V_{SS} so that the amplifier has the same quiescent conditions as the amplifier of Example 9.1 *Ans.* $V_{SS} = 4$ V, $V_{DD} = 16$ V

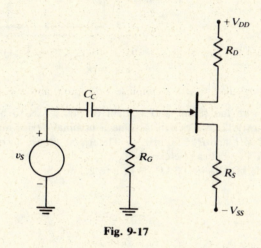

Fig. 9-17

9.18 Show that the transconductance of a JFET varies as the square root of drain current.
Ans. $g_m = \dfrac{2\sqrt{I_{DSS}}}{V_{p0}}\sqrt{i_D}$

9.19 For the drain-feedback-biased amplifier of Fig. 9-9(a), $V_{DD} = 15$ V, $R_F = 5$ MΩ, $I_{DQ} = 0.75$ mA, and $V_{GSQ} = 4.5$ V. Find (a) V_{DSQ} and (b) R_L. *Ans.* (a) 4.5 V; (b) 14 kΩ

9.20 For the amplifier of Fig. 9-13, $R_1 = 20$ kΩ, $R_2 = 100$ kΩ, $R_3 = 1$ MΩ, $r_{ds} = 30$ kΩ, $\mu = 150$ (see Problem 9.5), and $R_S = 1$ kΩ. Find (a) A_v ($= v_o/v_i$), (b) A_i ($= i_i/i_d$), and (c) Z_o.
Ans. (a) 0.829; (b) 843; (c) 198.7 Ω

9.21 Evaluate the voltage gain of the CG amplifier of Fig. 9-15(a).
Ans. $A_v = \dfrac{v_o}{v_i} = \dfrac{(\mu + 1)R_D}{R_D + r_{ds} + (\mu + 1)R_S}$

9.22 Find the voltage gain A_{v2} ($= v_2/v_i$) for the circuit of Fig. 9-18(a). Figure 9-18(b) is a small-signal equivalent circuit in which impedance reflection has been used for simplification.
Ans. $A_{v2} = -\mu R_D/[R_D + r_{ds} + (\mu + 1)R_S]$

9.23 If $R_D = R_S$ for the amplifier of Fig. 9-18(a), the circuit is commonly called a *phase-splitter*, since $v_2 = -v_1$ (the outputs are equal in magnitude but 180° out of phase). Find A_{v1} ($= v_1/v_i$) and by comparison with A_{v2} of Problem 9.22 verify that the circuit actually is a phase-splitter.
Ans. $A_{v1} = \mu R_S/[R_D + r_{ds} + (\mu + 1)R_S]$

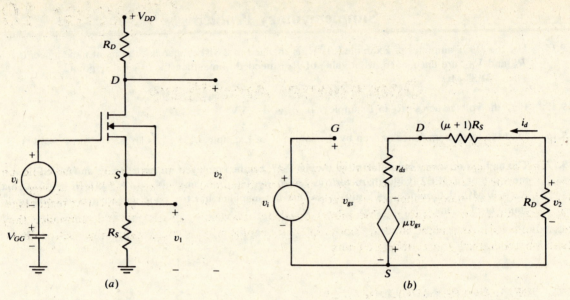

Fig. 9-18

9.24 A JFET amplifier with the circuit arrangement of Fig. 9-3 is to be manufactured using devices as described in Problem 9.12. For design, assume a nominal device and use $V_{DD} = 24$ V, $V_{DSQ} = 15$ V, $I_{DQ} = 2$ mA, $R_1 = 2$ MΩ, and $R_2 = 30$ MΩ. (a) Determine the values of R_S and R_D for the amplifier. (b) Predict the range of I_{DQ} that can be expected.

Ans.   (a) $R_S = 1.475$ kΩ, $R_D = 3.03$ kΩ; (b) 1.8 to 2.2 mA

Chapter 10

Operational Amplifiers

10.1 INTRODUCTION

The name *operational amplifier* (Op Amp) was originally given to an amplifier that could be easily modified by external circuitry to perform mathematical operations (addition, scaling, integration, etc.) in analog computer applications. However, with the advance of solid-state technology, Op Amps have become highly reliable, miniaturized, and consistently predictable in performance; they now figure as fundamental building blocks in basic amplification, signal conditioning, active filters, function generators, and switching circuits.

10.2 IDEAL AND PRACTICAL OP AMPS

An Op Amp amplifies the difference, $v_d \equiv v_1 - v_2$, between two input signals (see Fig. 10-1), exhibiting the open-loop voltage gain

$$A_{OL} \equiv \frac{v_o}{v_d} \qquad (10.1)$$

Terminal *1*, labeled with a minus sign, is the *inverting input*; signal v_1 is amplified in magnitude and appears phase-inverted at the output. Conversely, terminal *2*, labeled with a plus sign, is the *noninverting input*; output due to v_2 is phase-preserved.

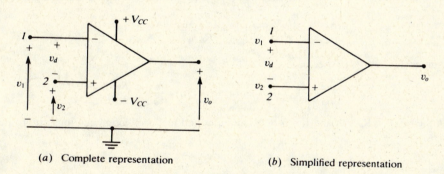

(*a*) Complete representation   (*b*) Simplified representation

Fig. 10-1. Operational Amplifier

In magnitude, open-loop voltage gain in Op Amps ranges from 10^4 to 10^7. The maximum magnitude of output voltage from an Op Amp is called *saturation voltage*; this voltage is approximately 2 V smaller than power supply voltage. In other words, the amplifier is linear over the range

$$-(V_{CC} - 2) \quad (V) < v_o < (V_{CC} - 2) \quad (V) \qquad (10.2)$$

The *ideal* Op Amp has three essential characteristics which serve as a standard to assess goodness of a *practical* Op Amp:

1. Open-loop voltage gain (A_{OL}) is negatively infinite.

2. Input impedance (R_d) between terminals *1* and *2* is infinitely large; thus, input current is zero.

3. Output impedance is zero; consequently, output voltage is independent of load.

153

Example 10.1 An Op Amp has saturation voltage (V_{osat}) 10 V, open-loop voltage gain -10^5, and input resistance 100 kΩ. Find (a) the value of v_d to just drive the amplifier to saturation, and (b) the Op Amp input current at saturation onset.

(a) By (10.1),
$$v_d = \frac{\pm V_{osat}}{A_{OL}} = \frac{\pm 10}{-10^5} = \mp 0.1 \text{ mV}$$

(b) Let i_{in} be the current into terminal 1 of Fig. 10-1(b); then
$$i_{in} = \frac{v_d}{R_d} = \frac{\mp 0.1 \times 10^{-3}}{100 \times 10^3} = \mp 1 \text{ nA}$$

In application, the operational amplifier is utilized with a large percentage of negative feedback, giving a circuit of which the characteristics depend almost entirely on circuit elements external to the basic Op Amp. Error due to treatment of the basic Op Amp as ideal tends to diminish in the presence of negative feedback.

10.3 INVERTING AMPLIFIER

The *inverting amplifier* of Fig. 10-2 has the noninverting input connected to ground or common. A signal is applied through input resistor R_1, and a negative current feedback (see Problem 10.1) is implemented by connection of *feedback resistor R_F*.

Example 10.2 For the inverting amplifier of Fig. 10-2, find the voltage gain v_o/v_S using only (a) characteristic 1, (b) characteristic 2, of the ideal Op Amp.

(a) By method of node voltages at the inverting input, the current balance is
$$\frac{v_S - v_d}{R_1} + \frac{v_o - v_d}{R_F} = i_{in} = \frac{v_d}{R_d} \tag{1}$$

where R_d is the differential input resistance. By (10.1), $v_d = v_o/A_{OL}$, which substituted into (1) gives
$$\frac{v_S - (v_o/A_{OL})}{R_1} + \frac{v_o - (v_o/A_{OL})}{R_F} = \frac{v_o/R_d}{A_{OL}} \tag{2}$$

In the limit as $A_{OL} \to -\infty$, (2) becomes
$$\frac{v_S}{R_1} + \frac{v_o}{R_F} = 0 \qquad \text{or} \qquad A_v \equiv \frac{v_o}{v_S} = -\frac{R_F}{R_1} \tag{3}$$

(b) If $i_{in} = 0$, then $v_d = i_{in} R_d = 0$, and $i_1 = i_F \equiv i$. The input and feedback loop equations are, respectively,
$$v_S = iR_1 \qquad \text{and} \qquad v_o = -iR_F$$

whence
$$A_v \equiv \frac{v_o}{v_S} = -\frac{R_F}{R_1} \tag{4}$$

in agreement with (3).

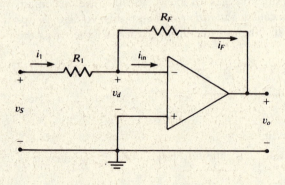

Fig. 10-2. Inverting Amplifier

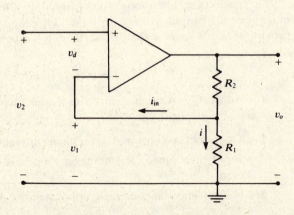

Fig. 10-3. Noninverting Amplifier

10.4 NONINVERTING AMPLIFIER

The *noninverting amplifier* of Fig. 10-3 is realized by grounding R_1 of Fig. 10-2 and applying the input signal at the noninverting Op Amp terminal. For a positive v_2, v_o is positive and current i is positive. Voltage $v_1 = iR_1$ then is applied to the inverting terminal, acting as negative voltage feedback.

Example 10.3 For the noninverting amplifier of Fig. 10-3, assume that the current into the inverting terminal of the Op Amp is zero, so that $v_d \approx 0$ and $v_1 \approx v_2$. Derive an expression for the voltage gain v_o/v_2.

With zero input current to the basic Op Amp, the currents through R_2 and R_1 must be identical, leading to

$$\frac{v_o - v_1}{R_2} = \frac{v_1}{R_1}$$

or

$$A_v \equiv \frac{v_o}{v_2} \approx \frac{v_o}{v_1} = 1 + \frac{R_2}{R_1}$$

10.5 SUMMER AMPLIFIER

The *inverting summer amplifier* (also called *inverting adder*) of Fig. 10-4 is formed by adding parallel inputs to the inverting amplifier of Fig. 10-2. The output is a weighted sum of the inputs, but inverted in polarity. For an ideal Op Amp, there is no limit to the number of inputs that can be used; however, the gain is reduced as inputs are added to a practical Op Amp (see Problem 10.18).

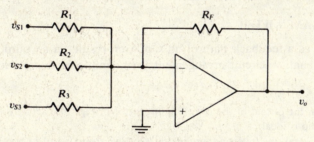

Fig. 10-4. Inverting Summer Amplifier

Example 10.4 Find an expression for the output of the inverting summer amplifier of Fig. 10-4, if the basic Op Amp is ideal.

The principle of superposition is used. With $v_{S2} = v_{S3} = 0$, the current in R_1 is not affected by the presence of R_2 and R_3, since the inverting node is a virtual ground (see Problem 10.1). Hence, the output voltage due to v_{S1} is determined by (3) of Example 10.2 as $v_{o1} = -(R_F/R_1)v_{S1}$. Similarly,

$$v_{o2} = -(R_F/R_2)v_{S2} \qquad v_{o3} = -(R_F/R_3)v_{S3}$$

Then, by superposition,

$$v_o = v_{o1} + v_{o2} + v_{o3} = -R_F\left(\frac{v_{S1}}{R_1} + \frac{v_{S2}}{R_2} + \frac{v_{S3}}{R_3}\right)$$

10.6 DIFFERENTIATING AMPLIFIER

Introduction of a capacitor into the input path of an Op Amp leads to time differentiation of the input signal. The circuit of Fig. 10-5 represents the simplest *inverting differentiator* involving an Op Amp. As such, the circuit finds limited practical use, since high-frequency noise can present a derivative of comparable magnitude to that of the signal. In practice, high-pass filtering is utilized to reduce the noise effects (see Problem 10.7).

Example 10.5 Find an expression for the output of the inverting differentiator of Fig. 10-5, if the basic Op Amp is ideal.

Since the Op Amp is ideal, $v_d \approx 0$, and the inverting terminal is a virtual ground. Consequently, v_S appears across capacitor C.

$$i_S = C \frac{dv_S}{dt}$$

But capacitor current is also the current through $R (i_{in} = 0)$.

$$v_o = -i_F R = -i_S R = -RC \frac{dv_S}{dt}$$

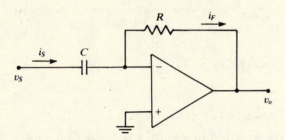

Fig. 10-5.   Differentiating Amplifier

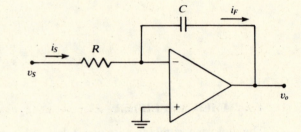

Fig. 10-6.   Integrating Amplifier

10.7   INTEGRATING AMPLIFIER

Use of a capacitor as a feedback path of an Op Amp results in an output signal that is a time integral of the input signal. A circuit arrangement for a simple *inverting integrator* is given in Fig. 10-6.

Example 10.6   Show that the output of the inverting integrator of Fig. 10-6 actually is the time integral of the input signal, if the Op Amp is ideal.

Since the Op Amp is ideal, the inverting terminal is a virtual ground, and v_S appears across R.

$$i_S = \frac{v_S}{R}$$

But, with negligible current into the Op Amp, the current through R must also flow through C.

$$v_o = -\frac{1}{C} \int i_F \, dt = -\frac{1}{C} \int i_S \, dt = -\frac{1}{RC} \int v_S \, dt$$

10.8   FILTER APPLICATIONS

The use of Op Amps in active RC filters has increased with the move to integrated circuits. Active filter realizations can eliminate the need for bulky inductors, which do not satisfactorily lend themselves to integrated circuitry. Further, active filters do not necessarily attenuate the signal over the band pass, as do the passive-element counterparts. A simple *inverting, first-order, low-pass filter* using an Op Amp as the active device is shown in Fig. 10-7(a).

Example 10.7   (a) For the low-pass filter whose s-domain (Laplace transform) representation is given in Fig. 10-7(a), find the transfer function $A(s) \equiv V_o(s)/V_S(s)$. ($b$) Make a frequency-domain plot of gain magnitude, showing that the filter passes signals of low frequency and attenuates high-frequency signals.

(a)   The feedback impedance $Z_F(s)$ and the input impedance $Z_1(s)$ are found as

$$Z_F(s) = \frac{R(1/sC)}{R + (1/sC)} = \frac{R}{sRC + 1} \qquad Z_1(s) = R_1 \qquad (1)$$

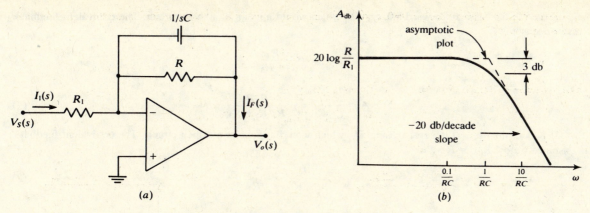

Fig. 10-7. First-Order, Low-Pass Filter

The resistive circuit analysis of Example 10.2 extends directly to the s-domain; thus,

$$A(s) = -\frac{Z_F(s)}{Z_1(s)} = -\frac{R/R_1}{sRC+1} \tag{2}$$

(b) It is customary to plot the gain magnitude, in decibels, against the (common) logarithm of radian frequency ω. Setting $s = 0 + j\omega$ in (2),

$$A_{db} \equiv 20\log|A(j\omega)| = 20\log R/R_1 - 20\log|j\omega RC + 1| \tag{3}$$

A plot of (3) is displayed in Fig. 10-7(b), in which the horizontal scale is logarithmic. The curve is essentially flat below $\omega = 0.1/RC$; thus, all frequencies below $0.1/RC$ are passed with the dc gain R/R_1. A 3-db reduction in gain is experienced at the "corner frequency," $1/RC$, and gain is attenuated by 20 db per decade of frequency change for frequencies greater than $10/RC$.

10.9 FUNCTION GENERATORS AND SIGNAL CONDITIONERS

Frequently in analog system design, the need arises to alter gain characteristics of signals, to compare signals with a generated reference, or to limit signals depending on their values. Such circuit applications can often be implemented using the high-input-/low-output-impedance and high-gain characteristics of the Op Amp. Possibilities of such circuits are boundless; however, the most common case is introduction of nonlinear elements (such as diodes or transistors) into negative feedback paths, while using linear elements in the input branches.

Example 10.8 The signal conditioning amplifier of Fig. 10-8 changes gain depending upon the polarity of v_S. Find the circuit voltage gains for positive and negative v_S if diode D_2 is ideal.

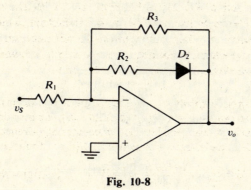

Fig. 10-8

If $v_S > 0$, then $v_o < 0$, forward-biasing D_2, which would appear as a short circuit. The equivalent feedback resistance is

$$R_{Feq} = \frac{R_2 R_3}{R_2 + R_3}$$

By (3) of Example 10.2,

$$A_v = -\frac{R_{Feq}}{R_1} = -\frac{R_2 R_3}{R_1(R_2 + R_3)} \tag{1}$$

If $v_S < 0$, then $v_o > 0$, reverse-biasing D_2, which would appear as an open circuit. The equivalent feedback resistance is now $R_{Feq} = R_3$, and

$$A_v = -\frac{R_{Feq}}{R_1} = -\frac{R_3}{R_1} \tag{2}$$

Solved Problems

10.1 For the inverting amplifier of Fig. 10-2, (a) show that as $A_{OL} \to -\infty$, $v_d \to 0$; thus, the inverting terminal remains nearly at ground potential (called a *virtual ground*). (b) Show that the current feedback is actually negative feedback.

(a)   Apply KVL around the outer loop.

$$v_S - v_o = i_1 R_1 + i_F R_F \tag{1}$$

Using (10.1) in (1), rearranging, and taking the limit gives

$$v_d = \lim_{A_{OL} \to -\infty} \left[\frac{i_1 R_1 + i_F R_F - v_S}{A_{OL}} \right] = 0 \tag{2}$$

(b)   The criterion for negative feedback is that i_F counteract i_1; that is, the two currents must have the same algebraic sign. Now, by two applications of KVL, with $v_d \approx 0$,

$$i_1 = \frac{v_S - v_d}{R_1} \approx \frac{v_S}{R_1} \qquad i_F = \frac{-v_o + v_d}{R_F} \approx \frac{-v_o}{R_F}$$

But, in an inverting amplifier, v_o and v_S have opposite signs; therefore, i_1 and i_F have like signs.

10.2 (a) Use (2) of Example 10.2 to derive an exact formula for the gain of a practical inverting Op Amp. (b) If $R_1 = 1 \text{ k}\Omega$, $R_F = 10 \text{ k}\Omega$, $R_d = 1 \text{ k}\Omega$, and $A_{OL} = -10^4$, evaluate the gain of this inverting amplifier. (c) Compare the result of (b) with the ideal Op Amp approximation given by (3) of Example 10.2.

(a)   Solving (2) of Example 10.2 for the voltage-gain ratio gives

$$A_v \equiv \frac{v_o}{v_S} = \frac{A_{OL}}{1 + \dfrac{R_1}{R_F}(1 - A_{OL}) + \dfrac{R_1}{R_d}} \tag{1}$$

(b)   Evaluating (1),

$$A_v = \frac{-10^4}{1 + \dfrac{1}{10}(1 + 10^4) + \dfrac{1}{1}} = -9.979 \tag{2}$$

(c)

$$A_{videal} = -\frac{R_F}{R_1} = -10$$

and the percent error is

$$\frac{-9.979 - (-10)}{-9.979} \times 100\% = -0.21\%$$

Note that R_d and A_{OL} are quite removed from the ideal, yet the error is small.

10.3   A *differential amplifier* (sometimes called a *subtractor*) responds to the difference of two input signals, removing any identical portions (often a bias or noise) of the input signals, in a process called *common-mode rejection*. Find an expression for v_o in Fig. 10-9 that shows this circuit to be a differential amplifier. Assume an ideal Op Amp.

Since the current into the ideal Op Amp is zero, a loop equation gives

$$v_1 = v_{S1} - Ri_1 = v_{S1} - R\left(\frac{v_{S1} - v_o}{R + R_1}\right)$$

By voltage division at the noninverting node,

$$v_2 = \frac{R_1}{R + R_1} v_{S2}$$

For the ideal Op Amp, $v_d = 0$, or $v_1 = v_2$, which leads to

$$v_o = \frac{R_1}{R}(v_{S1} - v_{S2})$$

Thus, the output voltage is directly proportional to the difference of the input voltages.

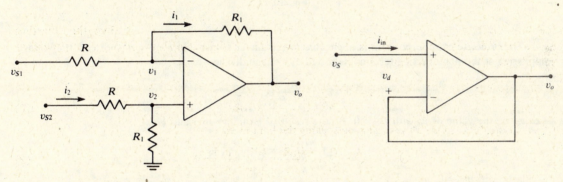

Fig. 10-9.   Differential Amplifier   **Fig. 10-10.**   Unity Follower

10.4   Find the amplifier input impedance (Z_1) of the inverting amplifier of Fig. 10-2, if the basic Op Amp is ideal.

Consider v_S a driving-point source. Since the Op Amp is ideal, the inverting terminal is a virtual ground, and a loop equation at the input leads to

$$v_S = i_1 R_1 + 0 \qquad \text{or} \qquad Z_1 = \frac{v_S}{i_1} = R_1$$

10.5   The *unity follower* amplifier of Fig. 10-10 has a voltage gain of 1, and the output is in phase with the input. It also has an extremely high input impedance, leading to use as an intermediate-stage (*buffer*) amplifier to prevent a small load impedance from loading a source. Assume a practical Op Amp having $A_{OL} = -10^6$ (typical value). (*a*) Show that $v_o \approx v_S$. (*b*) Find an expression for the amplifier input impedance, and evaluate for $R_d = 1$ mΩ (typical value).

(*a*)   Writing a loop equation and using (*10.1*),

$$v_S = v_o - v_d = v_o\left(1 - \frac{1}{A_{OL}}\right)$$

or $$v_o = \frac{v_S}{1 - (1/A_{OL})} = \frac{v_S}{1 + 10^{-6}} = 0.999999\, v_S \approx v_S$$

(*b*)   Considering v_S a driving-point source and using (*10.1*),

$$v_S = i_{in}R_d + v_o = i_{in}R_d - A_{OL}v_d = i_{in}R_d(1 - A_{OL})$$

or $$Z_{in} = \frac{v_S}{i_{in}} = R_d(1 - A_{OL}) \approx -A_{OL}R_d = -(-10^6)(10^6) = 1 \text{ TΩ}$$

10.6 Find an expression for the output (v_o) of the amplifier circuit of Fig. 10-11. Assume an ideal basic Op Amp. What mathematical operation does the circuit perform?

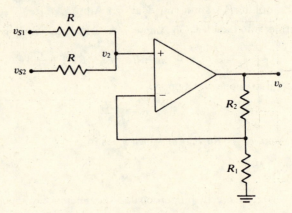

Fig. 10-11

The principle of superposition is applicable to this linear circuit. With $v_{S2} = 0$ (shorted), the voltage appearing at the noninverting terminal is found by voltage division to be

$$v_2 = \frac{R}{R+R} v_{S1} = \frac{v_{S1}}{2} \tag{1}$$

Let v_{o1} be the value of v_o with $v_{S2} = 0$. By the result of Example 10.3 and (1),

$$v_{o1} = \left(1 + \frac{R_2}{R_1}\right) v_2 = \left(1 + \frac{R_2}{R_1}\right) \frac{v_{S1}}{2} \tag{2}$$

Similarly, with $v_{S1} = 0$,

$$v_{o2} = \left(1 + \frac{R_2}{R_1}\right) \frac{v_{S2}}{2} \tag{3}$$

By superposition, the total output is

$$v_o = v_{o1} + v_{o2} = \frac{1}{2}\left(1 + \frac{R_2}{R_1}\right)(v_{S1} + v_{S2}) \tag{4}$$

The circuit is seen to be a noninverting adder.

10.7 The circuit of Fig. 10-12(a) (represented in the s-domain) is a more practical differentiator than that of Fig. 10-5, because it will attenuate high-frequency noise. (a) Find the s-domain transfer function relating V_o and V_S. (b) Sketch the gain-magnitude plot explaining how high-frequency noise effects are reduced.

(a) Assuming an ideal Op Amp, the inverting terminal is a virtual ground and $I_S(s) = -I_F(s)$. As in Example 10.7,

$$Z_F(s) = \frac{R}{sRC + 1}$$

and

$$I_F(s) = \frac{V_o(s)}{Z_F(s)} = \frac{sRC + 1}{R} V_o(s)$$

But

$$V_S(s) = I_S(s) Z_{in}(s) = -I_F(s) Z_{in}(s) = -\left[\frac{sRC + 1}{R} V_o(s)\right] \frac{sRC + 1}{sC}$$

whence

$$A(s) \equiv \frac{V_o(s)}{V_S(s)} = -\frac{sRC}{(sRC + 1)^2}$$

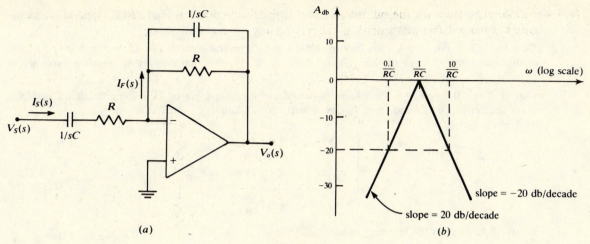

Fig. 10-12

(b)   From (a),

$$A_{db} \equiv 20 \log |A(j\omega)| = 20 \log \omega RC - 40 \log |j\omega RC + 1|$$

$$\approx \begin{cases} 20 \log \omega RC & (\omega RC < 1) \\ -20 \log \omega RC & (\omega RC > 1) \end{cases}$$

Figure 10-12(b) is a plot of this approximate (asymptotic) expression for A_{db}. For a true differentiator,

$$v_o = K \frac{dv_S}{dt} \qquad \text{or} \qquad V_o = sKV_S$$

which leads to $A_{db} = 20 \log \omega K$. Thus it is seen that the circuit indeed differentiates components of the signal whose frequency is less than the "break frequency," $f_1 \equiv 1/2\pi RC$ (Hz). Spectral components above the break frequency will be attenuated—the higher the frequency, the greater the attenuation.

10.8   In analog signal processing, the need often arises to introduce a *level clamp* (linear amplification to a desired output level or value, then no further increase in output value as input is increased). One such circuit, shown in Fig. 10-13(a), uses series Zener diodes in a negative feedback path. Assuming ideal Zeners and Op Amp, find the relationship between v_o and v_S. Sketch the results on a transfer characteristic.

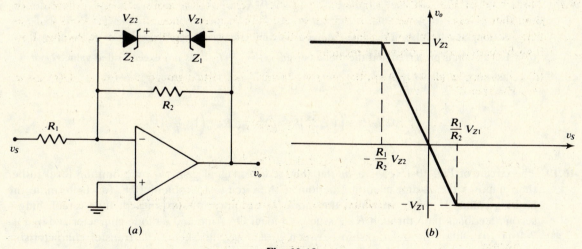

Fig. 10-13

Since the Op Amp is ideal, the inverting terminal is a virtual ground and v_o appears across the parallel-connected feedback paths.

Case I: $v_S > 0$. For $v_o < 0$, Z_2 is forward-biased and Z_1 reverse-biased. The Zener feedback path is an open circuit until $v_o = -V_{Z1}$; then Z_1 will limit v_o at $-V_{Z1}$ so that no further negative excursion is possible.

Case II: $v_S < 0$. For $v_o > 0$, Z_1 is forward-biased and Z_2 reverse-biased. The Zener feedback path acts as an open circuit until v_o reaches V_{Z2}, at which point Z_2 limits v_o to that value. In summary,

$$v_o = \begin{cases} V_{Z2} & \left(v_S < -\dfrac{R_1}{R_2} V_{Z2} \right) \\[2mm] -\dfrac{R_2}{R_1} v_S & \left(-\dfrac{R_1}{R_2} V_{Z2} \le v_S \le \dfrac{R_1}{R_2} V_{Z1} \right) \\[2mm] -V_{Z1} & \left(v_S > \dfrac{R_1}{R_2} V_{Z1} \right) \end{cases}$$

Figure 10-13(b) gives the transfer characteristic.

10.9   The circuit of Fig. 10-14 is an *adjustable-output voltage regulator.* Assume that the basic Op Amp is ideal. Regulation of the Zener is preserved if $i_Z \ge 0.1 I_Z$ (Section 7.9). (*a*) Find the regulated output v_o in terms of V_Z. (*b*) Given a specific Zener diode and the values of R_S and R_1, what is the range of V_S over which there is no loss of regulation?

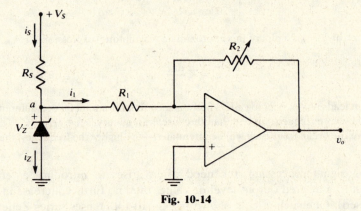

Fig. 10-14

(*a*)   Since V_Z is the voltage at node a, (3) of Example 10.2 gives

$$v_o = -\frac{R_2}{R_1} V_Z$$

So long as $i_Z \ge 0.1 I_Z$, a regulated value of v_o can be achieved by adjustment of R_2.

(*b*)   Regulation is preserved and the diode current, $i_Z = i_S - i_1$, does not exceed rated value (I_Z) if

$$0.1 I_Z \le i_S - i_1 \le I_Z \qquad \text{or} \qquad 0.1 I_Z \le \frac{V_S - V_Z}{R_S} - \frac{V_Z}{R_1} \le I_Z$$

or

$$0.1 I_Z R_S + \left(1 + \frac{R_S}{R_1} \right) V_Z \le V_S \le I_Z R_S + \left(1 + \frac{R_S}{R_1} \right) V_Z$$

10.10   The circuit of Fig. 10-15(*a*) is a *limiter* that reduces signal gain to some limiting level rather than imposes the abrupt clamping action of the circuit of Problem 10.8. (*a*) Determine the limiting value, V_ℓ, of v_o at which the diode D becomes forward-biased, thus establishing a second feedback path through R_3. Assume an ideal Op Amp and a diode characterized by Fig. 7-4(*a*). (*b*) Obtain the relationship between v_o and v_S, and sketch the transfer characteristic.

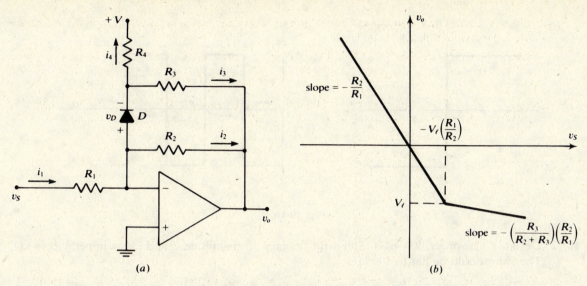

Fig. 10-15

(a) The diode voltage, v_D, is found by writing a loop equation. Since the inverting input is a virtual ground, v_o appears across R_2, and

$$v_D = -v_o - i_3 R_3 = -v_o - \left(\frac{V - v_o}{R_3 + R_4}\right) R_3 \tag{1}$$

When $v_D = 0$, $v_o = V_\ell$, and (1) gives

$$V_\ell = -\frac{R_3}{R_4} V \tag{2}$$

(b) For $v_o > V_\ell$, the diode blocks and R_2 constitutes the only feedback path. From $i_1 = i_2$,

$$\frac{v_S}{R_1} = -\frac{v_o}{R_2} \tag{3}$$

For $v_o \le V_\ell$, the diode conducts and the parallel combination of R_2 and R_3 forms the feedback path. From $i_1 = i_2 + i_3 + i_4$,

$$\frac{v_S}{R_1} = -\left(\frac{v_o}{R_2} + \frac{v_o}{R_3} + \frac{V}{R_4}\right) \tag{4}$$

It follows from (2), (3), and (4) that

$$v_o = \begin{cases} -\dfrac{R_2}{R_1} v_S & (v_o > V_\ell) \\[2ex] -\left(\dfrac{R_3}{R_2 + R_3}\right)\left(\dfrac{R_2}{R_1}\right) v_S + \dfrac{R_2}{R_2 + R_3} V_\ell & (v_o \le V_\ell) \end{cases}$$

This transfer characteristic is plotted in Fig. 10-15(b).

10.11 Under what modifications and specifications will the circuit of Fig. 10-13(a) become a 3-V square-wave generator, if $v_S = 0.02 \sin \omega t$ (V)? Sketch the circuit transfer characteristic and the input and output waveforms.

Modify the circuit of Fig. 10-13(a) by removing R_2. Specify Zener diodes such that $V_{Z1} = V_{Z2} = 3$ V. The transfer characteristic of Fig. 10-13(b) will change to that of Fig. 10-16(a). The resulting time relationship between v_S and v_o is displayed in Fig. 10-16(b).

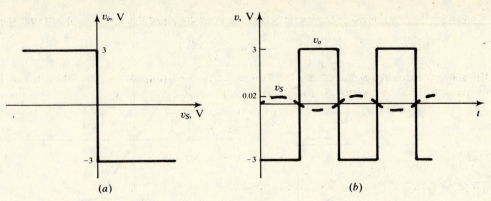

Fig. 10-16

10.12 Design a first-order, low-pass filter with dc gain of magnitude 2 and input impedance 5 kΩ. The gain should be flat to 100 Hz.

   The filter is shown in Fig. 10-7. For an ideal Op Amp, Problem 10.4 gives $Z_{in} = R_1 = 5$ kΩ. The dc gain is given by (2) of Example 1.07 as

$$A(0) = -\frac{R}{R_1}$$

whence $R = 2R_1 = 10$ kΩ. From Fig. 10-7(b) it is seen that gain magnitude is flat to $\omega = 0.1/RC$, or the capacitor must be sized such that

$$C = \frac{0.1}{2\pi f R} = \frac{0.1}{2\pi (100)(10 \times 10^3)} = 15.9 \text{ nF}$$

10.13 The analog computer utilizes operational amplifiers to solve differential equations. Devise an analog solution for $i(t)$, $t > 0$, in the circuit of Fig. 10-17(a). Assume that you have available an inverting integrator with unity gain ($R_1 C_1 = 1$), inverting amplifiers, a variable dc source, and a switch.

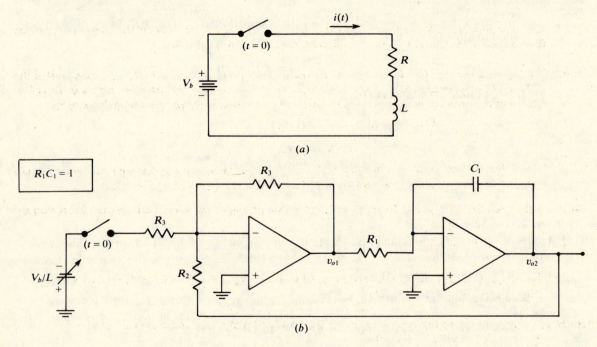

Fig. 10-17

For $t > 0$, the governing differential equation for the circuit of Fig. 10-17(a) may be written as

$$-\frac{di}{dt} = -\frac{V_b}{L} + \frac{R}{L}i \qquad (1)$$

The summation on the right of (1) can be simulated by the left-hand inverting adder of Fig. 10-17(b), where $v_{o1} = -di/dt$ and where R_2 and R_3 are chosen such that

$$\frac{R_3}{R_2} = \frac{R}{L}$$

Then

$$v_{o2} = -\int v_{o1}\, dt$$

will be an analog of $i(t)$, on a scale of 1 A per volt.

Supplementary Problems

10.14 For the noninverting amplifier of Fig. 10-3, (a) find an exact expression for the voltage-gain ratio, and (b) evaluate for $R_1 = 1$ kΩ, $R_2 = 10$ kΩ, $R_d = 1$ kΩ, and $A_{OL} = -10^4$. (c) Compare the result of (b) with the value found using the ideal Op Amp expression derived in Example 10.3.
   Ans. (a) $A_v = \dfrac{R_1 + R_2}{R_1 - \dfrac{R_1 R_2}{A_{OL} R_d} - \dfrac{R_1 + R_2}{A_{OL}}}$; ($b$) 10.977; ($c$) $A_{videal} = 11$, or $+0.21\%$ error

10.15 For the first-order, low-pass filter of Example 10.7, $R = 10$ kΩ, $R_1 = 1$ kΩ, and $C = 0.1$ μF. Find (a) the gain for dc signals, (b) the corner or break frequency (f_1) at which the gain drops off by 3 db, (c) the frequency (f_u) at which the gain has dropped to unity (called the *unity-gain bandwidth*).
   Ans. (a) -10; (b) 159.2 Hz; (c) 1583.6 Hz

10.16 The noninverting amplifier circuit of Fig. 10-3 has an infinite input impedance if the basic Op Amp is ideal. If the Op Amp is not ideal, but instead $R_d = 1$ MΩ and $A_{OL} = -10^6$, find the input impedance. Let $R_2 = 10$ kΩ and $R_1 = 1$ kΩ.   *Ans.* 1 TΩ

10.17 Let $R_1 = R_2 = R_3 = 3R_F$ in the inverting summer amplifier of Fig. 10-4. What mathematical operation does this circuit perform?   *Ans.* negative instantaneous average value

10.18 An inverting summer (see Fig. 10-4) has n inputs with $R_1 = R_2 = R_3 = \cdots = R_n = R$. Assume that the open-loop basic Op Amp gain (A_{OL}) is finite, but that the inverting terminal input current is negligible. Derive a relationship that shows how gain magnitude is reduced in the presence of multiple inputs.
   Ans. $A_n \equiv \dfrac{v_o}{v_{S1} + v_{S2} + \cdots + v_{Sn}} = -\dfrac{R_F/R}{1 - \dfrac{nR_F}{(R+1)A_{OL}}}$

   For a single input, v_{S1}, the gain is A_1. For the same input v_{S1}, together with ($n - 1$) zero inputs $v_{S2} = \cdots = v_{Sn} = 0$, then gain is A_n. But, since $A_{OL} < 0$, $|A_n| < |A_1|$ for $n > 1$.

10.19 If the basic Op Amp in Fig. 10-18 is ideal, find v_o and determine the mathematical operation performed by the amplifier circuit.
   Ans. $v_o = \left(1 + \dfrac{R_2}{R_1}\right)(v_{S2} - v_{S1})$ (a subtractor)

10.20 Describe the transfer characteristic of the level-clamp circuit of Fig. 10-13(a) if diode Z_2 is shorted.
   Ans. Let $V_{Z2} \to \infty$ in Fig. 10-13(b).

10.21 Find the gain of the inverting amplifier of Fig. 10-19, if the Op Amp and diodes are ideal.
   Ans. $A_v = \begin{cases} -R_2/R_1 & (v_S > 0) \\ -R_3/R_1 & (v_S \leq 0) \end{cases}$

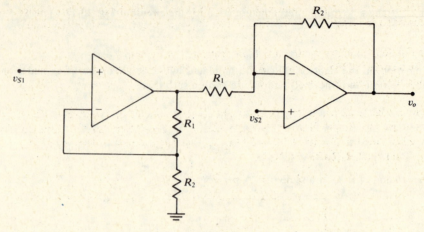

Fig. 10-18

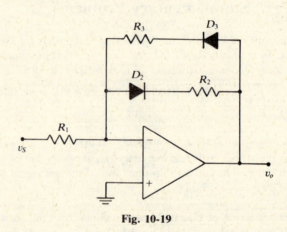

Fig. 10-19

10.22 If the Op Amp of the circuit in Fig. 10-20 is ideal, find an expression for v_o in terms of v_S and infer the function of the circuit.

Ans. $v_o = \dfrac{2}{R_1 C} \displaystyle\int v_S \, dt$ (a noninverting integrator)

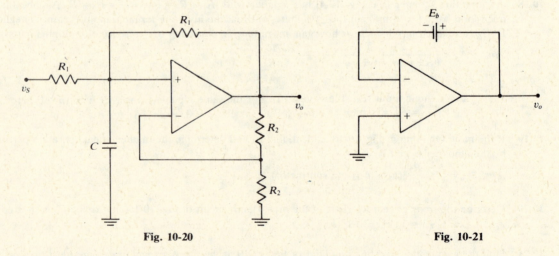

Fig. 10-20   **Fig. 10-21**

10.23 If the nonideal Op Amp of the circuit of Fig. 10-21 has an open-loop gain $A_{OL} = -10^4$, find v_o.
     Ans. $0.9999 E_b$

10.24   How can the square-wave generator of Problem 10.11 be used to make a triangular-wave generator?
   Ans.   Cascade the integrator of Fig. 10-6 to the output of the square-wave generator.

10.25   Describe Op Amp circuitry to simulate the equation $3v_1 + 2v_2 + v_3 = v_o$.
   Ans.   The summer of Fig. 10-4, with $R_F/R_1 = 3$, $R_F/R_2 = 2$, and $R_F/R_3 = 1$, cascaded into the inverting
   amplifier of Fig. 10-2, with $R_F/R_1 = 1$.

10.26   The circuit of Fig. 10-22 (called a *gyrator*) can be used to simulate an inductor in active *RC* filter design.
   Assuming ideal Op Amps, find (*a*) the *s*-domain input impedance, $Z(s)$; (*b*) the equivalent value of
   inductance simulated if $C = 10$ nF, $R_1 = 20$ kΩ, $R_2 = 100$ kΩ, and $R_3 = R_4 = 10$ kΩ.
   Ans.   (*a*) $Z(s) = sR_1R_2R_3C/R_4$; (*b*) 2 H

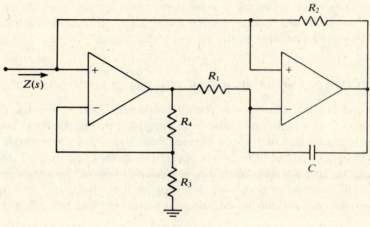

Fig. 10-22

Chapter 11

Electron Tubes: The Vacuum Triode

11.1 INTRODUCTION

The dominant electron tube is the *vacuum tube*, an evacuated enclosure containing (1) a *cathode* which emits electrons, with a *heater* acting to enhance the process; (2) an *anode* or *plate* which attracts the emitted electrons when operated at a positive potential relative to the cathode; and (3) one or more intermediate electrodes (called *grids*) which alter the emission-attraction process. In this chapter attention will be restricted to the case of a single grid.

11.2 CONSTRUCTION AND SYMBOLS OF THE VACUUM TRIODE

The single grid of the *vacuum triode* is called the *control grid*; it is made of small-diameter wire and inserted between the plate and cathode as suggested in Fig. 11-1(*a*). The mesh of the grid is sufficiently coarse so as not to impede current flow from plate to cathode through collision of electrons with the grid wire, and the grid is placed physically close to the cathode so that the grid's electric field can exert considerable control over electron emission from the cathode surface. Symbols for the total instantaneous currents and voltages of the triode are established in Fig. 11-1(*b*); component, average, rms, and maximum values are symbolized as in Table 8-1.

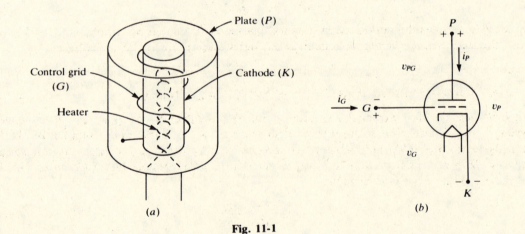

(a) (b)

Fig. 11-1

11.3 TRIODE TERMINAL CHARACTERISTICS

The v-i characteristics of the triode are experimentally determined with the cathode sharing a common connection between the input and output ports. If plate voltage (v_P) and grid voltage (v_G) are taken as independent variables, and grid current (i_G) as the dependent variable, then the *input characteristics* (*grid characteristics*) have the form

$$i_G = f_1(v_P, v_G) \qquad (11.1)$$

of which Fig. 11-2(*a*) is a typical experimentally determined plot. Similarly, taking v_P and v_G as independent variables, the plate current (i_P) becomes the dependent variable of the *output characteristics* (*plate characteristics*)

$$i_P = f_2(v_P, v_G) \tag{11.2}$$

with a corresponding typical plot displayed in Fig. 11-2(*b*).

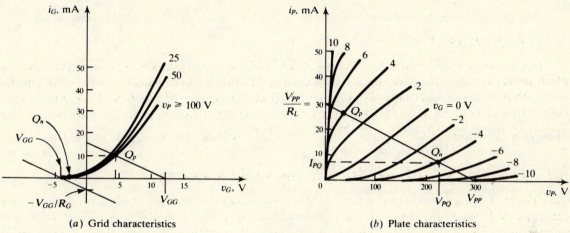

(*a*) Grid characteristics

(*b*) Plate characteristics

Fig. 11-2

The triode input characteristics of Fig. 11-2(*a*) show that operation with a positive grid voltage results in flow of grid current; however, with a negative grid voltage (the common application), negligible grid current flows and the plate characteristics are reasonably approximated by the three-halves-power law (Section 7.10) involving a linear combination of plate and grid voltages:

$$i_P = \kappa(v_P + \mu v_G)^{3/2} \tag{11.3}$$

where κ again denotes the perveance (a constant that depends upon mechanical design of the tube) and μ is the *amplification factor*, a constant whose significance is elucidated in Problem 11.1.

11.4 BIAS AND GRAPHICAL ANALYSIS OF TRIODE AMPLIFIERS

In order to establish a range of triode operation favorable to the signal to be amplified, a quiescent or Q-point must be determined by dc bias circuitry. The basic triode amplifier of Fig. 11-3 has a grid power supply (V_{GG}) of such polarity as to maintain v_G negative (the more common mode of operation). With no signal present ($v_S = 0$), application of KVL around the grid loop of Fig. 11-3 yields the equation of the *grid bias line*,

$$i_G = -\frac{V_{GG}}{R_G} - \frac{v_G}{R_G} \tag{11.4}$$

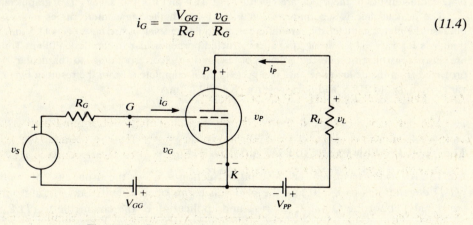

Fig. 11-3. Basic Triode Amplifier

which can be solved simultaneously with (11.1), or plotted as indicated on Fig. 11-2(a) to determine the quiescent values I_{GQ} and V_{GQ}. If V_{GG} is of the polarity indicated in Fig. 11-3, the grid is negatively biased, giving the Q-point labeled Q_n; here $I_{GQ} \approx 0$ and $V_{GQ} \approx -V_{GG}$, and these approximate solutions suffice in the case of negative grid bias. However, if the polarity of V_{GG} were reversed, the grid would have a positive bias, and the quiescent point Q_p would give $I_{GQ} > 0$ and $V_{GQ} < V_{GG}$.

Voltage summation around the plate circuit of Fig. 11-3 leads to the *dc load line* equation

$$i_P = \frac{V_{PP}}{R_L} - \frac{v_P}{R_L} \qquad (11.5)$$

which when plotted on the plate characteristics of Fig. 11-2(b) yields the quiescent values V_{PQ} and I_{PQ} at its intersection with the curve $v_G = V_{GQ}$.

Example 11.1 For the triode amplifier of Fig. 11-3, $V_{GG} = 4$ V, $V_{PP} = 300$ V, $R_L = 10$ kΩ, and $R_G = 2$ kΩ. Plate characteristics of the triode are given by Fig. 11-2(b). (a) Draw the dc load line, and determine the quiescent values (b) I_{GQ}, (c) V_{GQ}, (d) I_{PQ}, (e) V_{PQ}.

(a) For the given values, the dc load line, (11.5), has the i_P-intercept

$$\frac{V_{PP}}{R_L} = \frac{300}{10 \times 10^3} = 30 \text{ mA}$$

and the v_P-intercept $V_{PP} = 300$ V. These intercepts have been utilized to draw the dc load line on the plate characteristics of Fig. 11-2(b).

(b) Since the polarity of V_{GG} is such that v_G is negative, negligible grid current will flow ($I_{GQ} \approx 0$).

(c) For negligible grid current, (11.4) evaluated at the Q-point yields $V_{GQ} = -V_{GG} = -4$ V.

(d) Quiescent plate current is read from the projection of Q_n on the i_P-axis of Fig. 11-2(b) as $I_{PQ} = 8$ mA.

(e) Projection of Q_n on the v_P-axis of Fig. 11-2(b) gives $V_{PQ} = 220$ V.

Application of a time-varying v_S to the amplifier of Fig. 11-3 results in grid voltage with a time-varying component,

$$v_G = V_{GQ} + v_g$$

It is usual practice to assure that $v_G \geq 0$ by proper selection of bias and signal combination. Thus, $i_G = 0$, and the operating point must move along the dc load line from the Q-point in accordance with the variation of v_g, giving instantaneously values of v_P and i_P that simultaneously satisfy (11.2) and (11.5).

Example 11.2 The triode amplifier of Fig. 11-3 has V_{GG}, V_{PP}, R_G, and R_L as given in Example 11.1. If the plate characteristics of the triode are given by Fig. 11-4 and if $v_S = 2 \sin \omega t$ (V), graphically find v_P and i_P.

The dc load line is superimposed on Fig. 11-4 with the same intercepts as in Example 11.1; however, because the plate characteristics are different, the quiescent values are now $I_{PQ} = 11.3$ mA and $V_{PQ} = 186$ V. A time axis for $v_G = -4 + 2 \sin \omega t$ (V) is constructed perpendicular to the dc load line. Time axes for i_P and v_P are also constructed, and the values of i_P and v_P corresponding to particular values of $v_G(t)$ are projected from the dc load line until one cycle of v_G is completed. From the result in Fig. 11-4, it is seen that v_P varies from 152 to 218 V and i_P ranges from 8.1 to 14.7 mA.

11.5 EQUIVALENT CIRCUIT OF A TRIODE

The following treatment echoes that of Section 9.5. For the usual case of negligible grid current, (11.1) degenerates to $i_G = 0$ and the grid acts as an open circuit. Considering only small excursions (ac signals) about the Q-point, $\Delta i_P = i_p$, and application of the chain rule to (11.2) leads to

$$i_p = \Delta i_P \approx di_P = \frac{1}{r_p} v_p + g_m v_g \qquad (11.6)$$

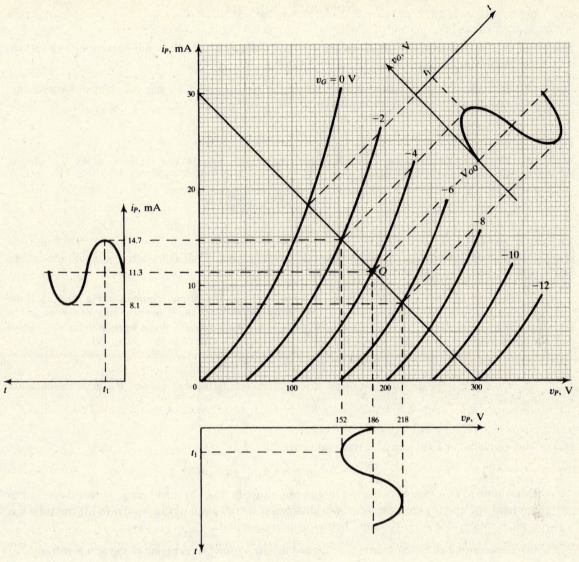

Fig. 11-4

In (11.6) we have defined

$$\text{plate resistance}\quad r_p \equiv \left.\frac{\partial v_P}{\partial i_P}\right|_Q \approx \left.\frac{\Delta v_P}{\Delta i_P}\right|_Q \tag{11.7}$$

$$\text{transconductance}\quad g_m \equiv \left.\frac{\partial i_P}{\partial v_G}\right|_Q \approx \left.\frac{\Delta i_P}{\Delta v_G}\right|_Q \tag{11.8}$$

Equation (11.6), under the condition $i_G = 0$, is simulated by the current-source equivalent circuit of Fig. 11-5(a). The frequently used voltage-source model of Fig. 11-5(b) is developed in Prob. 11.2.

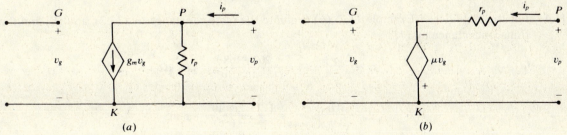

Fig. 11-5.   Triode Small-Signal Equivalent Circuits

Solved Problems

11.1 For the triode with plate characteristics given by Fig. 11-4, find (a) the perveance (κ) and (b) the amplification factor (μ).

(a) Perveance can be evaluated at any point on the $v_G = 0$ curve. Choosing the point with coordinates $i_P = 15$ mA, $v_P = 100$ V, we have from (11.3)

$$\kappa = \frac{i_P}{(v_P)^{3/2}} = \frac{15 \times 10^{-3}}{(100)^{3/2}} = 15 \ \mu A/V^{3/2}$$

(b) The amplification factor is easiest evaluated along the v_P-axis. Use the point $i_P = 0$, $v_P = 100$ V, $v_G = -4$ V in (11.3) to obtain

$$\mu = -\frac{v_P}{v_G} = -\frac{100}{-4} = 25$$

11.2 Use the current-source, small-signal triode model of Fig. 11-5(a) to derive the voltage-source model of Fig. 11-5(b).

Find the Thévenin equivalent looking to the left through the output terminals of Fig. 11-5(a). If the independent source is deactivated, then $v_g = 0$; thus, $g_m v_g = 0$, and the dependent current source acts as an open circuit. The Thévenin resistance is $R_{Th} = r_p$. Open-circuit voltage appearing at the output terminals is

$$v_{Th} = -g_m v_g r_p \equiv -\mu v_g$$

where $\mu \equiv g_m r_p$ is the *amplification factor*. Proper series arrangement of v_{Th} and R_{Th} gives the circuit of Fig. 11-5(b).

11.3 The amplifier of Example 11.1 has plate current

$$i_P = I_P + i_p = 8 + 1 \cos \omega t \quad \text{(mA)}$$

Determine (a) the power delivered by the plate supply V_{PP}, (b) the average power delivered to the load R_L, and (c) the average power dissipated by the plate of the triode. (d) If the tube has a plate rating of 2 W, is the tube being properly applied?

(a) Power supplied by the source V_{PP} is found by integration over a period of the ac waveform.

$$P_{PP} = \frac{1}{T} \int_0^T V_{PP} i_P \, dt = V_{PP} I_P = (300)(8 \times 10^{-3}) = 2.4 \text{ W}$$

(b) $$P_L = \frac{1}{T} \int_0^T i_P^2 R_L \, dt = R_L(I_P^2 + I_p^2) = (10 \times 10^3)\left[(8 \times 10^{-3})^2 + \left(\frac{1 \times 10^{-3}}{\sqrt{2}}\right)^2\right] = 0.645 \text{ W}$$

(c) Average power dissipated by the plate is found as

$$P_P = P_{PP} - P_L = 2.4 - 0.645 = 1.755 \text{ W}$$

(d) The tube is not properly applied. If the signal is removed ($i_p = 0$), then the plate dissipation must increase ($P_P = P_{PP}$) to 2.4 W, which exceeds the power rating.

11.4 For the amplifier of Example 11.2, (a) evaluate the plate resistance by (11.7), and (b) find the transconductance by (11.8).

(a) $$r_p \approx \left.\frac{\Delta v_P}{\Delta i_P}\right|_{v_G = -4} = \frac{202 - 168}{(15 - 8) \times 10^{-3}} = 4.86 \text{ k}\Omega$$

(b) $$g_m \approx \left.\frac{\Delta i_P}{\Delta v_G}\right|_{v_P = 186} = \frac{(17 - 7) \times 10^{-3}}{-3 - (-5)} = 5 \text{ mS}$$

11.5   Find an expression for the voltage gain, $A_v = v_p/v_g$, of the basic triode amplifier of Fig. 11-3, using an ac equivalent circuit.

The equivalent circuit of Fig. 11-5(b) is applicable if R_L is connected from P to K. By voltage division in the plate circuit,

$$v_p = \frac{R_L}{R_L + r_p}(-\mu v_g) \qquad \text{or} \qquad A_v = \frac{v_p}{v_g} = \frac{-\mu R_L}{R_L + r_p}$$

11.6   *Plate efficiency* of a tube amplifier is defined as the ratio of ac signal power to the load (P_{Lac}) to plate supply power (P_{PP}). (*a*) Calculate the efficiency of the amplifier of Problem 11.3. (*b*) What is the maximum efficiency possible for this amplifier without changing the Q-point or clipping the signal?

(*a*) $$\eta = \frac{P_{Lac}}{P_{PP}} \times 100\% = \frac{I_p^2 R_L}{V_{PP}I_P} \times 100\% = \frac{(10^{-3}/\sqrt{2})^2(10 \times 10^3)}{2.4} \times 100\% = 0.208\%$$

(*b*)   Ideally, the input signal could be increased until i_p swings ± 8 mA; thus,

$$\eta_{max} = (8/1)^2(0.208\%) = 13.31\%$$

11.7   The triode amplifier of Fig. 11-6 utilizes *cathode bias* to eliminate the need for a grid power supply. The very large resistance R_G provides a path to ground for stray charge collected by the grid; this current is so small, however, that the voltage drop across R_G is negligible. It follows that the grid is maintained at a negative bias, or

$$v_G = -R_K i_P \qquad\qquad (1)$$

A plot of (*1*) on the plate characteristics is called the *grid bias line*, and its intersection with the dc load line determines the Q-point. Let $R_L = 11.6$ kΩ, $R_K = 400$ Ω, $R_G = 1$ MΩ, and $V_{PP} = 300$ V. If the plate characteristics of the triode are given by Fig. 11-7, (*a*) draw the dc load line, (*b*) sketch the grid bias line, and (*c*) determine the Q-point quantities.

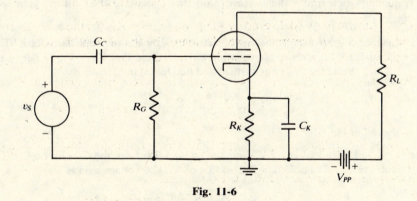

Fig. 11-6

(*a*)   The dc load line has horizontal intercept $V_{PP} = 300$ V and vertical intercept

$$\frac{V_{PP}}{R_{dc}} = \frac{V_{PP}}{R_L + R_K} = \frac{300}{(11.6 + 0.4) \times 10^3} = 25 \text{ mA}$$

on the plate characteristics of Fig. 11-7.

(*b*)   Points for the plot of (*1*) are determined by selecting values of i_P and calculating the corresponding values of v_G. For example, if $i_P = 5$ mA, $v_G = -(400)(5 \times 10^{-3}) = -2$ V, which plots as point 1 of the grid bias line in Fig. 11-7. Note that this is not a straight line.

(*c*)   From the intersection of the grid bias line with the dc load line, $I_{PQ} = 10$ mA, $V_{PQ} = 180$ V, and $V_{GQ} = -4$ V.

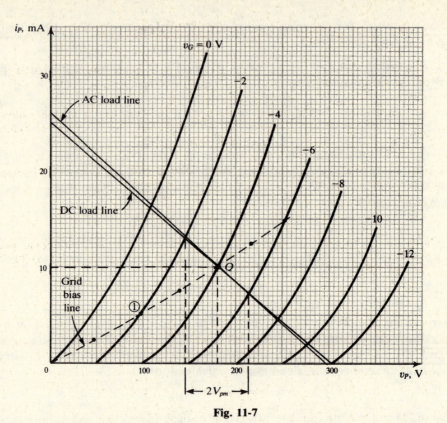

Fig. 11-7

11.8   For the amplifier of Problem 11.7, let $v_S = 2 \cos \omega t$ (V). (a) Draw the ac load line on Fig. 11-7. (b) Graphically determine the voltage gain. (c) Calculate the voltage gain using small-signal analysis.

(a)   If capacitor C_K appears as a short circuit to ac signals, then application of KVL around the plate circuit of Fig. 11-6 gives as the equation of the ac load line $V_{PP} + V_{GQ} = i_P R_L + v_P$.

Thus, the ac load line has vertical and horizontal intercepts

$$\frac{V_{PP} + V_{GQ}}{R_L} = \frac{300 - 4}{11.6 \times 10^3} = 25.5 \text{ mA} \qquad V_{PP} + V_{GQ} = 296 \text{ V}$$

as shown on Fig. 11-7.

(b) $v_g = v_S$, and as v_g swings ± 2 V along the ac load line from the Q-point on Fig. 11-7, it is seen that the total swing of v_p is $2V_{pm} = 213 - 145 = 68$ V. The voltage gain is

$$A_v = -\frac{2V_{pm}}{2V_{gm}} = -\frac{68}{4} = -17$$

where the negative sign is inserted to account for the phase reversal between v_p and v_g.

(c)   Applying (11.7) and (11.8) at the Q-point of Fig. 11-7,

$$r_p = \left. \frac{\Delta v_P}{\Delta i_P} \right|_{v_G = -4} = \frac{202 - 168}{(15 - 8) \times 10^{-3}} = 4.86 \text{ k}\Omega$$

$$g_m = \left. \frac{\Delta i_P}{\Delta v_G} \right|_{v_P = 180} = \frac{(15.5 - 6.5) \times 10^{-3}}{-3 - (-5)} = 4.5 \text{ mS}$$

Then, $\mu \equiv g_m r_p = 21.87$, and Problem 11.5 yields

$$A_v = -\frac{\mu R_L}{R_L + r_p} = -\frac{(21.87)(11.6 \times 10^3)}{(11.6 + 4.86) \times 10^3} = -15.41$$

Supplementary Problems

11.9 If the amplifier of Problem 11.3 has plate resistance (r_p) 20 kΩ and $v_S = 1 \cos \omega t$ (V), find its amplification factor (μ) using the small-signal voltage-source model of Fig. 11-5(b). *Ans.* 30

11.10 If the bypass capacitor C_K is removed from the amplifier of Fig. 11-6, find (a) an expression for the voltage gain, and (b) the percentage deviation of the voltage gain from the result of Problem 11.8. *Ans.* (a) $A_v = -\mu R_L/[R_L + r_p + (\mu + 1)R_K]$; ($b$) 35.7% decrease

11.11 Two triodes are parallel-connected plate to plate, grid to grid, and cathode to cathode. Find the equivalent amplification factor (μ_{eq}) and plate resistance (r_{peq}) for the combination.

Ans. $\mu_{eq} = \dfrac{\mu_1 r_{p2} + \mu_2 r_{p1}}{r_{p1} + r_{p2}}, \quad r_{peq} = \dfrac{r_{p1} r_{p2}}{r_{p1} + r_{p2}}$

11.12 The circuit of Fig. 11-8 is a *cathode follower*, so called because v_o is in phase with v_S and nearly equal to it in magnitude. Find a voltage-source equivalent circuit of the form of Fig. 11-5(b) to model the cathode follower. *Ans.* See Fig. 11-9.

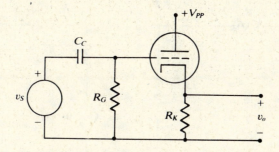

Fig. 11-8

11.13 For the cathode follower of Fig. 11-8, $r_p = 5$ kΩ, $\mu = 25$, and $R_K = 15$ kΩ. (a) Use the equivalent circuit, Fig. 11-9, to find a formula for the voltage gain. (b) Evaluate the voltage gain. *Ans.* (a) $A_v = \mu R_K/[r_p + (\mu + 1)R_K]$; ($b$) 0.95

11.14 The cathode follower is frequently used as a final-stage amplifier to effect an impedance match with a low-impedance load for maximum power transfer (see Problem 3.10). In such a case, the load (resistor R_L) is capacitor-coupled to the right of R_K in Fig. 11-9. Find an expression for the internal impedance (output impedance) of the cathode follower as seen by the load. *Ans.* $R_o = R_K r_p/[r_p + (\mu + 1)R_K]$

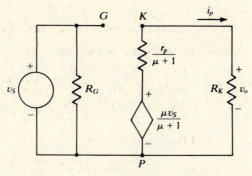

Fig. 11-9

11.15 The amplifier of Fig. 11-10 is a *common-grid* amplifier. By taking a Thévenin equivalent looking to the right through GK and another equivalent looking to the left through R_P, verify that the small-signal circuit of Fig. 11-11 is valid. Then, (a) find an expression for the voltage gain; (b) evaluate the voltage gain for the typical values $\mu = 20$, $r_p = 5$ kΩ, $R_K = 1$ kΩ, and $R_P = 15$ kΩ; (c) find R_{in}, the input resistance; and (d) find R_o, the output resistance.

Ans.   (a) $A_v = (\mu + 1)R_P/[R_P + r_p + (\mu + 1)R_K]$; (b) 7.7; (c) $R_{in} = 1.95$ kΩ; (d) $R_o = 26$ kΩ

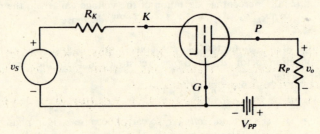

Fig. 11-10

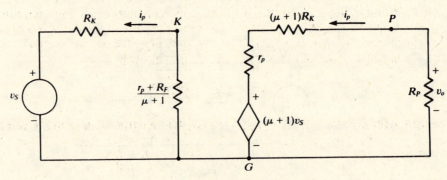

Fig. 11-11

Chapter 12

Magnetic Circuits

12.1 BASIC CONCEPTS

By a *magnetic circuit* is meant a path for magnetic flux, just as an electric circuit provides a path for the flow of electric current. Transformers, electric machines, and numerous other electromechanical devices utilize magnetic circuits.

We define the *magnetic induction* or *magnetic flux density*, *B*, by the force equation

$$F = B\ell I \tag{12.1}$$

where F (N) is the force experienced by a straight conductor of length ℓ (m) which carries a current I (A) and is oriented at right angles to a uniform magnetic field of flux density B (T). In other words, if a conductor is 1 meter long, carries 1 ampere current, and experiences 1 newton force when located at right angles to parallel magnetic flux lines, then the magnetic flux density is 1 tesla. Actually, the magnetic induction is a vector, **B**, of magnitude B and tangential to the flux lines.

The *magnetic flux*, ϕ, through a given surface S is defined by

$$\phi = \int_S \mathbf{B} \cdot d\mathbf{S} \tag{12.2}$$

If **B** is uniform over a surface of area A and is everywhere perpendicular to the surface, (*12.2*) gives

$$\phi = BA \qquad \text{or} \qquad B = \frac{\phi}{A} \tag{12.3}$$

The unit of magnetic flux is the *weber* (Wb), and (*12.3*) shows that $1\ T = 1\ Wb/m^2$.

The source of magnetic flux is either a permanent magnet or an electric current. To measure the effectiveness of electric current in producing a magnetic field (or flux) we introduce the concept of *magnetomotive force* (mmf), $\mathscr{F}$, defined as

$$\mathscr{F} \equiv NI \tag{12.4}$$

where I is the current flowing in an N-turn coil. The unit of $\mathscr{F}$ is the *ampere turn* (At). Schematically, a magnetic circuit with an mmf and magnetic flux is shown in Fig. 12-1.

The mutual relationship between an electric current I and the corresponding *magnetic field intensity*, **H**, is given by *Ampère's circuital law*,

$$\oint \mathbf{H} \cdot d\mathbf{l} = I \tag{12.5}$$

In (*12.5*), I is the total current through a given open surface, and the line integral is extended around the boundary of that surface. If the current cuts the surface N times (as it does the central rectangle in Fig. 12-1), (*12.5*) becomes

$$\oint \mathbf{H} \cdot d\mathbf{l} = NI \equiv \mathscr{F} \tag{12.6}$$

Ampère's circuital law implies the unit A/m for **H** (or H).

The core material of a magnetic circuit (constituting a transformer or an electric machine) is generally ferromagnetic, and the variation of B with H is depicted by a *saturation curve*, as shown in Fig. 12-2. For region II, where the slope of the curve is nearly constant, we may write

$$B = \mu H \tag{12.7}$$

177

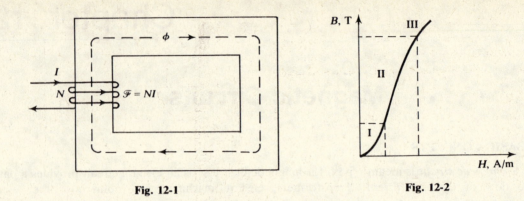

Fig. 12-1   Fig. 12-2

where μ (H/m) is defined as the *permeability* of the material. For free space (or air), $\mu = \mu_0 = 4\pi \times 10^{-7}$ H/m. Sometimes (12.7) is written in terms of μ_0 as

$$B = \mu_r \mu_0 H \qquad\qquad (12.8)$$

where $\mu_r \equiv \mu/\mu_0$ is called the *relative permeability*. In regions I and III, the relationship between B and H is nonlinear. Figure 12-3 shows actual saturation curves of some common materials; note that the scale of H is logarithmic.

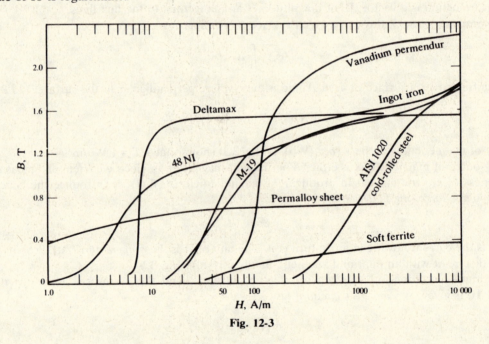

Fig. 12-3

12.2 LAWS GOVERNING MAGNETIC CIRCUITS

Magnetic circuits obey the analog of Ohm's law: the complete analogy between magnetic and dc resistive circuits is displayed in Table 12-1.

In Table 12-1, ℓ is the length and A is the cross-sectional area of the path for the current in the electric circuit, or for the flux in the magnetic circuit. By the analogy, the laws of resistances in series or parallel also hold for reluctances.

The *differences* between a dc resistive circuit and a magnetic circuit are: (1) there is an I^2R-loss in a resistance but no $\phi^2 \mathcal{R}$-loss in a reluctance; (2) magnetic fluxes take *leakage* paths, as ϕ_ℓ in Fig. 12-4, whereas electric currents (flowing through resistances) do not; and (3) in magnetic circuits with air gaps we encounter *fringing* (Fig. 12-4) of flux lines, but we do not have fringing of currents in electric circuits. Fringing increases with the length of the air gap, and increases the effective cross-sectional area of the air gap.

Table 12-1

DC Resistive Circuit	Magnetic Circuit
current, I	flux, ϕ
voltage, V	mmf, $\mathcal{F}$
conductivity, σ	permeability, μ
Ohm's law, $I = V/R$	$\phi = \mathcal{F}/\mathcal{R}$
resistance, $R = \ell/\sigma A$	reluctance, $\mathcal{R} = \ell/\mu A$
conductance, $G = 1/R$	permeance, $\mathcal{P} = 1/\mathcal{R}$

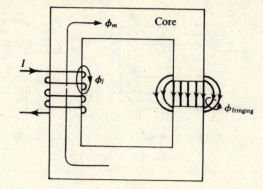

Fig. 12-4

12.3 AC LOSSES IN MAGNETIC CIRCUITS

If the mmf acting in a magnetic circuit is ac, then the *B-H* curve takes the form of a *hysteresis loop*, as shown in Fig. 12-5. The area within the loop is proportional to the energy loss (as heat) per cycle; this energy loss is known as *hysteresis loss*; it is proportional to the frequency f of the exciting mmf. In addition, *eddy-current loss*, proportional to f^2, arises due to the eddy currents induced in the core material.

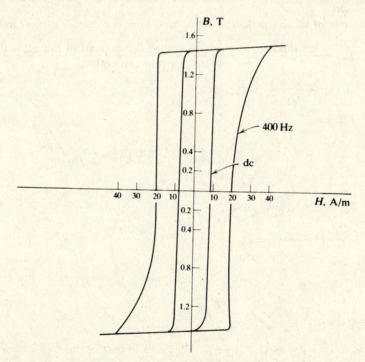

Fig. 12-5.   Deltamax Tape-Wound Core, 0.002-in Strip Hysteresis Loop

The hysteresis-loss component of the core loss in a magnetic circuit is reduced by using as the core material 'good' quality electrical steel having a narrow hysteresis loop. To reduce eddy-current loss, the core is made of laminations, or thin sheets, with very thin layers of insulation alternating with the laminations. The laminations are oriented parallel to the direction of flux [Fig. 12-6(*b*)]. Laminating a core increases its cross-sectional area, and hence the volume. The ratio of the volume actually occupied by the magnetic material to the total volume of the core is called the *stacking factor*. Table 12-2 gives some values of the stacking factor.

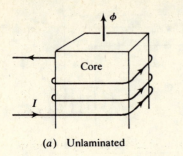

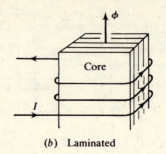

(a) Unlaminated   (b) Laminated

Fig. 12-6

Table 12-2

Lamination Thickness, mm	Stacking Factor
0.0127	0.50
$0.0254 = 10^{-3}$ in	0.75
0.0508	0.85
0.10 to 0.25	0.90
0.27 to 0.36	0.95

12.4 INDUCTANCE AND MAGNETIC ENERGY

In Section 2.2, we introduced inductance, L, as an electric circuit element. Alternatively, we may relate inductance to magnetic circuit (or field) quantities. Thus, we redefine the *inductance* of an N-turn coil as the flux linkage, λ, per unit current:

$$L \equiv \frac{\lambda}{i} \equiv \frac{N\phi}{i} \qquad (12.9)$$

In (12.9), ϕ is the flux through each of the turns (or else, the mean flux per turn). The unit of inductance is the *henry* (H).

For a magnetic circuit having n distinct coils, Fig. 12-7, we may define n^2 inductances by

$$L_{pq} \equiv \frac{\text{flux linking the } p\text{th coil due to the current in the } q\text{th coil}}{\text{current in the } q\text{th coil}}$$

$$\equiv \frac{N_p(k_{pq}\phi_q)}{i_q} \qquad (p, q = 1, 2, \dots, n) \qquad (12.10)$$

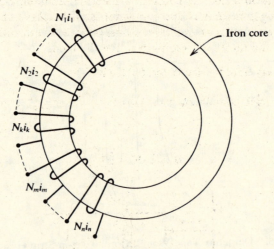

$N_1 i_1$

$N_2 i_2$

$N_k i_k$

$N_m i_m$

$N_n i_n$

Iron core

Fig. 12-7

where N_p is the number of turns in the pth coil and where k_{pq}, the *coupling coefficient* between the pth and qth coils, is defined as the fraction of the flux produced by the qth that links the pth coil. For $p \neq q$, L_{pq} is called the *mutual inductance* between coils p and q; for $p = q$, L_{pp} is the *self-inductance* of coil p. It can be shown that, for all p and q,

$$k_{pq} = k_{qp} \qquad \text{and} \qquad L_{pq} = L_{qp}$$

To express L_{pq} in terms of the magnetic circuit parameters, substitute

$$\phi_q = \frac{N_q i_q}{\mathcal{R}} = N_q i_q \frac{\mu A}{\ell}$$

in (12.10), obtaining

$$L_{pq} = \frac{k_{pq} N_p N_q}{\mathcal{R}} = k_{pq} N_p N_q \frac{\mu A}{\ell} \tag{12.11}$$

If the n coils of Fig. 12-7 carry currents $i_1, i_2, \ldots, i_n$, it may be shown that the energy stored in the inductances is given by

$$W_m = \frac{1}{2} \sum_{p=1}^{n} \sum_{q=1}^{n} L_{pq} i_p i_q \tag{12.12}$$

Further, it may be shown that the same energy may be expressed, in terms of magnetic field quantities **B** and **H**, as

$$W_m = \frac{1}{2} \int_v \mathbf{B} \cdot \mathbf{H} \, dv = \frac{1}{2} \mu \int_v H^2 \, dv = \frac{1}{2\mu} \int_v B^2 \, dv \tag{12.13}$$

in which the volume integration may, for practical purposes, be restricted to the common core of the coils. Thus, to evaluate an inductance L_{pq} we may either directly use (12.11) or equate (12.12) and (12.13).

Solved Problems

12.1 A square loop, $2a$ (m) on a side, is placed in the field of an infinitely long, straight conductor (in free space) carrying a current I. The conductor is in the plane of the square, is parallel to two sides of the square, and is at a distance b(m) from the center of the square, as shown in Fig. 12-8. Determine the total flux passing through the loop.

At a point r (m) away from the conductor, (12.5) gives

$$\oint \mathbf{H} \cdot d\mathbf{l} = 2\pi r H_\varphi = I \qquad \text{or} \qquad H_\varphi = \frac{I}{2\pi r} \quad \text{(A/m)}$$

Thus

$$B_\varphi = \mu_0 H_\varphi = \frac{\mu_0 I}{2\pi r} \quad \text{(T)}$$

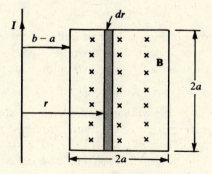

Fig. 12-8

μ_0 being the permeability of free space. The flux $d\phi$ through the elementary area $dA = 2a\,dr$ is given by

$$d\phi = B_\varphi\,dA = \left(\frac{\mu_0 Ia}{\pi}\right)\frac{dr}{r}$$

Hence

$$\phi = \frac{\mu_0 Ia}{\pi}\int_{b-a}^{b+a}\frac{dr}{r} = \frac{\mu_0 Ia}{\pi}\ln\frac{b+a}{b-a}\quad\text{(Wb)}$$

12.2   For the magnetic circuit shown in Fig. 12-9, $N = 70$ turns, $\ell_g = 0.2$ mm, $\ell_m = 150$ mm. The core material is 0.2-mm-thick M-19 laminations. (The saturation characteristic of M-19 is shown in Fig. 12-3.) Calculate I required to produce a 1-T flux density in the air gap. Neglect fringing and leakage.

For the air gap,

$$H_g = \frac{B_g}{\mu_0} = \frac{1}{4\pi \times 10^{-7}} = 7.95 \times 10^5 \text{ A/m}$$
$$\mathscr{F}_g = H_g\ell_g = (7.95 \times 10^5)(2 \times 10^{-4}) = 159 \text{ At}$$

For the core, the stacking factor is 0.9 (from Table 12-2), and

$$B_m = \frac{B_g}{\text{stacking factor}} = \frac{1}{0.9} = 1.11 \text{ T}$$

From Fig. 12-3, at 1.11 T, we have

$$H_m = 130 \text{ A/m} \qquad \text{and} \qquad \mathscr{F}_m = (130)(150 \times 10^{-3}) = 19.5 \text{ At}$$

Then the total required mmf is $\mathscr{F}_{\text{total}} = \mathscr{F}_g + \mathscr{F}_m = 159 + 19.5 = 178.5$ At, whence

$$I = \frac{178.5}{70} = 2.55 \text{ A}$$

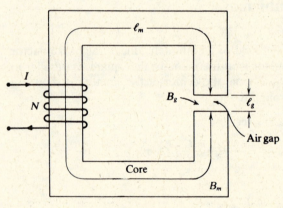

Fig. 12-9

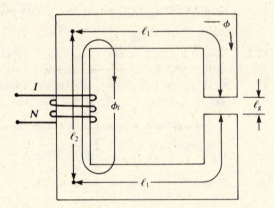

Fig. 12-10

12.3   The magnetic circuit shown in Fig. 12-10 is of uniform cross section, the air-gap flux density is 1 T, and there is 10 percent leakage flux having the path shown. The core material is M-19. Given: $N = 60$ turns; $\ell_g = 0.2$ mm; $\ell_1 = \ell_2 = 200$ mm. Calculate the coil current.

As in Problem 12.2, $\mathscr{F}_g = 159$ At. For the core, Fig. 12-3 gives, at $B_{\ell_1} = 1$ T, $H_{\ell_1} = 90$ A/m; thus,

$$\mathscr{F}_{\ell_1} = (90)(2 \times 200 \times 10^{-3}) = 36 \text{ At}$$

Including leakage flux, at $B_{\ell_2} = 1.1$ T, $H_{\ell_2} = 130$ A/m, and

$$\mathscr{F}_{\ell_2} = (130)(200 \times 10^{-3}) = 26 \text{ At}$$

Hence, $\mathscr{F}_{\text{total}} = NI = 60I = 159 + 36 + 26 = 221$ At, and $I = 221/60 = 3.68$ A.

12.4   From the data of Problem 12.3, determine (a) the permeability of the portion of the core marked ℓ_1 in Fig. 12-10, (b) the effective permeability of the composite portion of the core consisting of the air gap and the lengths ℓ_1.

(a)   From Fig. 12-3, at $B = 1$ T, $H = 90$ A/m. Consequently,

$$\mu = \frac{B}{H} = \frac{1}{90} = 0.0111 \text{ H/m}$$

(b)   By Ampère's law, $H\ell = NI$, or

$$\mu = \frac{B\ell}{NI} = \frac{(1)(400.2 \times 10^{-3})}{221} = 0.00181 \text{ H/m}$$

12.5   The core of a magnetic circuit is of mean length 40 cm and uniform cross-sectional area 4 cm². The relative permeability of the core material is 1000. An air gap of 1 mm is cut in the core, and 1000 turns are wound on the core. Determine the inductance of the coil if fringing is negligible.

Using $\mathscr{R} = \ell/\mu A$,

$$\mathscr{R}_{\text{core}} = \frac{(400-1) \times 10^{-3}}{1000(4\pi \times 10^{-7})(4 \times 10^{-4})} = 0.793 \times 10^{6} \text{ H}^{-1} \qquad \mathscr{R}_{\text{air}} = \frac{1 \times 10^{-3}}{(4\pi \times 10^{-7})(4 \times 10^{-4})} = 1.990 \times 10^{6} \text{ H}^{-1}$$

Therefore, $\mathscr{R}_{\text{total}} = 2.783 \times 10^{6}$ H⁻¹, and, by (12.11) ($k_{11} = 1$),

$$L = \frac{N^2}{\mathscr{R}_{\text{total}}} = \frac{10^{6}}{2.783 \times 10^{6}} = 0.359 \text{ H}$$

12.6   The magnetic circuit of Fig. 12-11 is composed of laminations stacked to make a net core thickness of 2 cm. The relative permeability of the core is 1500. Find the self- and mutual inductances of the pair of coils.

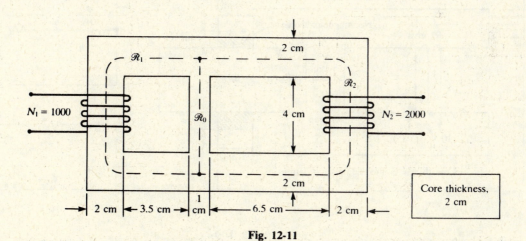

Fig. 12-11

Defining $\mathscr{R}_0$, $\mathscr{R}_1$, and $\mathscr{R}_2$ as shown in Fig. 12-11, we have

$$\mathscr{R}_0 = \frac{6 \times 10^{-2}}{1500(4\pi \times 10^{-7})(2 \times 10^{-4})} = 159\,000 \text{ H}^{-1}$$

$$\mathscr{R}_1 = \frac{16 \times 10^{-2}}{1500(4\pi \times 10^{-7})(4 \times 10^{-4})} = 212\,000 \text{ H}^{-1}$$

$$\mathscr{R}_2 = \frac{22 \times 10^{-2}}{1500(4\pi \times 10^{-7})(4 \times 10^{-4})} = 292\,000 \text{ H}^{-1}$$

The net reluctance $\mathscr{R}'$ seen by the coil $N_1 = 1000$ is $\mathscr{R}_1$ in series with the parallel combination of $\mathscr{R}_0$ and $\mathscr{R}_2$. Thus,

$$\mathscr{R}' = \mathscr{R}_1 + \frac{\mathscr{R}_0\mathscr{R}_2}{\mathscr{R}_0 + \mathscr{R}_2} = \frac{\mathscr{R}_0\mathscr{R}_1 + \mathscr{R}_1\mathscr{R}_2 + \mathscr{R}_0\mathscr{R}_2}{\mathscr{R}_0 + \mathscr{R}_2} \equiv \frac{\Delta}{\mathscr{R}_0 + \mathscr{R}_2}$$

Similarly, the coil $N_2 = 2000$ sees net reluctance

$$\mathscr{R}'' = \frac{\Delta}{\mathscr{R}_0 + \mathscr{R}_1}$$

It follows that

$$L_{11} = \frac{N_1^2}{\mathscr{R}'} = \frac{N_1^2}{\Delta}(\mathscr{R}_0 + \mathscr{R}_2) \qquad L_{22} = \frac{N_2^2}{\mathscr{R}''} = \frac{N_2^2}{\Delta}(\mathscr{R}_0 + \mathscr{R}_1)$$

The total flux of coil N_1, $N_1I_1/\mathscr{R}'$, is divided between the limbs $\mathscr{R}_0$ and $\mathscr{R}_2$, with the latter receiving the amount

$$\phi_{21} = \frac{\mathscr{R}_0}{\mathscr{R}_0 + \mathscr{R}_2}\left(\frac{N_1I_1}{\mathscr{R}'}\right) = \frac{\mathscr{R}_0 N_1 I_1}{\Delta}$$

Then, by (12.10),

$$L_{12} = L_{21} = \frac{N_2\phi_{21}}{I_1} = \frac{N_1 N_2}{\Delta}\mathscr{R}_0$$

Substituting numerical values yields

$$L_{11} = 3.18 \text{ H} \qquad L_{22} = 10.44 \text{ H} \qquad L_{12} = L_{21} = 2.24 \text{ H}$$

12.7 Assuming an ideal core ($\mu_i \to \infty$), calculate the flux density in the air gap of the magnetic circuit shown in Fig. 12-12(a).

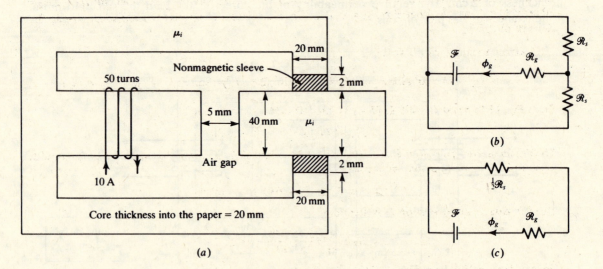

Fig. 12-12

The electrical analog. Fig. 12-12(b), may be reduced to the simpler Fig. 12-12(c). From the latter, we obtain (with $\mu_0 = 4\pi \times 10^{-7}$ H/m):

$$\mathscr{R}_g \equiv \text{air-gap reluctance} = \frac{5 \times 10^{-3}}{\mu_0(20 \times 40 \times 10^{-6})} = \frac{50}{8\mu_0} \quad (\text{H}^{-1})$$

$$\mathscr{R}_s \equiv \text{sleeve reluctance} = \frac{2 \times 10^{-3}}{\mu_0(20 \times 20 \times 10^{-6})} = \frac{20}{4\mu_0} \quad (\text{H}^{-1})$$

$$\mathscr{R}_t \equiv \text{total reluctance} = \mathscr{R}_g + \frac{1}{2}\mathscr{R}_s = \frac{70}{8\mu_0} \quad (\text{H}^{-1})$$

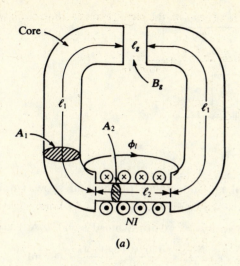

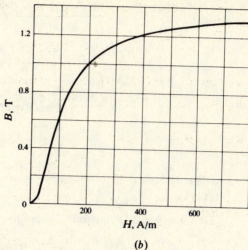

Fig. 12-13

$$\phi_g \equiv \text{air-gap flux} = \frac{\mathscr{F}}{\mathscr{R}_t} = \frac{(50)(10)}{70/8\mu_0} = \frac{400\mu_0}{7} \quad \text{(Wb)}$$

$$B_g \equiv \text{air-gap flux density} = \frac{\phi_g}{A_g} = \frac{400\mu_0/7}{20 \times 40 \times 10^{-6}} = 90 \text{ mT}$$

or 900 gauss.

12.8 A composite magnetic circuit of varying cross section is shown in Fig. 12-13(a); the iron portion has the B-H characteristic of Fig. 12-13(b). Given: $N = 100$ turns; $\ell_1 = 4\ell_2 = 40$ cm; $A_1 = 2A_2 = 10$ cm^2; $\ell_g = 2$ mm; leakage flux, $\phi_l = 0.01$ mWb. Calculate I required to establish an air-gap flux density of 0.6 T.

Corresponding to $B_g = 0.6$ T,

$$H_g = \frac{0.6}{\mu_0} = 4.78 \times 10^5 \text{ A/m} \qquad \mathscr{F}_g = (4.78 \times 10^5)(2 \times 10^{-3}) = 956 \text{ At} \qquad B_{\ell_1} = B_g = 0.6 \text{ T}$$

From Fig. 12-13(b), at $B = 0.6$ T, $H = 100$ A/m. Thus, for the two lengths ℓ_1,

$$\mathscr{F}_{\ell_1} = (100)(0.40 + 0.40) = 80 \text{ At}$$

The flux in the air gap is $\phi_g = B_g A_1 = (0.6)(10 \times 10^{-4}) = 0.6$ mWb. The total flux produced by the coil, ϕ_c, is the sum of the air-gap flux and the leakage flux:

$$\phi_c = \phi_g + \phi_l = 0.6 + 0.01 = 0.61 \text{ mWb}$$

The flux density in the portion ℓ_2 is therefore

$$B_2 = \frac{\phi_c}{A_2} = \frac{0.61 \times 10^{-3}}{5 \times 10^{-4}} = 1.22 \text{ T}$$

For this flux density, from Fig. 12-13(b), $H = 410$ A/m and

$$\mathscr{F}_{\ell_2} = (410)(0.10) = 41 \text{ At}$$

The total required mmf is thus $\mathscr{F}_g + \mathscr{F}_{\ell_1} + \mathscr{F}_{\ell_2} = 956 + 80 + 41 = 1077$ At, and, for $N = 100$ turns, the required current is

$$I = \frac{1077}{100} = 10.77 \text{ A}$$

12.9 Draw an electrical analog for the magnetic circuit shown in Fig. 12-13(a).

See Fig. 12-14.

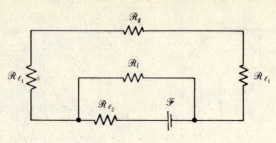

Fig. 12-14

12.10 Calculate the total self-inductance and the leakage inductance of the coil shown in Fig. 12-13(a).

From Problem 12.8, the total flux produced by the coil is $\phi_c = 0.61$ mWb and $I = 10.77$ A. Hence,

$$L = \frac{N\phi_c}{I} = \frac{(100)(0.61 \times 10^{-3})}{10.77} = 5.66 \text{ mH} \qquad L_l = \frac{N\phi_l}{I} = \frac{(100)(0.01 \times 10^{-3})}{10.77} = 0.093 \text{ mH}$$

12.11 Determine the magnetic energy stored in the iron and in the air gap of the magnetic circuit of Fig. 12-13(a).

From (*12.13*),

$$W_{\text{air}} = \frac{1}{2\mu_0} B_g^2 \times \text{vol}_{\text{gap}} = \frac{(0.6)^2}{2\mu_0}(10 \times 10^{-4})(2 \times 10^{-3}) = 0.286 \text{ J}$$

From (*12.12*) and Problem 12.10,

$$W_{\text{iron}} = \frac{1}{2}LI^2 - W_{\text{air}} = \frac{1}{2}N\phi_c I - W_{\text{air}} = 0.328 - 0.286 = 0.042 \text{ J}$$

Notice that $W_{\text{air}} \gg W_{\text{iron}}$.

12.12 A toroid of rectangular cross section is shown in Fig. 12-15. The mean diameter is large compared to the core thickness in the radial direction, so that the core flux density is essentially uniform. Derive an expression for the inductance of the toroid, and evaluate it if $r_1 = 80$ mm, $r_2 = 100$ mm, $a = 20$ mm, and $N = 200$ turns. The core relative permeability is 900.

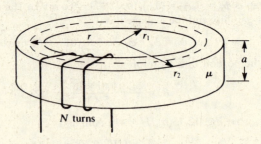

Fig. 12-15

$$\phi = \frac{Ni}{\mathcal{R}} = \frac{\mu A N_i}{2\pi r} = \frac{\mu a (r_2 - r_1) N_i}{\pi(r_2 + r_1)}$$

since $A = a(r_2 - r_1)$ and $r = (r_2 + r_1)/2$. Hence,

$$L = \frac{N\phi}{i} = \frac{\mu a (r_2 - r_1) N^2}{\pi(r_2 + r_1)} = \frac{(900)(4\pi \times 10^{-7})(20 \times 10^{-3})(20 \times 10^{-3})(200)^2}{\pi(180 \times 10^{-3})} = 0.32 \text{ H}$$

12.13 A magnetic circuit, made up of laminations of sheet steel, is of the dimensions shown in Fig. 12-16(a). The net core (stack) thickness is 5 cm. For a 320-At mmf on limb B, calculate the fluxes in the limbs A, B, and C. See Fig. 12-16(b) for the B-H characteristics of sheet steel.

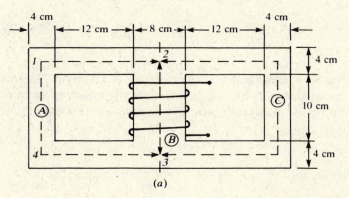

(a)

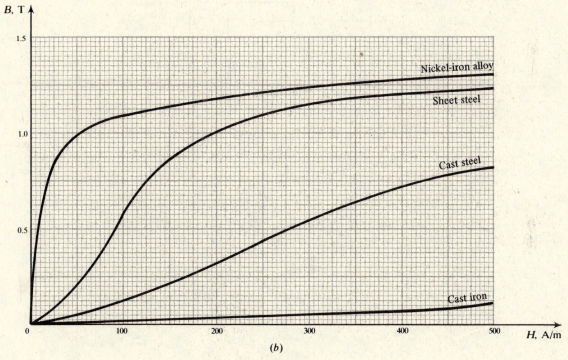

(b)

Fig. 12-16

From Ampère's law, (*12.6*), around the closed path *12341*,

$$H[(18 + 14 + 18 + 14) \times 10^{-2}] = 320 \quad \text{or} \quad H = \frac{320}{0.64} = 500 \text{ A/m}$$

Correspondingly, from Fig. 12-16(b), $B = 1.25$ T. Therefore,

$$\phi_A = (1.25)(4 \times 5 \times 10^{-4}) = 2.5 \text{ mWb} \qquad \phi_B = (1.25)(8 \times 5 \times 10^{-4}) = 5.0 \text{ mWb}$$

and, by symmetry, $\phi_C = \phi_A = 2.5$ mWb. (*Check*: $\phi_B = \phi_A + \phi_C$.)

12.14 If the coil on limb B in Fig. 12-16(a) has 100 turns and carries 3.2 A current, find (a) the magnetic energy density in the core, and (b) the relative permeability at the operating point.

(a)   magnetic energy density $\equiv w_m = \frac{1}{2}\mathbf{B}\cdot\mathbf{H} = \frac{1}{2}BH$

assuming that **B** and **H** are parallel. From Problem 12.13, $B = 1.25$ T and $H = 500$ A/m. Thus,

$$w_m = \frac{1}{2}(1.25)(500) = 312.5 \text{ J/m}^3$$

(b) $\mu = \dfrac{B}{H} = \dfrac{1.25}{500}$ H/m   and $\mu_r = \dfrac{\mu}{\mu_0} = \dfrac{1.25/500}{4\pi \times 10^{-7}} = 1990$

12.15 An inductor, made of a highly permeable material, has N turns. The dimensions of the core and the coil are as shown in Fig. 12-17. Calculate the input power to the coil to establish a given flux density, B, in the air gap. The *winding space-factor* of the coil is k_s and its conductivity is σ.

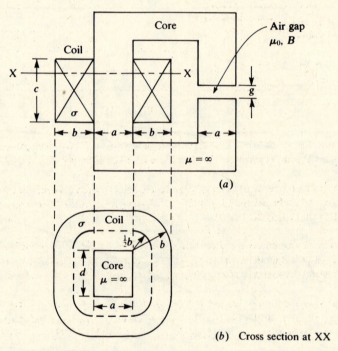

Fig. 12-17

From Fig. 12-17(b), the mean length per turn is

$$\ell = 2a + 2d + 4\left(\frac{1}{4}\right)\left(2\pi\frac{b}{2}\right) = 2\left(a + d + \pi\frac{b}{2}\right)$$

and the total length of wire making the coil is ℓN. Let A_c denote the cross-sectional area of the wire; then its resistance is $R = \ell N/\sigma A_c$, giving an input power

$$P_i = I^2 R = \frac{I^2 \ell N}{\sigma A_c} \tag{1}$$

But $\mathscr{F} = NI = \phi\mathscr{R} = BA\dfrac{g}{\mu_0 A} = \dfrac{Bg}{\mu_0}$ or $I = \dfrac{Bg}{\mu_0 N}$

and the total volume of the wire is $(bc\ell)k_s = \ell N A_c$. Hence, substituting for I and A_c in (1),

$$P_i = \frac{B^2 g^2 \ell}{\mu_0^2 \sigma b c k_s}$$

Supplementary Problems

12.16 From Fig. 12-16(b), determine the relative permeabilities of (a) nickel-iron alloy and (b) sheet steel, at an operating flux density of 1 T.   *Ans.*   (a) 16 000; (b) 4000

12.17 A magnetic circuit has mean length 20 cm and cross-sectional area 10 cm^2. If the relative permeability of the material is 2000, what is the reluctance of the circuit?   *Ans.* 7.96×10^4 H^{-1}

12.18 The magnetic circuit of Fig. 12-18 is made of sheet steel having the *B-H* characteristic shown in Fig. 12-16(b). Calculate the mmf of the coil to establish a 1-T flux density in the air gap.   *Ans.*   902 At

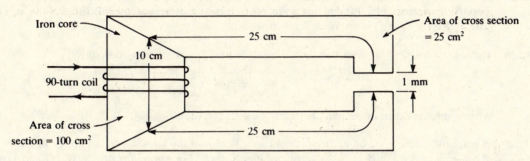

Fig. 12-18

12.19 The coil of Fig. 12-18 has 90 turns. For the data of Problem 12.18, determine the energy stored (a) in the coil, (b) in the air gap, (c) in iron.   *Ans.*   (a) 1.13 J; (b) 0.995 J; (c) 0.135 J

12.20 Calculate the inductance of the coil of Fig. 12-18, (a) excluding the effect of the iron core (i.e. assuming the core to be infinitely permeable) and (b) including the effect of the iron core.
Ans.   (a) 25.45 mH; (b) 22.45 mH

12.21 The reluctance of the core of a magnetic circuit is 3.18×10^5 H^{-1}. (a) If the circuit carries a 1000-turn coil, determine its inductance. (b) Let a current of 0.1 A flow through the coil. Assuming the magnetic circuit to be linear, calculate the core flux.   *Ans.*   (a) 3.14 H; (b) 0.314 mWb

12.22 A system of three coils on an ideal core is shown in Fig. 12-19, where $N_1 = N_3 = 2N_2 = 500$ turns, $g_1 = 2g_2 = 2g_3 = 4$ mm, and $A = 1000$ mm^2. Calculate (a) the self-inductance of coil N_1 and (b) the mutual inductance between coils N_2 and N_3.   *Ans.*   (a) 5.2 mH; (b) 0.314 mH

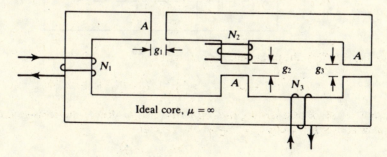

Fig. 12-19

12.23 If gap g_1 (Fig. 12-19) were closed, what would be the mutual inductances between (a) N_1 and N_2, (b) N_2 and N_3, and (c) N_3 and N_1?   *Ans.*   (a) 1.57 mH; (b) 0; (c) 3.14 mH

12.24 The coils of Problem 12.23 are connected in series (all mutual inductances being positive) and carry a current of 10 A. What is the total energy stored in the entire magnetic circuit?   *Ans.*   12.25 J

12.25 A toroid has a core of square cross section, 2500 mm² in area, and a mean diameter of 250 mm. The core material is of relative permeability 1000. (a) Calculate the number of turns to be wound on the core to obtain an inductance of 1 H. (b) If the coil thus wound carries 1 A, what are the values of H and B at the mean radius of the core?   Ans. (a) 500 turns; (b) 636 A/m, 0.8 T

12.26 A toroid is constructed of 48 NI alloy (Fig. 12-3). The mean length of the toroid is 250 mm and its cross-sectional area is 200 mm². If the toroid is to be used in an application requiring a flux of 0.2 mWb, (a) what mmf must be applied to the toroid? (b) It is desired that the coil have an inductance of 10 mH when the flux is 0.2 mWb. Determine the number of turns in the coil.
Ans. (a) 3.75 At; (b) 14 turns

12.27 For the square loop of Problem 12.1, we have: $a = 0.2$ m, $b = 0.25$ m, and $I = 5$ A (rms) at 60 Hz. If the loop has a resistance of 2.0 Ω and has negligible inductance, determine the rms value of the loop current. [Hint: Refer to (13.5).]   Ans. 0.166 mA

12.28 The mutual coupling between two coils is measured by the *coefficient of coupling*,

$$k \equiv \frac{L_{12}}{\sqrt{L_{11}L_{22}}} = \frac{k_{12}}{\sqrt{k_{11}k_{22}}}$$

What is the coefficient of coupling between the coils shown in Fig. 12-11?   Ans. 0.389

12.29 A magnetic circuit made of a 5-cm stack of laminations is of the form shown in Fig. 12-20. The material is sheet steel, having the B-H characteristic of Fig. 12-16(b). For a core flux density of 1 T and a current of 2 A in the 500-turn coil, determine the current in the 1000-turn coil.   Ans. 1.2 A or 0.8 A

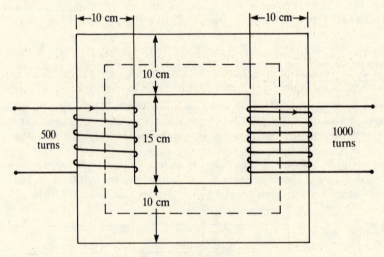

Fig. 12-20

Chapter 13

Transformers

13.1 PRINCIPLE OF OPERATION

A *transformer* is an electromagnetic device having two or more mutually coupled windings. Figure 13-1 shows a two-winding, *ideal* transformer; the transformer is ideal in the sense that its core is lossless and is infinitely permeable, it has no leakage fluxes, and the windings have no losses.

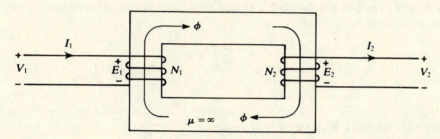

Fig. 13-1

In Figure 13-1, the basic components are the core, the *primary winding*, N_1, and the *secondary winding*, N_2. If ϕ is the mutual (or core) flux threading each turn of N_1 and N_2, then according to *Faraday's law* of electromagnetic induction, emfs e_1 and e_2 are induced in N_1 and N_2, given by

$$e_1 = N_1 \frac{d\phi}{dt} \qquad e_2 = N_2 \frac{d\phi}{dt} \tag{13.1}$$

The direction of e is such as to produce a current that gives rise to a flux which opposes the flux change $d\phi$ (*Lenz's law*). From (*13.1*), $e_1/e_2 = N_1/N_2$, or, in terms of rms values,

$$\frac{E_1}{E_2} = \frac{N_1}{N_2} \equiv a \tag{13.2}$$

where a is the *turns ratio*.

The transformer being ideal, $e_1 = v_1$ and $e_2 = v_2$ in Fig. 13-1; hence the flux and voltage are related by

$$\phi = \frac{1}{N_1} \int v_1 \, dt = \frac{1}{N_2} \int v_2 \, dt \tag{13.3}$$

If the flux varies sinusoidally, $\phi = \phi_m \sin \omega t$, then the corresponding induced voltage, e, in an N-turn winding is given by

$$e = \omega N \phi_m \cos \omega t \tag{13.4}$$

and the rms value of this induced voltage is

$$E = \frac{\omega N \phi_m}{\sqrt{2}} = 4.44 f N \phi_m \tag{13.5}$$

which is known as the *emf equation*. In (*13.5*), $f = \omega/2\pi$ is the flux frequency in Hz.

13.2 VOLTAGE, CURRENT, AND IMPEDANCE TRANSFORMATIONS

Transformers are used to effect voltage, current, and impedance transformations, and to provide isolation (that is, to eliminate direct connections between electrical circuits). The voltage transformation property of an ideal transformer was developed in Section 13.1:

$$\frac{V_1}{V_2} = \frac{E_1}{E_2} = a \qquad (13.6)$$

where the subscripts 1 and 2 respectively correspond to the primary and the secondary sides.

For an ideal transformer, the net mmf around its magnetic circuit must be zero. Thus, $N_1 I_1 - N_2 I_2 = 0$, or

$$\frac{I_2}{I_1} = \frac{N_1}{N_2} \equiv a \qquad (13.7)$$

From (13.6) and (13.7), it can be shown that if an impedance Z_2 is connected to the secondary, the impedance Z_1 seen at the primary satisfies

$$\frac{Z_1}{Z_2} = \left(\frac{N_1}{N_2}\right)^2 \equiv a^2 \qquad (13.8)$$

13.3 NONIDEAL TRANSFORMERS

A *nonideal* (or *actual*) transformer differs from an ideal transformer in that the former has hysteresis and eddy-current (core) losses, and has resistive ($I^2 R$) losses in its primary and secondary windings. Furthermore, the core of a nonideal transformer is not perfectly permeable, and requires a finite mmf for its magnetization. Also, because of leakage, not all fluxes link with the primary and the secondary windings simultaneously in a nonideal transformer.

13.4 EQUIVALENT CIRCUITS

An equivalent circuit of an ideal transformer is shown in Fig. 13-2(a). When the (nonideal) effects of winding resistances, leakage reactances, magnetizing reactance, and core losses are included, the circuit of Fig. 13-2(b) results, in which the primary and the secondary are coupled by an ideal transformer. By using (13.6), (13.7), and (13.8), the ideal transformer may be removed from Fig. 13-2(b) and the entire equivalent circuit may be referred either to the primary, as in Fig. 13-3(a), or to the secondary, as in Fig. 13-3(b).

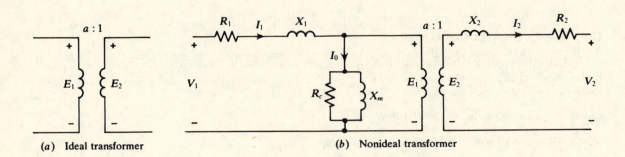

(a) Ideal transformer   (b) Nonideal transformer

Fig. 13-2

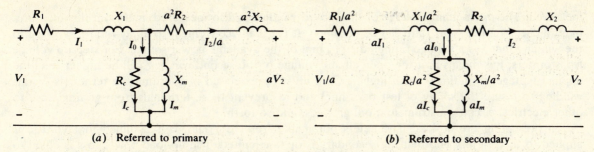

(a) Referred to primary   (b) Referred to secondary

Fig. 13-3. Equivalent Circuits of a Nonideal Transformer

A phasor diagram for the circuit of Fig. 13-3(a), in the case of lagging power factor, is shown in Fig. 13-4. In Figs. 13-2, 13-3, and 13-4, the various symbols are:

$a \equiv$ turns ratio
$E_1 \equiv$ primary induced voltage
$E_2 \equiv$ secondary induced voltage
$V_1 \equiv$ primary terminal voltage
$V_2 \equiv$ secondary terminal voltage
$I_1 \equiv$ primary current
$I_2 \equiv$ secondary current
$I_0 \equiv$ no-load (primary) current
$R_1 \equiv$ resistance of the primary winding
$R_2 \equiv$ resistance of the secondary winding
$X_1 \equiv$ primary leakage reactance
$X_2 \equiv$ secondary leakage reactance
$I_m, X_m \equiv$ magnetizing current and reactance
$I_c, R_c \equiv$ current and resistance accounting for the core losses

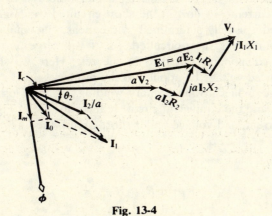

Fig. 13-4

13.5   TESTS ON TRANSFORMERS

Transformer performance characteristics can be obtained from the equivalent circuits of Section 13.4. The circuit parameters are determined either from design data or from test data. The two common tests are as follows.

Open-Circuit (No-Load) Test

Here one winding is open-circuited and a voltage—usually the rated voltage at rated frequency—is applied to the other winding. Voltage, current, and power at the terminals of this winding are measured. The open-circuit voltage of the second winding is also measured, and from this measurement a check on the turns ratio can be obtained. It is usually convenient to apply the test voltage to the winding that has a voltage rating equal to that of the available power source. For step-up voltage transformers, this means that the open-circuit voltage of the second winding will be higher than the applied voltage, sometimes much higher. Care must be exercised in guarding the terminals of this winding to insure the safety of test personnel and to prevent these terminals from getting close to other electrical circuits, instrumentation, grounds, and so forth.

In presenting the no-load parameters obtainable from test data, it is assumed that voltage is applied to the primary and that the secondary is open-circuited. The no-load power loss is equal to the wattmeter reading in this test; core loss is found by subtracting the ohmic loss in the primary, which is usually small and may be neglected in some cases. Thus, if P_0, I_0, and V_0 are the input power, current, and voltage, then the core loss is given by

$$P_c = P_0 - I_0^2 R_1 \tag{13.9}$$

The primary induced voltage is, in phasor form,

$$\mathbf{E_1} = V_0\underline{/0°} - (I_0\underline{/\theta_0})(R_1 + jX_1) \tag{13.10}$$

where $\theta_0 \equiv$ no-load power-factor angle $= \cos^{-1}(P_0/V_0 I_0) < 0$. Other circuit quantities are found from:

$$R_c = \frac{E_1^2}{P_c} \qquad I_c = \frac{P_c}{E_1} \qquad I_m = \sqrt{I_0^2 - I_c^2} \qquad X_m = \frac{E_1}{I_m} \qquad a \approx \frac{V_0}{E_2} \tag{13.11}$$

Short-Circuit Test

In this test, one winding is short-circuited across its terminals, and a reduced voltage is applied to the other winding. This reduced voltage is of such a magnitude as to cause a specific value of current—usually the rated current—to flow in the short-circuited winding. Again, the choice of the winding to be short-circuited is usually determined by the measuring equipment available. Care must be taken to note which winding is short-circuited, for it will serve as the reference winding for expressing the impedance components obtained in the test. Here we shall let the secondary be short-circuited and the reduced voltage be applied to the primary.

With a very low voltage applied to the primary winding, the core-loss current and magnetizing current become very small, and the equivalent circuit reduces to that of Fig. 13-5. Thus, if P_s, I_s, and V_s are the input power, current, and voltage under short circuit, then, referred to the primary,

$$Z_s = \frac{V_s}{I_s} \tag{13.12}$$

$$R_1 + a^2 R_2 \equiv R_s = \frac{P_s}{I_s^2} \tag{13.13}$$

$$X_1 + a^2 X_2 \equiv X_s = \sqrt{Z_s^2 - R_s^2} \tag{13.14}$$

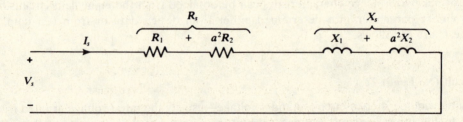

Fig. 13-5

Given R_1 and a, R_2 can be found from (13.13). In (13.14) it is usually assumed that the leakage reactance is divided equally between the primary and the secondary; that is,

$$X_1 = a^2 X_2 = \frac{1}{2} X_s \qquad\qquad (13.15)$$

13.6 TRANSFORMER CONNECTIONS

Of the eight types of transformer connection shown in Table 13-1, the first six are for purposes of voltage transformation and the last two are for changing the number of phases. (Not included is the single-phase voltage transformer.) Each line segment in the diagrams corresponds to one winding of a transformer.

Table 13-1

Type of Connection	Primary	Secondary
Two-phase	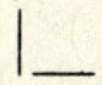 ⌐	⌐ or ⌐
Three-phase, delta-delta	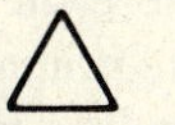 △	 △
Three-phase, delta-wye	 △	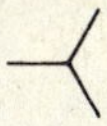 ⅄
Three-phase, wye-wye	 Y	 Y
Three-phase, open-delta	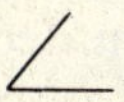 ∠	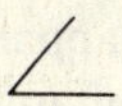 ∠
Three-phase, tee	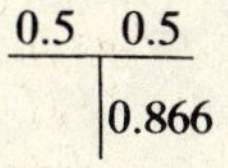 0.5 0.5 / 0.866	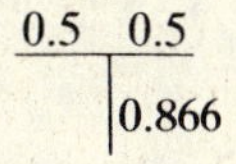 0.5 0.5 / 0.866
Two-phase-to-three-phase (Scott)	⌐	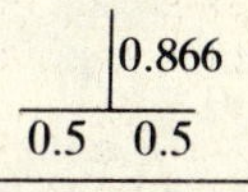 0.866 / 0.5 0.5
Three-phase-to-six-phase (diametrical)	 Y or  ▷	 ✳

It is important to observe the polarity markings in polyphase transformer connections, and for the sake of illustration the connection of three identical transformers in delta-wye is shown in some detail in Fig. 13-6.

13.7 AUTOTRANSFORMERS

An *autotransformer* is a single-winding transformer; it is a very useful device for some applications because of its simplicity and relatively low cost compared to multiwinding transformers. However, it does not provide electrical isolation and therefore cannot be used where this feature is required. The autotransformer circuit, Fig. 13-7, can be developed from a two-winding transformer

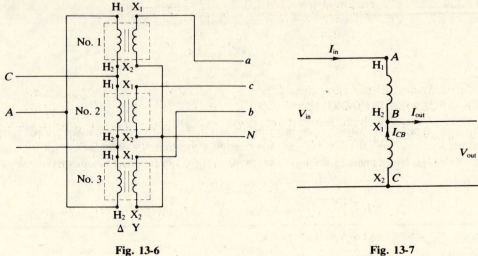

Fig. 13-6   Fig. 13-7

by connecting the two windings electrically in series so that the polarities are additive. Assume that this has been done in the circuit of Fig. 13-7, where the primary of the two-winding transformer is winding AB and the secondary is winding BC. The primary of the autotransformer is now the sum of these two windings, AC, and the secondary is winding BC. Hence, the autotransformer voltage and turns ratio is

$$a' = \frac{E_{AB} + E_{BC}}{E_{BC}} = \frac{N_{AB} + N_{BC}}{N_{BC}} = a + 1 \qquad (13.16)$$

where a is the voltage and turns ratio of the original two-winding transformer. Besides furnishing a greater transformation ratio, a pair of windings connected as an autotransformer can also deliver more voltamperes (apparent power) than when connected as a two-winding transformer. The reason is that the transfer of apparent power from primary to secondary in an autotransformer is not only by induction, as in a two-winding transformer, but by conduction as well.

Solved Problems

13.1   It is desired to have a 4.13 mWb maximum core flux in a transformer operating at 110 V and 60 Hz. Determine the required number of turns on the primary.

From the emf equation, (13.5),

$$N_1 = \frac{E_1}{4.44\phi_m f} = \frac{110}{(4.44)(4.13 \times 10^{-3})(60)} = 100 \text{ turns}$$

13.2   A 50-Hz transformer takes 75 W in power at 1.5 A and 120 V. If the primary winding resistance is 0.4 Ω, find (a) the core loss and (b) the no-load power factor.

(a) $P_c = 75 - (1.5)^2(0.4) = 74.1 \text{ W}$

(b) $\cos \theta_0 = \dfrac{75}{(120)(1.5)} = 0.417 \text{ lagging}$

13.3   Represent the core of the transformer of Problem 13.2 by a parallel combination of R_c and X_m, and obtain their numerical values. Neglect entirely the effect of the primary winding.

$$R_c = \frac{(120)^2}{75} = 192 \ \Omega \qquad I_m = \sqrt{(1.5)^2 - (0.625)^2} = 1.36 \ \text{A}$$

$$I_c = \frac{120}{192} = 0.625 \ \text{A} \qquad X_m = \frac{120}{1.36} = 88 \ \Omega$$

13.4 The parameters of the equivalent circuit of a 150-kVA, 2400-V/240-V transformer, shown in Fig. 13-3, are $R_1 = 0.2 \ \Omega$, $R_2 = 2 \ \text{m}\Omega$, $X_1 = 0.45 \ \Omega$, $X_2 = 4.5 \ \text{m}\Omega$, $R_c = 10 \ \text{k}\Omega$, and $X_m = 1.55 \ \text{k}\Omega$. Using the circuit referred to the primary, determine the (*a*) *voltage regulation* and (*b*) *efficiency* of the transformer, operating at rated load with 0.8 lagging power factor.

See Figs. 13-3(*a*) and 13-4. Given $V_2 = 240 \ \text{V}$, $a = 10$, and $\theta_2 = \cos^{-1} 0.8 = -36.8°$,

$$aV_2 = 2400\underline{/0°} \ \text{V}$$

$$I_2 = \frac{150 \times 10^3}{240} = 625 \ \text{A} \qquad \frac{I_2}{a} = 62.5\underline{/-36.8°} = 50 - j37.5 \ \text{A}$$

Also, $a^2R_2 = 0.2 \ \Omega$ and $a^2X_2 = 0.45 \ \Omega$, so that

$$\mathbf{E}_1 = (2400 + j0) + (50 - j37.5)(0.2 + j0.45) = 2427 + j15 = 2427\underline{/0.35°} \ \text{V}$$

$$\mathbf{I}_m = \frac{2427\underline{/0.35°}}{1550\underline{/90°}} = 1.56\underline{/-89.65°} = 0.0095 - j1.56 \ \text{A}$$

$$\mathbf{I}_c = \frac{2427 + j15}{10 \times 10^3} \approx 0.2427 + j0 \ \text{A}$$

Therefore

$$\mathbf{I}_0 = \mathbf{I}_c + \mathbf{I}_m = 0.25 - j1.56 \ \text{A}$$

$$\mathbf{I}_1 = \mathbf{I}_0 + (\mathbf{I}_2/a) = 50.25 - j39.06 = 63.65\underline{/-37.85°} \ \text{A}$$

$$\mathbf{V}_1 = (2427 + j15) + (50.25 - j39.06)(0.2 + j0.45) = 2455 + j30 = 2455\underline{/0.7°} \ \text{V}$$

(*a*)   percent regulation $\equiv \dfrac{V_{\text{no-load}} - V_{\text{load}}}{V_{\text{load}}} \times 100\%$

$$= \frac{V_1 - aV_2}{aV_2} \times 100\% = \frac{2455 - 2400}{2400} \times 100\% = 2.3\%$$

(*b*)   efficiency $\equiv \dfrac{\text{output}}{\text{input}} = \dfrac{\text{output}}{\text{output} + \text{losses}}$

Now   output $= (150 \times 10^3)(0.8) = 120 \ \text{kW}$

$$\text{losses} = I_1^2 R_1 + I_c^2 R_c + I_2^2 R_2$$

$$= (63.65)^2(0.2) + (0.2427)^2(10 \times 10^3) + (625)^2(2 \times 10^{-3}) = 2.18 \ \text{kW}$$

Hence   efficiency $= \dfrac{120}{122.18} = 0.982 = 98.2\%$

13.5 An approximate equivalent circuit of a transformer is shown in Fig. 13-8. Using this circuit, repeat the calculations of Problem 13.4 and compare the results. Draw a phasor diagram showing all the voltages and currents.

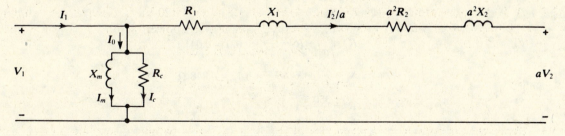

Fig. 13-8

From Problem 13.4,

$$a\mathbf{V}_2 = 2400\underline{/0°}\ \text{V} \qquad \frac{\mathbf{I}_2}{a} = 50 - j37.5\ \text{A} \qquad R_1 + a^2 R_2 = 0.4\ \Omega \qquad X_1 + a^2 X_2 = 0.9\ \Omega$$

Hence,

$$\mathbf{V}_1 = (2400 + j0) + (50 - j37.5)(0.4 + j0.9)$$
$$= 2453 + j30 = 2453\underline{/0.7°}\ \text{V}$$

$$\mathbf{I}_c = \frac{2453\underline{/0.7°}}{10 \times 10^3} = 0.2453\underline{/0.7°}\ \text{A}$$

$$\mathbf{I}_m = \frac{2453\underline{/0.7°}}{1550\underline{/90°}} = 1.58\underline{/-89.3°}\ \text{A}$$

$$\mathbf{I}_0 = 0.2453 - j1.58\ \text{A}$$

$$\mathbf{I}_1 = 50.25 - j39.08 = 63.66\underline{/-37.9°}\ \text{A}$$

The phasor diagram is given in Fig. 13-9.

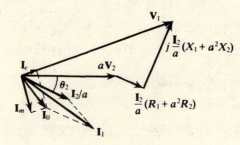

Fig. 13-9

(a) $$\text{percent regulation} = \frac{2453 - 2400}{2400} \times 100\% = 2.2\%$$

(b) $$\text{efficiency} = \frac{120 \times 10^3}{(120 \times 10^3) + (63.66)^2(0.4) + (0.2453)^2(10 \times 10^3)} = 0.982 = 98.2\%$$

Notice that the approximate circuit yields results that are sufficiently accurate.

13.6 The ohmic values of the circuit parameters of a transformer, having a turns ratio of 5, are $R_1 = 0.5\ \Omega$; $R_2 = 0.021\ \Omega$; $X_1 = 3.2\ \Omega$; $X_2 = 0.12\ \Omega$; $R_c = 350\ \Omega$, referred to the primary; and $X_m = 98\ \Omega$, referred to the primary. Draw the approximate equivalent circuits of the transformer, referred to (a) the primary and (b) the secondary. Show the numerical values of the circuit parameters.

The circuits are respectively shown in Fig. 13-10(a) and Fig. 13-10(b). The calculations are as follows:

(a)
$$R' \equiv R_1 + a^2 R_2 = 0.5 + (5)^2(0.021) = 1.025\ \Omega$$
$$X' \equiv X_1 + a^2 X_2 = 3.2 + (5)^2(0.12) = 6.2\ \Omega$$
$$R'_c = 350\ \Omega$$
$$X'_m = 98\ \Omega$$

(b)
$$R'' \equiv \frac{R_1}{a^2} + R_2 = \frac{0.5}{25} + 0.021 = 0.041\ \Omega$$

$$X'' \equiv \frac{X_1}{a^2} + X_2 = \frac{3.2}{25} + 0.12 = 0.248\ \Omega$$

$$R''_c = \frac{350}{25} = 14\ \Omega$$

$$X''_m = \frac{98}{25} = 3.92\ \Omega$$

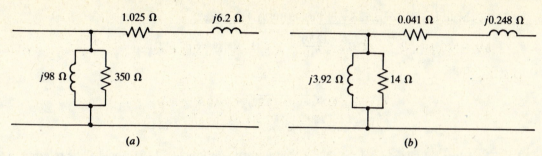

Fig. 13-10

13.7 The results of open-circuit and short-circuit tests on a 25-kVA, 440-V/220-V, 60-Hz transformer are as follows:

> **Open-circuit test.** Primary open-circuited, with instrumentation on the low-voltage side. Input voltage, 220 V; input current, 9.6 A; input power, 710 W.

> **Short-circuit test.** Secondary short-circuited, with instrumentation on the high-voltage side. Input voltage, 42 V; input current, 57 A; input power, 1030 W.

Obtain the parameters of the exact equivalent circuit (Fig. 13-3), referred to the high-voltage side. Assume that $R_1 = a^2 R_2$ and $X_1 = a^2 X_2$.

From the short-circuit test:

$$Z_{s1} = \frac{42}{57} = 0.737\ \Omega \qquad R_{s1} = \frac{1030}{(57)^2} = 0.317\ \Omega \qquad X_{s1} = \sqrt{(0.737)^2 - (0.317)^2} = 0.665\ \Omega$$

Consequently,

$$R_1 = a^2 R_2 = 0.158\ \Omega \qquad R_2 = 0.0395\ \Omega$$
$$X_1 = a^2 X_2 = 0.333\ \Omega \qquad X_2 = 0.0832\ \Omega$$

From the open-circuit test:

$$\theta_0 = \cos^{-1}\frac{710}{(9.6)(220)} = \cos^{-1} 0.336 = -70° \qquad I_{c2} = \frac{219}{67.5} = 3.24\ \text{A}$$

$$\mathbf{E}_2 = 220\underline{/0°} - (9.6\underline{/-70°})(0.0395 + j0.0832) \qquad I_{m2} = \sqrt{(9.6)^2 - (3.24)^2} = 9.03\ \text{A}$$

$$\approx 219\underline{/0°}\ \text{V}$$

$$P_{c2} = 710 - (9.6)^2(0.0395) \approx 710\ \text{W} \qquad X_{m2} = \frac{219}{9.03} = 24.24\ \Omega$$

$$R_{c2} = \frac{(219)^2}{710} = 67.5\ \Omega \qquad X_{m1} = a^2 X_{m2} = 97\ \Omega$$

$$R_{c1} = a^2 R_{c2} = 270\ \Omega$$

Thus, the equivalent circuit has the parameters as labeled in Fig. 13-11.

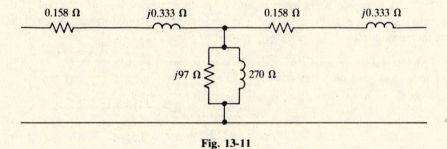

Fig. 13-11

13.8 From the test data of Problem 13.7, obtain the values of the circuit constants for the approximate equivalent circuit referred to the low-voltage side.

The circuit has the appearance of Fig. 13-10(b), but now:

$$R_{c2} = \frac{(220)^2}{710} = 68.2 \ \Omega \quad \text{(in place of 14 } \Omega\text{)}$$

$$I_{c2} = \frac{220}{68.2} = 3.22 \ \text{A}$$

$$I_{m2} = \sqrt{(9.6)^2 - (3.22)^2} = 9.04 \ \text{A}$$

$$X_{m2} = \frac{220}{9.04} = 24.33 \ \Omega \quad \text{(in place of 3.92 } \Omega\text{)}$$

The values calculated from the short-circuit test data (in Problem 13.7) should all be referred to the secondary. Thus,

$$R_{s2} = \frac{0.317}{4} = 0.079 \ \Omega \quad \text{(in place of 0.041 } \Omega\text{)}$$

$$X_{s2} = \frac{0.665}{4} = 0.166 \ \Omega \quad \text{(in place of 0.248 } \Omega\text{)}$$

13.9 A 220-V/110-V, 60-Hz transformer has a total no-load loss of 800 W while drawing 3.5 A current at 220 V. The primary winding resistance is 0.54 Ω. From the manufacturer's core-loss data, hysteresis loss at 60 Hz was found to be 520 W. If the operating voltage and frequency are doubled, calculate the new core losses.

Core losses in a transformer are given, on a per-unit-mass basis, by the approximate formulas

$$\text{eddy-current loss} \equiv p_e = k_e f^2 B_m^2 \quad \text{(W/kg)}$$
$$\text{hysteresis loss} \equiv p_h = k_h f B_m^\nu \quad \text{(W/kg)}$$

in which k_e and k_h are material constants, f is the frequency of the ac flux, B_m is the maximum core flux density, and the exponent ν ranges from 1.5 to 2.5. Now, the emf equation, (13.5), implies that B_m is proportional to E/f and thus remains constant when the operating voltage and frequency are both increased by the same factor (in particular, by the factor 2). At 220 V and 60 Hz, we are given that

$$\text{no-load } I^2R\text{-loss} = (3.5)^2(0.54) = 6.6 \ \text{W}$$
$$P_h = 520 \ \text{W}$$
$$P_e = 800 - (520 + 6.6) = 273.4 \ \text{W}$$

Hence, at 440 V and 120 Hz,

$$P_h = 2(520) = 1040 \ \text{W} \qquad P_e = 2^2(273.4) = 1094 \ \text{W}$$

and the new core losses are $1040 + 1094 = 2134$ W.

13.10 Two transformers, with equivalent impedances $\mathbf{Z}_e'$ and $\mathbf{Z}_e''$ referred to the respective primaries, operate in parallel at a secondary terminal voltage $\mathbf{V}_t$ and a primary terminal voltage $\mathbf{V}_1$ [Fig. 13-2(a)]. The transformers have a' and a'' as their respective turns ratios. If the total primary current is $\mathbf{I}_1$, determine how the load is shared by the two transformers. Neglect the core losses and magnetizing current.

An equivalent circuit for two transformers in parallel is shown in Fig. 13-12(b), from which we obtain the following phasor relations:

$$\mathbf{V}_1 = \mathbf{V}_1' = a'\mathbf{V}_t + \mathbf{I}_1'\mathbf{Z}_e'$$
$$\mathbf{V}_1 = \mathbf{V}_1'' = a''\mathbf{V}_t + \mathbf{I}_1''\mathbf{Z}_e''$$
$$\mathbf{I}_1 = \mathbf{I}_1' + \mathbf{I}_1''$$

Subtracting the first two equations, and then solving simultaneously with the third, we obtain for the two load currents:

$$\mathbf{I}_1' = \frac{-\mathbf{V}_t(a' - a'') + \mathbf{I}_1\mathbf{Z}_e''}{\mathbf{Z}_e' + \mathbf{Z}_e''} \qquad \mathbf{I}_1'' = \frac{\mathbf{V}_t(a' - a'') + \mathbf{I}_1\mathbf{Z}_e'}{\mathbf{Z}_e' + \mathbf{Z}_e''}$$

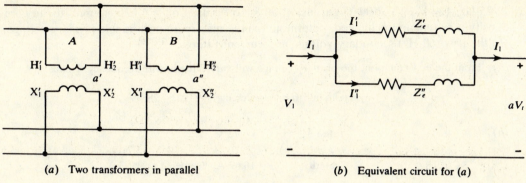

(a) Two transformers in parallel   (b) Equivalent circuit for (a)

Fig. 13-12

13.11 Two transformers, each rated 100 kVA, 11 000 V/2300 V, 60 Hz, are connected in open-delta on primary and secondary sides. (a) What is the total load that can be supplied from this transformer bank? (b) A 120-kVA, 2300-V, 0.866-lagging-power-factor, three-phase, delta-connected load is connected to the transformer bank. What is the line current on the high-voltage side?

(a)   load on open-delta $= \sqrt{3} \times$ (kVA-rating of each transformer) $= \sqrt{3} \times 100 = 173.2$ kVA

The circuit and phasor diagrams are shown in Fig. 13-13.

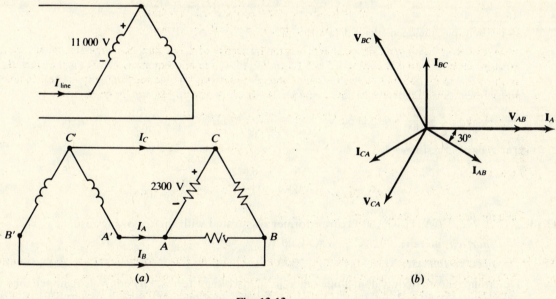

(a)   (b)

Fig. 13-13

(b) For the delta-connected load,

$$I_{AB} = I_{BC} = I_{CA} = \frac{1}{3}\left(\frac{120 \times 10^3}{2300}\right) = 17.4 \text{ A}$$

From the phasor diagram, Fig. 13-13(b),

$$\mathbf{I}_A = \mathbf{I}_{AB} - \mathbf{I}_{CA} = \left(2 \times \frac{\sqrt{3}}{2} \times 17.4\right)\underline{/0°} = 30.12\underline{/0°} \text{ A}$$

$$\text{transformation ratio } a = \frac{11000}{2300} = 4.78$$

$$I_{line} \equiv \text{current in the 11 000-V winding} = \frac{30.12}{4.78} = 6.3 \text{ A}$$

13.12 A 25-Hz, 120-V/30-V, 500-VA transformer is to be used on a 60-Hz source. If the core flux density is to remain unchanged, determine (a) the maximum permissible primary voltage, and (b) the new (60-Hz) secondary voltage and current ratings.

(a) From the emf equation, the primary voltage will vary directly with frequency. Hence,

$$\text{maximum primary voltage} = \frac{60}{25}(120) = 288 \text{ V}$$

(b)
$$\text{rated } V_2 = \frac{60}{25}(30) = 72 \text{ V}$$

$$\text{rated } I_2 = \frac{500}{30} = 16.67 \text{ A}\quad\text{(same as at 25 Hz)}$$

13.13 A quantity is expressed in *per unit* if it has been divided by a chosen *base quantity* (having the same physical dimension). Suppose that for a 10-kVA, 2400-V/240-V transformer we choose

$$P_{\text{base}} = 10 \text{ kW} \qquad V_{1,\text{base}} = 2400 \text{ V} \qquad V_{2,\text{base}} = 240 \text{ V}$$

This transformer has the following test data:

    open-circuit test (on low-voltage side): 240 V, 0.8 A, 80 W

    short-circuit test (on high-voltage side): 80 V, 5.1 A, 220 W

Convert all test data into per-unit values and find the series equivalent resistance in per unit.

$$I_{1,\text{base}} = \frac{10 \times 10^3}{2400} = 4.17 \text{ A} \qquad I_{2,\text{base}} = 41.7 \text{ A}$$

In per unit (pu), the open-circuit data are

$$V_0 = \frac{240}{240} = 1 \text{ pu} \qquad I_0 = \frac{0.8}{41.7} = 0.019 \text{ pu} \qquad P_0 = \frac{80}{10 \times 10^3} = 0.008 \text{ pu}$$

and the short-circuit data are

$$V_s = \frac{80}{2400} = 0.0333 \text{ pu} \qquad I_s = \frac{5.1}{4.17} = 1.22 \text{ pu} \qquad P_s = \frac{220}{10 \times 10^3} = 0.022 \text{ pu}$$

The equivalent resistance is

$$R_e = \frac{P_s(\text{pu})}{[I_s\,(\text{pu})]^2} = \frac{0.022}{(1.22)^2} = 0.0148 \text{ pu}$$

13.14 A 75-kVA, 230-V/115-V, 60-Hz transformer was tested with the following results:

    short-circuit test: 9.5 V, 326 A, 1200 W

    open-circuit test: 115 V, 16.3 A, 750 W

Determine the (a) equivalent impedance in high-voltage terms; (b) equivalent impedance in per unit; (c) voltage regulation at rated load, 0.8 power factor lagging; (d) efficiency at rated load, 0.8 power factor lagging, and at 1/2 load, unity power factor.

(a)
$$Z_s = \frac{9.5}{326} = 0.029 \ \Omega$$

(b) Proceeding as in Problem 13.13,

$$\text{per-unit } V_s = \frac{9.5}{230} = 0.0413 \text{ pu}$$

$$\text{per-unit } I_s = \frac{326}{326} = 1 \text{ pu}$$

$$\text{per-unit } Z_s = \frac{V_s\,(\text{pu})}{I_s\,(\text{pu})} = \frac{0.0413}{1} = 0.0413 \text{ pu}$$

(c)
$$\text{per-unit } P_s = \frac{1200}{75 \times 10^3} = 0.016 \text{ pu} = [I_s \text{ (pu)}]^2[R_s \text{ (pu)}]$$

Thus,

$$R_s = 0.016 \text{ pu} \qquad X_s = \sqrt{(0.0413)^2 - (0.016)^2} = 0.0384 \text{ pu}$$
$$\mathbf{V}_0 = \mathbf{V} + \mathbf{IZ} = 1 + (0.8 - j0.6)(0.016 + j0.0384)$$

whence $V_0 = 1.036$ pu. Then,

$$\text{voltage regulation} = \frac{V_0 - V_2}{V_2} = \frac{1.036 - 1}{1} = 0.036 \text{ pu} = 3.6\%$$

(d)
$$\eta_{\text{ rated load}} = \frac{(75 \times 10^3)(0.8)}{(60 \times 10^3) + 1200 + 750} = 96.85\%$$

$$\eta_{\text{ 1/2 rated load}} = \frac{(37.5 \times 10^3)(1)}{(37.5 \times 10^3) + 300 + 750} = 97.27\%$$

13.15 The core of a transformer has reluctance $\mathscr{R}$, and a resistance R is connected across the secondary. The turns ratio is N_1/N_2. The core flux is sinusoidal, of angular frequency ω. Obtain an expression for the instantaneous primary current in terms of N_1, N_2, ω, R, $\mathscr{R}$, and the induced primary voltage E_1.

Write $\phi = \phi_m \sin \omega t$; then

$$e_1 = \omega N_1 \phi_m \cos \omega t \qquad e_2 = \omega N_2 \phi_m \cos \omega t \qquad (1)$$

The secondary current, i_2, is given by

$$i_2 = \frac{e_2}{R} \qquad (2)$$

For the magnetic circuit we have $\mathscr{F} = \mathscr{R}\phi$, or

$$N_1 i_1 - N_2 i_2 = \mathscr{R}\phi_m \sin \omega t \qquad (3)$$

From (1), (2), and (3), after some manipulation,

$$i_1 = \left(\frac{N_2}{N_1}\right)^2 \frac{E_1}{R} \cos \omega t + \frac{E_1 \mathscr{R}}{\omega N_1^2} \sin \omega t$$

13.16 For an autotransformer, such as the one shown in Fig. 13-7, we have $V_{\text{in}} = V_1 = 220$ V, $V_{\text{out}} = V_2 = 110$ V, and $I_{\text{out}} = I_2 = 10$ A. Compare this autotransformer with a 220-V/110-V, two-winding transformer, supplying 10 A in current at the secondary, in terms of the copper required for the windings of the two transformers.

Transformer windings are generally operated at a specified current density. This makes the weight of copper in a winding proportional to the ampere turns (the mmf). Thus, for a two-winding transformer,

$$\text{(weight of copper)}_{\text{trans}} = k(N_1 I_1 + N_2 I_2) \qquad (1)$$

where k is a constant.

In the autotransformer, we have, from Fig. 13-7:

$$\text{number of turns from } A \text{ to } C = N_1 \qquad \text{current in } AB = I_{\text{in}} = I_1$$
$$\text{number of turns from } B \text{ to } C = N_2 \qquad \text{current in } BC = I_1 - I_2$$
$$\text{number of turns from } A \text{ to } B = N_1 - N_2$$

and so

$$\text{(weight of copper)}_{\text{auto}} = k[(N_1 - N_2)I_1 + N_2(I_2 - I_1)] \qquad (2)$$

From (*1*) and (*2*),

$$\frac{\text{weight}_{auto}}{\text{weight}_{trans}} = \frac{(N_1 - N_2)I_1 + N_2(I_2 - I_1)}{N_1 I_1 + N_2 I_2} = 1 - \frac{2N_2/N_1}{1 + (N_2/N_1)(I_2/I_1)}$$

$$= 1 - \frac{2/a}{2} = 1 - \frac{1}{a}$$

since $N_1/N_2 = I_2/I_1 = a$. Hence,

$$\text{saving in copper} = \text{weight}_{trans} - \text{weight}_{auto} = \frac{1}{a}\text{weight}_{trans}$$

Considering the numerical values, we have

$$\frac{1}{a} = \frac{N_2}{N_1} = \frac{100}{220} = 0.5$$

or a 50% saving in copper.

Supplementary Problems

13.17 The *B-H* curve of the core of a transformer is as shown in Fig. 12-16(*b*), and the maximum flux density is 1.2 T for a sinusoidal input voltage. Show qualitatively that the exciting current is nonsinusoidal.

13.18 A flux, $\phi = 2\sin 377t + 0.08\sin 1885t$ (mWb), completely links a 500-turn coil. Calculate the (*a*) instantaneous and (*b*) rms induced voltage in the coil.
Ans. (*a*) $v = 377\cos 377t + 75.4\cos 1885t$ (V); (*b*) $V = 271.86$ V

13.19 A 100-kVA, 60-Hz, 2200-V/220-V transformer is designed to operate at a maximum flux density of 1 T and an induced voltage of 15 volts per turn. Determine the number of turns of (*a*) the primary winding, (*b*) the secondary winding. (*c*) What is the cross-sectional area of the core?
Ans. (*a*) 147 turns; (*b*) 15 turns; (*c*) 0.0563 m^2

13.20 A transformer has a turns ratio of 4. (*a*) If a 50-Ω resistor is connected across the secondary, what is its resistance referred to the primary? (*b*) If the same resistor is instead connected across the primary, what is its resistance referred to the secondary?   *Ans.* (*a*) 800 Ω; (*b*) 3.125 Ω

13.21 Refer to Fig. 13-3. For a 110-kVA, 2200-V/110-V transformer the ohmic values of the circuit parameters are $R_1 = 0.22$ Ω, $R_2 = 0.5$ mΩ, $X_1 = 2.0$ Ω, $X_2 = 5$ mΩ, $R_c = 5494.5$ Ω, and $X_m = 1099$ Ω. Calculate (*a*) the voltage regulation and (*b*) the efficiency of the transformer, at full-load and unity power factor. (*c*) Find the core losses of the transformer.   *Ans.* (*a*) 1.55%; (*b*) 98.6%; (*c*) 1346.3 W

13.22 Open-circuit and short-circuit tests are performed on a 10-kVA, 220-V/110-V, 60-Hz transformer. Both tests are performed with instrumentation on the high-voltage side, and the following data are obtained:

    open-circuit test: input power, 500 W; input voltage, 220 V; input current, 3.16 A
    short-circuit test: input power, 400 W; input voltage, 65 V; input current, 10 A

Determine the parameters of the approximate equivalent circuit referred to the (*a*) primary and (*b*) secondary.
Ans. (*a*) $R_c = 96.8$ Ω, $X_m = 100$ Ω, $R_1 + a^2R_2 = 4$ Ω, $X_1 + a^2X_2 = 5.1$ Ω; (*b*) $R_c = 24.2$ Ω, $X_m = 25$ Ω, $R_2 + (R_1/a^2) = 1$ Ω, $X_2 + (X_1/a^2) = 1.28$ Ω

13.23 The core loss of a transformer is P_c. The equivalent resistance referred to the secondary is R_{e2}. At what secondary current I' will the transformer efficiency be maximum? Use the approximate equivalent circuit for the analysis.   *Ans.* $I' = \sqrt{P_c/R_{e2}}$

13.24 A 110-kVA, 2200-V/110-V, 60-Hz transformer has the following circuit parameters: $R_1 = 0.22\ \Omega$, $R_2 = 0.5\ m\Omega$, $X_1 = 2.0\ \Omega$, $X_2 = 5\ m\Omega$, $R_c = 5.5\ k\Omega$, and $X_m = 1.1\ k\Omega$. During a 24-hour (all-day) period, the transformer had the following load cycle: 4 h on no-load, 8 h on $\frac{1}{4}$ full-load at 0.8 power factor, 8 h on $\frac{1}{2}$ full-load at unity power factor, and 4 h on full-load at unity power factor. During the 24-h period the core losses remain constant at 1346 W. Determine the all-day energy efficiency of the transformer, defined as the ratio of energy output to energy input over 24 h.   *Ans.*   96.1%

13.25 A 1.5-kVA, 230-V/115-V transformer has the following circuit parameters: $R_1 = 4R_2 = 0.4\ \Omega$, $X_1 = 4X_2 = 1.2\ \Omega$, $X_m = 1300\ \Omega$, and $R_c = 1800\ \Omega$. Using the exact equivalent circuit, at full-load and unity power factor compute (*a*) the transformer efficiency and (*b*) the ratio of the terminal voltages, V_1/V_2.   *Ans.*   (*a*) 95.8%; (*b*) 2.045

13.26 Repeat Problem 13.25 using the approximate equivalent circuit.   *Ans.*   (*a*) 95.8%; (*b*) 2.04

13.27 Repeat Problem 13.25 if the power factor is 0.8 lagging.   *Ans.*   (*a*) 94.8%; (*b*) 2.12

13.28 If an autotransformer is made from a two-winding transformer having a turns ratio $N_1/N_2 = a$, show that:

$$\frac{\text{magnetizing current as an autotransformer}}{\text{magnetizing current as a 2-winding transformer}} = \frac{a-1}{a}$$

$$\frac{\text{short-circuit current as an autotransformer}}{\text{short-circuit current as a 2-winding transformer}} = \frac{a}{a-1}$$

13.29 Two transformers, operating in parallel, deliver a 230-V, 400-kVA load at 0.8 lagging power factor. One transformer is rated 2300 V/230 V and has impedance $1.84\underline{/84.2°}\ \Omega$, referred to the primary; corresponding data for the second transformer are 2300 V/225 V and $0.77\underline{/82.5°}\ \Omega$. Calculate (*a*) the current and (*b*) the power, delivered by each transformer.
Ans.   (*a*) 657 A, 1112 A; (*b*) 96 kW, 224 kW

13.30 The high-voltage windings of three 100-kVA, 19000-V/220-V transformers are connected in delta. The phase windings carry rated current at 0.866 lagging power factor. Determine the primary line and phase voltages and currents.
Ans. $\mathbf{V}_{AB} = 19\,000\underline{/0°}$ V, $\mathbf{V}_{BC} = 19\,000\underline{/120°}$ V, $\mathbf{V}_{CA} = 19\,000\underline{/240°}$ V, $\mathbf{I}_{AB} = 5.26\underline{/-30°}$ A, $\mathbf{I}_A = 9.1\underline{/0°}$ A

13.31 The flux linking a 500-turn coil is given by $\phi = 8t^2$ (Wb), where t is in s. Plot the induced voltage versus t. Compute the induced voltage at $t = 2$ s and at $t = 4$ s.   *Ans.*   16 kV, 32 kV

13.32 Bearing in mind the equivalent circuit of a transformer referred to the primary, as shown in Fig. 13-3(*a*), answer the following questions. (*a*) What experimental tests are used in obtaining the impedances shown? (*b*) What approximations are involved in relating the results of the experimental tests to the impedances shown? (*c*) Where does leakage flux in the transformer core show up in the diagram? (*d*) Which impedances result in energy losses? (*e*) What is the expression for the equivalent series impedance of the transformer in terms of symbols of the above diagram? (*f*) If the lamination thickness of the core material in the transformer were doubled, which impedances in the equivalent circuit would be affected? Explain.
Ans.   (*e*) To a good approximation, the shunt branch is moved to the extreme left, as in Fig. 13-8. The series impedance is then $(R_1 + a^2R_c) + j(X_1 + a^2X_2)$. (*f*) Doubling the lamination thickness quadruples the eddy-current loss; thus, the value of R_c would decrease.

13.33 Which transformer would you expect to be the heavier, a 25-Hz unit or a 60-Hz unit of identical voltampere rating and identical secondary voltage rating? Explain.
Ans.   From (*13.5*) it follows that the 25-Hz transformer requires the greater flux. This implies a greater core cross section for the 25-Hz transformer (for the same B_{max}). Hence the 25-Hz transformer is heavier.

13.34   An *ideal* transformer is rated 2400 V/240 V. A certain load of 50 A, unity power factor, is to be connected to the low-voltage winding. This load must have exactly 200 V across it. With 2400 V applied to the high-voltage winding, what resistance must be added in series with the transformer, if located (*a*) in the low-voltage winding, (*b*) in the high-voltage winding?   *Ans.*   (*a*) 0.8 Ω; (*b*) 80 Ω

13.35   A 50-kVA, 2300-V/230-V, 60-Hz transformer takes 200 W and 0.30 A at no-load when 2300 V is applied to the high-voltage side. The primary resistance is 3.5 Ω. Neglecting the leakage reactance drop, determine (*a*) the no-load power factor, (*b*) the primary induced voltage, (*c*) the magnetizing current, (*d*) the core-loss current component.   *Ans.*   (*a*) 0.29; (*b*) ≈2300 V; (*c*) 0.286 A; (*d*) 0.088 A

Chapter 14

Electromechanics and Electric Machines

14.1 BASIC PRINCIPLES

Electromechanical energy conversion is the transformation of electrical energy into mechanical energy, and vice versa. A black-box diagram of a rotary energy converter, having no losses, is given in Fig. 14-1; the corresponding energy conversion equation is

$$T\omega_m = vi \qquad (14.1)$$

where T and ω_m are the torque (in N·m) and speed of rotation (in rad/s) at the mechanical port, and v and i are the voltage and current at the electrical port.

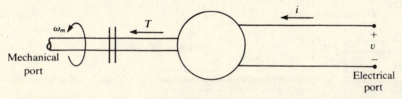

Fig. 14-1

Equation (14.1) is simply a statement of equivalence of mechanical and electrical power and does not give an insight into the process of energy conversion. *Electric generators*, which convert mechanical energy into electrical energy, operate on the basis of Faraday's law of electromagnetic induction (see Section 13.1). According to Faraday's law, a voltage is induced in a conductor when it "cuts" magnetic flux lines, or in a circuit when the flux linking the circuit changes in time (as in a transformer, Section 13.1). These two versions of Faraday's law can be mathematically expressed as:

$$e = \ell uB \sin \theta = B\ell u_\perp \qquad (14.2)$$

$$e = N \frac{d\phi}{dt} = \frac{d}{dt}(Li) \qquad (14.3)$$

Equation (14.2) pertains to the situation indicated in Fig. 14-2 (we need not consider the more general case). Here, a straight conductor, of length ℓ, moves with velocity **u** through a uniform

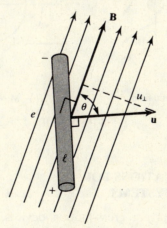

Fig. 14-2

magnetic field **B**, cutting the lines of **B** at right angles. Only the velocity component normal to the field is effective in inducing voltage e, whose polarity is as shown. Note that if $\theta = 90°$, (14.2) becomes $e = B\ell u$ (the $B\ell u$-rule). The first equality in (14.3) represents the most general form of Faraday's law; it implies (14.2) when applied to the situation of Fig. 14-2. The second equality in (14.3) follows from the definition of inductance, (12.9).

Electric motors and *electromechanical transducers*, which convert electrical energy into mechanical energy, operate either on the principle of (1) alignment of flux, or (2) interaction between magnetic fields and current-carrying conductors (Ampère's law). Two examples of "alignment of flux" are illustrated in Fig. 14-3. In Fig. 14-3(*a*), the force on the ferromagnetic pieces causes them to align with the flux lines, thus shortening the magnetic flux path and reducing the reluctance. Figure 14-3(*b*) shows the alignment of two current-carrying coils. Examples of "interaction between current-carrying conductors and magnetic fields" are given in Fig. 14-4. Thus, in Fig. 14-4(*b*), a force is produced by the interaction of flux lines and coil current, resulting in a torque on the coil.

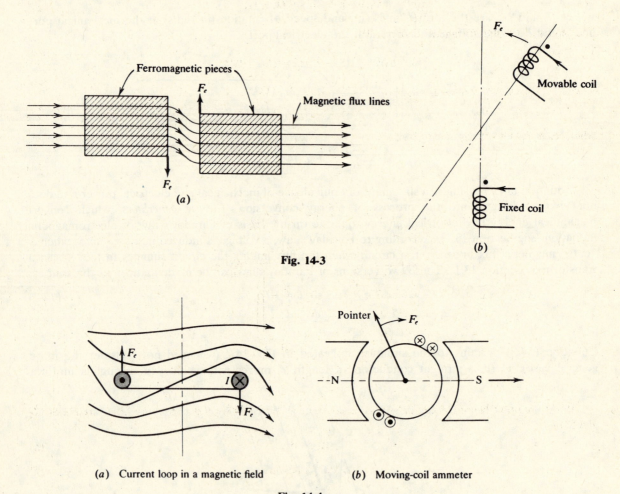

Fig. 14-3

(*a*) Current loop in a magnetic field   (*b*) Moving-coil ammeter

Fig. 14-4

14.2 FORCE AND TORQUE EQUATIONS FOR INCREMENTAL-MOTION SYSTEMS

Whereas motors and generators are gross-motion devices, electromechanical transducers are incremental-motion devices in the sense that their motion is restricted (to small displacements). For instance, the displacement of the diaphragm of a loudspeaker (processing electrical energy and converting it to mechanical form) is small compared to the motion of an electric motor. An electromagnetic solenoid also belongs to the category of incremental-motion devices. Mechanical forces of electrical origin in such devices can be determined from the principle of energy conservation, as demonstrated by the following example.

Example 14.1 In any lossless (conservative) electromechanical system,

$$\begin{pmatrix} \text{electrical} \\ \text{energy input} \end{pmatrix} = \begin{pmatrix} \text{mechanical} \\ \text{work done} \\ \text{by system} \end{pmatrix} + \begin{pmatrix} \text{increase in} \\ \text{stored energy} \end{pmatrix} \tag{1}$$

In particular, consider an electromagnet that attracts an iron mass, as shown in Fig. 14-5(a), where ① and ② indicate respectively the initial and final positions of the iron mass, which undergoes a displacement $-dx$ (against the positive x-direction). If the coil current stays constant at $i = I_0$ during the motion from ① to ②, while $\lambda(=N\phi)$ changes from λ_1 to λ_2, then the input electrical energy, dW_e, from the current source is given by Faraday's law (14.3) as

$$dW_e = I_0 e \, dt = I_0(\lambda_2 - \lambda_1) \tag{2}$$

The increase in stored (magnetic) energy, dW_m, is

$$dW_m = \frac{1}{2}(L_2 - L_1)I_0^2 = \frac{1}{2}(\lambda_2 - \lambda_1)I_0 \tag{3}$$

where we have assumed a linear magnetic circuit, $L = \lambda/i$. By (1),

$$dW_e = (-F_e)(-dx) + dW_m \tag{4}$$

where F_e is the electrically-caused force. Thus, from (2), (3), and (4),

$$F_e \, dx = \frac{1}{2}(\lambda_2 - \lambda_1)I_0 = dW_m \tag{5}$$

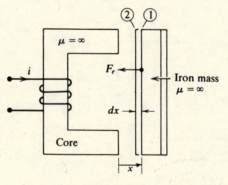

Fig. 14-5

If, on the other hand, the flux linkage stays constant at $\lambda = \lambda_0$ during the motion, we have, instead of (2) and (3),

$$dW_e = 0 \tag{6}$$

$$dW_m = \frac{1}{2}\lambda_0(i_2 - i_1) \tag{7}$$

which, together with (4), yield

$$F_e \, dx = -\frac{1}{2}\lambda_0(i_2 - i_1) = -dW_m \tag{8}$$

We may restate (5) and (8) of Example 14.1 as

$$\text{for current excitation:} \quad F_e = \frac{\partial W_m(i, x)}{\partial x} \tag{14.4a}$$

$$\text{for voltage excitation:} \quad F_e = -\frac{\partial W_m(\lambda, x)}{\partial x} \tag{14.4b}$$

These are the two forms of the force equation, giving the value of the mechanical force of electrical origin. For rotary-motion systems (gross- as well as incremental-motion), the analogous expressions for torque are

$$\text{for current excitation:} \qquad T_e = \frac{\partial W_m(i, \theta)}{\partial \theta} \tag{14.5a}$$

$$\text{for voltage excitation:} \qquad T_e = -\frac{\partial W_m(\lambda, \theta)}{\partial \theta} \tag{14.5b}$$

Equations (*14.4a*) and (*14.4b*)—or (*14.5a*) and (*14.5b*)—may be used interchangeably for a *linear* magnetic circuit.

14.3   DC MACHINES

Generators and Motors

As discussed in Section 14.1, generator action is based on Faraday's law of electromagnetic induction. Applying either (*14.2*) or (*14.3*) to an N-turn rectangular coil, of axial length ℓ and radius r, that rotates at a constant angular velocity ω in a uniform magnetic field B [Fig. 14-6(a)], one obtains

$$e = 2BN\ell r\omega \sin \omega t = BNA\omega \sin \omega t \tag{14.6}$$

The second form of (*14.6*) holds for an arbitrary planar coil of area A. A voltage $v = e$ is available at the slip rings (or brushes), as shown in Fig. 14-6. The direction of the induced voltage is often determined by use of the *right-hand rule*, Fig. 14-7(*a*), which agrees with Fig. 14-2.

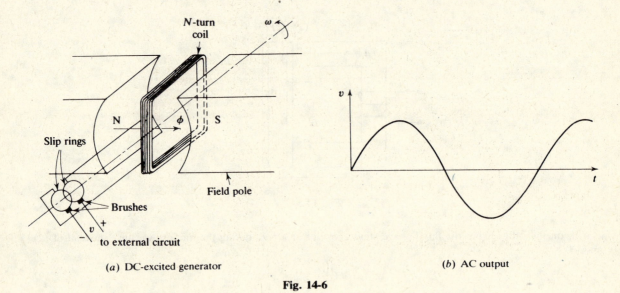

(*a*) DC-excited generator   (*b*) AC output

Fig. 14-6

Motor action is based on Ampère's law, which we rewrite as the *Bℓi-rule*:

$$F = B(\ell i)_\perp \tag{14.7}$$

Here, F is the magnitude of the force on a conductor carrying a directed current element $i\mathbf{l}$ whose component normal to the uniform magnetic field $\mathbf{B}$ is $(\ell i)_\perp$. The direction of the force may be obtained from the left-hand rule, Fig. 14-7(*b*).

Just as an ac sinusoidal voltage is produced at the terminals of a generator, the torque produced by the coil fed at the brushes from a dc source would be alternating in nature, with a zero time-average value.

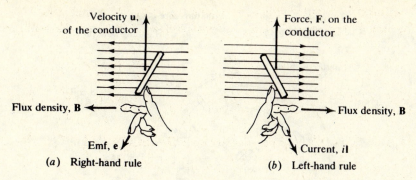

(a) Right-hand rule   (b) Left-hand rule

Fig. 14-7

Commutator Action

In order to get a unidirectional polarity at a brush, or to obtain a unidirectional torque from a coil in a magnetic field, the slip-ring/brush mechanism of Fig. 14-6(a) is modified to the one shown in Fig. 14-8(a). Notice that instead of two slip rings we now have one ring split into two halves that are insulated from each other. The brushes slide on these halves, known as *commutator segments*. It can be readily verified by applying the right-hand rule that such a commutator/brush system results in the brushes' having definite polarities, corresponding to the output voltage waveform of Fig. 14-8(b). This dc output at the brushes has a nonzero time-average value.

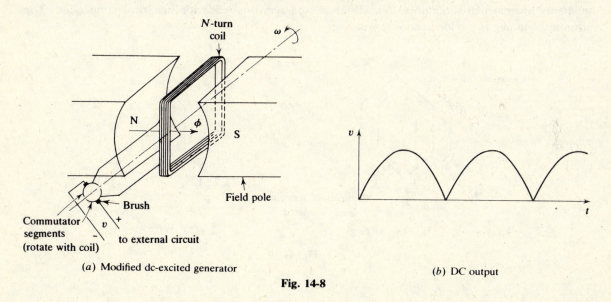

(a) Modified dc-excited generator   (b) DC output

Fig. 14-8

It may also be verified, by applying the left-hand rule, that if the coil connected to the commutator/brush system is fed from a dc source, the resulting torque is unidirectional.

Armature Windings and Physical Features

Figure 14-9 shows some of the important parts and physical features of a dc machine. The *field poles*, which produce the needed flux, are mounted on the *stator* and carry windings called *field windings* or *field coils*. Some machines carry several sets of field windings on the same *pole core*. To facilitate their assembly, the cores of the poles are built of sheet-steel laminations. Because the field windings carry direct current, it is not electrically necessary to have the cores laminated. It is, however, necessary for the *pole faces* to be laminated, because of their proximity to the *armature windings*.

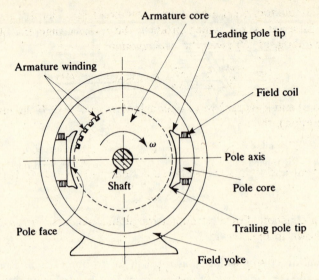

Fig. 14-9

The *armature core*, which carries the armature windings, is generally on the *rotor* and is made of sheet-steel laminations. The *commutator* is made of hard-drawn copper segments insulated from one another by mica. As shown in Fig. 14-10, the armature windings are connected to the commutator segments, over which the carbon *brushes* slide and serve as leads for electrical connection. The armature winding is the load-carrying winding.

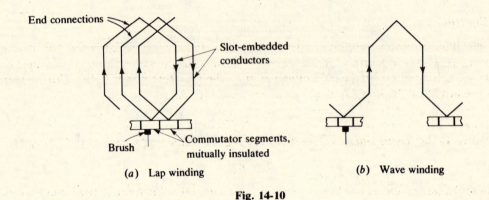

(a) Lap winding   (b) Wave winding

Fig. 14-10

The armature winding may be a *lap winding* [Fig. 14-10(a)] or a *wave winding* [Fig. 14-10(b)], and the various coils forming the armature winding may be connected in a series-parallel combination. It is found that in a lap winding the number of paths in parallel, a, is equal to the number of poles, p; whereas in a wave winding the number of parallel paths is always 2.

EMF Equation

Consider a conductor rotating at n rpm in the field of p poles having a flux ϕ per pole. The total flux cut by the conductor in n revolutions is $p\phi n$; hence, the flux cut per second, giving the induced voltage e, is

$$e = \frac{p\phi n}{60} \quad (V) \tag{14.8}$$

If there is a total of z conductors on the armature, connected in a parallel paths, then the effective number of conductors in series is z/a, which produce the total voltage E in the armature winding. Hence, for the entire winding, (14.8) gives the *emf equation*

$$E = \frac{p\phi n}{60}\frac{z}{a} = \frac{zp}{2\pi a}\,\phi\omega_m = k_a\phi\omega_m \quad \text{(V)} \tag{14.9}$$

where $\omega_m \equiv 2\pi n/60$ (rad/s) and $k_a \equiv zp/2\pi a$ (a dimensionless constant). If the magnetic circuit is linear (i.e., if there is no saturation), then

$$\phi = k_f i_f \tag{14.10}$$

where i_f is the field current and k_f is a proportionality constant; and (14.9) becomes

$$E = k i_f \omega_m \tag{14.11}$$

where $k \equiv k_f k_a$ ($\Omega \cdot \text{s}$), a constant. For a nonlinear magnetic circuit, E versus I_f is a nonlinear curve for a given speed, as shown in Fig. 14-11.

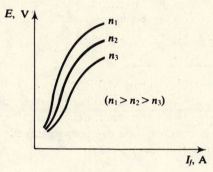

Fig. 14-11

Torque Equation

The mechanical power developed by the armature is $T_e\omega_m$, where T_e is the (electromagnetic) torque and ω_m is the armature's angular velocity. If this torque is developed while the armature current is i_a at an armature (induced) voltage E, then the armature power is Ei_a. Thus, ignoring any losses in the armature, $T_e\omega_m = Ei_a$, which becomes, from (14.9),

$$T_e = k_a\phi i_a \tag{14.12}$$

This is known as the *torque equation*. For a linear magnetic circuit, (14.10) and (14.12) yield

$$T_e = k i_f i_a \tag{14.13}$$

where $k \equiv k_f k_a$, as in (14.11). Thus, k may be termed the *electromechanical energy-conversion constant*. Notice that in (14.10) through (14.13) lowercase letters have been used, designating instantaneous values; however, these equations are equally valid under steady state.

Speed Equation

The armature of a dc motor may be schematically represented as in Fig. 14-12. Under steady state,

$$V - E = I_a R_a \tag{14.14}$$

Substituting (14.19) in (14.14) yields

$$\omega_m = \frac{V - I_a R_a}{k_a\phi} \tag{14.15}$$

which, for a linear magnetic circuit, becomes

$$\omega_m = \frac{V - I_a R_a}{k I_f} \tag{14.16}$$

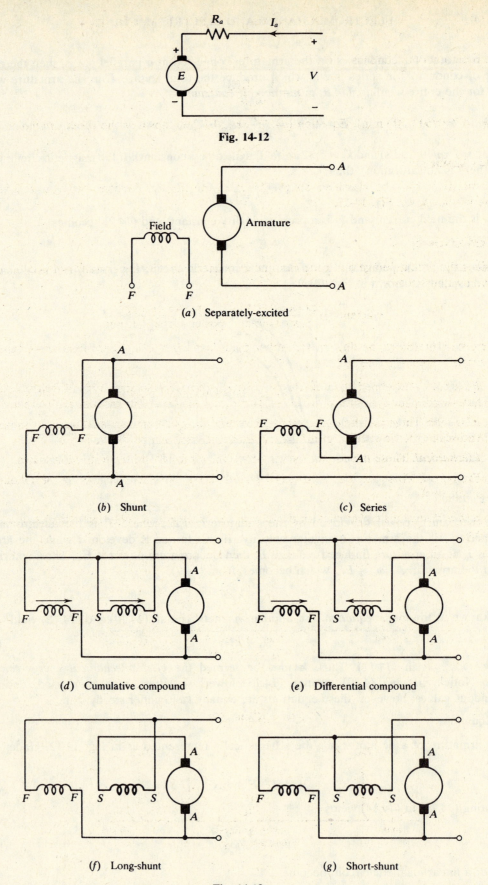

Fig. 14-12

(a) Separately-excited

(b) Shunt

(c) Series

(d) Cumulative compound

(e) Differential compound

(f) Long-shunt

(g) Short-shunt

Fig. 14-13

An alternate form of (*14.16*) is

$$n = \frac{V - I_a R_a}{k_m I_f} \quad \text{(rpm)} \tag{14.17}$$

where $k_m \equiv 2\pi k/60$ ($\Omega \cdot \text{min}$). Equation (*14.16*) or (*14.17*) is known as the *speed equation*.

Machine Classification

DC machines may be classified on the basis of the interconnections between the field and armature windings; see Fig. 14-13.

Losses and Efficiency

Besides the voltamperage and speed-torque characteristics, the performance of a dc machine is measured by its *efficiency*:

$$\text{efficiency} \equiv \frac{\text{power output}}{\text{power input}} = \frac{\text{power output}}{\text{power output} + \text{losses}} \tag{14.18}$$

Efficiency may, therefore, be determined either from load tests or by determination of losses. The various losses are classified as follows:

Electrical. (1) Copper losses in various windings, such as the armature winding and different field windings. (2) Loss due to the contact resistance of the brush (with the commutator).

Magnetic. These are the iron losses and include the hysteresis and eddy-current losses in the various magnetic circuits, primarily the armature core and pole faces.

Mechanical. These include the bearing-friction, windage, and brush-friction losses.

Stray-load. These are load losses not covered above. They are taken as 1% of the output (as a rule of thumb).

The power flow in a dc generator or motor is represented in Fig. 14-14, in which T_s denotes the shaft torque.

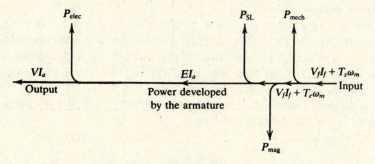

(*a*) DC generator

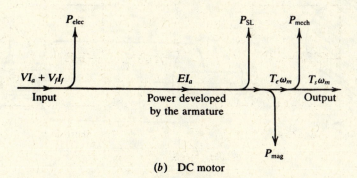

(*b*) DC motor

Fig. 14-14

Motor and Generator Characteristics

Load characteristics of motors and generators are usually of greatest interest in determining potential applications of these machines. In some cases (as in Fig. 14-12), no-load characteristics are also of importance. Typical load characteristics of dc generators are shown in Fig. 14-15, and Fig. 14-16 shows torque-speed characteristics of dc motors.

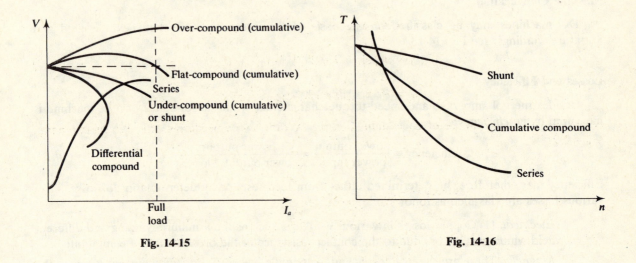

Fig. 14-15   Fig. 14-16

14.4   THREE-PHASE INDUCTION MOTORS

The induction motor is probably the most common of all motors. Like the dc machine, the induction motor consists of a stator and a rotor, the latter mounted on bearings and separated from the stator by an air gap. The stator core, made up of punchings (or laminations), carries slot-embedded conductors. These conductors are interconnected in a predetermined fashion and constitute the armature windings.

Alternating current is supplied to the stator windings, and the currents in the rotor windings are induced by the magnetic field of the stator currents. The rotor of the induction machine is cylindrical and carries either (1) conducting bars short-circuited at both ends by conducting rings, as in a *cage-type machine* [Fig. 14-17(a)]; or (2) a polyphase winding with terminals brought out to slip rings for external connections, as in a *wound-rotor machine* [Fig. 14-17(b)]. A wound-rotor winding is similar to that of the stator. Sometimes the cage-type machine is called a *brushless machine* and the wound-rotor machine termed a *slip-ring machine*.

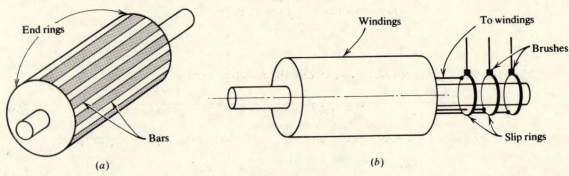

(a)   (b)

Fig. 14-17

An induction motor operates on the basis of interaction of induced rotor currents and the air-gap field. If the rotor is allowed to run under the torque developed by this interaction, the machine will operate as a motor. On the other hand, the rotor may be driven by an external agency at so high a speed that the machine begins to deliver electric power; it then operates as an induction generator.

The armature winding of a three-phase induction motor is appropriately distributed over the stator periphery, as in Fig. 14-18. We assume that the mmf produced by each phase is sinusoidal in space. These mmfs are displaced from each other by 120° (electrical) in space:

$$\mathscr{F}_A = \mathscr{F}_m \sin \omega t \cos \frac{\pi x}{\tau}$$

$$\mathscr{F}_B = \mathscr{F}_m \sin (\omega t - 120°) \cos \left(\frac{\pi x}{\tau} - 120° \right) \qquad (14.19)$$

$$\mathscr{F}_C = \mathscr{F}_m \sin (\omega t + 120°) \cos \left(\frac{\pi x}{\tau} + 120° \right)$$

where $\mathscr{F}_m$ is the amplitude of each mmf. For the N-turn coil, considering only the fundamental, $\mathscr{F}_m = 0.9\,NI$.

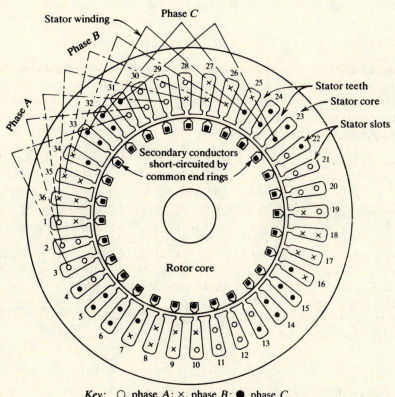

Key: ○, phase A; ×, phase B; ●, phase C.

Fig. 14-18

Adding the three mmfs of (14.19), we obtain the resultant mmf as

$$\mathscr{F}(x, t) = 1.5 \mathscr{F}_m \sin \left(\omega t - \frac{\pi x}{\tau} \right) \qquad (14.20)$$

It is seen that $\mathscr{F}(x, t)$ is a wave, of amplitude $1.5 \mathscr{F}_m$, that travels circumferentially at speed

$$v_s \equiv \frac{\tau \omega}{\pi} \quad \text{(m/s)} \qquad (14.21)$$

relative to the stator. We call v_s the *synchronous velocity*. Note that the wavelength is

$$\lambda = \frac{2\pi v_s}{\omega} = 2\tau \quad \text{(m)} \tag{14.22}$$

If the machine has p poles, (14.21) may be rewritten in the form

$$n_s \equiv \text{synchronous speed} = \frac{120 f_1}{p} \quad \text{(rpm)} \tag{14.23}$$

where $f_1 = \omega/2\pi$ is the stator-current (and mmf-rotational) frequency.

Equation (14.20) describes the rotating magnetic field produced by the stator of the induction motor. This field cuts the rotor conductors, and thereby voltages are induced in these conductors. The induced voltages give rise to rotor currents, which interact with the air-gap field to produce a torque, which is maintained as long as the rotating magnetic field and induced rotor currents exist. Consequently, the motor starts rotating at a speed $n < n_s$, in the direction of the rotating field.

Slip; Machine Equivalent Circuits

The actual speed, n, of the rotor is often related to the synchronous speed, n_s, via the *slip*:

$$s \equiv \frac{n_s - n}{n_s} \tag{14.24}$$

or the *percent slip*, $100s$.

At standstill ($s = 1$), the rotating magnetic field produced by the stator has the same speed with respect to the rotor windings as with respect to the stator windings. Thus, the frequency of the rotor currents, f_2, is the same as the frequency of the stator currents, f_1. At synchronous speed ($s = 0$), there is no relative motion between the rotating field and the rotor, and the frequency of rotor current is zero. (Indeed, the rotor current is zero.) At intermediate speeds the rotor-current frequency is proportional to the slip,

$$f_2 = s f_1 \tag{14.25}$$

wherefore f_2 is known as the *slip frequency*. Noting that the rotor currents are of slip frequency, we have the rotor equivalent circuit (on a per-phase basis) of Fig. 14-19(a), which gives the rotor current, I_2, as

$$I_2 = \frac{sE_2}{\sqrt{R_2^2 + (sX_2)^2}}$$

Here, E_2 is the induced rotor emf at standstill; X_2 is the rotor leakage reactance per phase at standstill; and R_2 is the rotor resistance per phase. This may also be written as

$$I_2 = \frac{E_2}{\sqrt{(R_2/s)^2 + X_2^2}} \tag{14.26}$$

For (14.26), we redraw the circuit of Fig. 14-19(a) as Fig. 14-19(b).

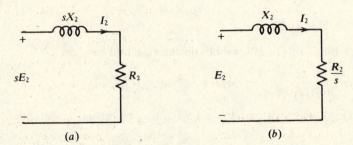

Fig. 14-19

To include the stator circuit, the induction motor may be viewed as a transformer with an air gap, having a variable resistance in the secondary [see (14.26)]. Thus, the primary of the transformer corresponds to the stator of the induction motor, whereas the secondary corresponds to the rotor on a per-phase basis. Because of the air gap, however, the value of the magnetizing reactance, X_m, tends to be low as compared to that of a true transformer. As in a transformer, we have a mutual flux linking both the stator and rotor, represented by the magnetizing reactance and various leakage fluxes. For instance, the total rotor leakage flux is denoted by X_2 in Fig. 14-19(b).

Considering the rotor to be coupled to the stator as the secondary of a transformer is coupled to its primary, we may draw the circuit shown in Fig. 14-20. To develop this circuit further, we express the rotor quantities as referred to the stator (as in a transformer), obtaining the exact equivalent circuit (per phase) shown in Fig. 14-21(a). For reasons that will immediately become clear, we split R_2'/s as

$$\frac{R_2'}{s} \equiv R_2' + \frac{R_2'}{s}(1-s)$$

to obtain the circuit shown in Fig. 14-21(b). Here, R_2' is simply the per-phase standstill rotor resistance referred to the stator, and $R_2'(1-s)/s$ is a per-phase dynamic resistance that depends on the rotor speed and corresponds to the load on the motor. Notice that all the parameters shown in Fig. 14-21 are standstill values.

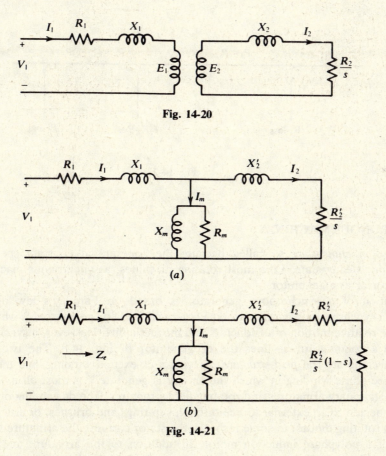

Fig. 14-20

(a)

(b)

Fig. 14-21

Calculations from Equivalent Circuits

The major usefulness of an equivalent circuit of an induction motor is in the calculation of its performance. All calculations are made on a per-phase basis, assuming a balanced operation of the machine; total quantities are then obtained by using the appropriate multiplying factor.

Figure 14-22(a) is Fig. 14-21(b) with R_m omitted. (Core losses, most of which are in the stator, will be included only in efficiency calculations.) In Fig. 14-22(b) are shown approximately the power flow and various power losses in one phase of the machine. Here:

$P_i \equiv$ input power

$P_g \equiv$ power crossing the air gap

$P_d \equiv$ developed electromagnetic power
    $\equiv$ power in the load $R_2'(1-s)/s$

$P_r \equiv$ rotational (mechanical) loss

$P_o \equiv$ shaft output power

The efficiency of the motor is $\eta \equiv P_o/P_i$.

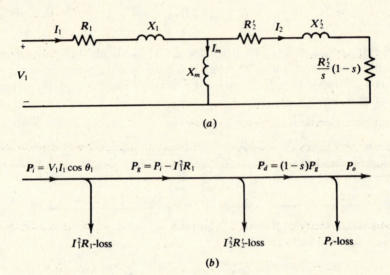

(a)

(b)

Fig. 14-22

14.5 SYNCHRONOUS MACHINES

Synchronous machines are so called because they operate at constant speeds and constant frequencies under steady state. Like most rotating machines, a synchronous machine can function either as a motor or as a generator.

The operation of a synchronous generator is based on Faraday's law of electromagnetic induction, and a synchronous generator works very much like a dc generator, in which the generation of emf is by the relative motion of conductors and magnetic flux. However, a synchronous generator does not have a commutator as does the dc generator of Fig. 14-8. The two basic parts of a synchronous machine are the *field structure*, carrying a dc-excited winding, and the *armature*, which often has a three-phase winding in which the ac emf is generated. Almost all modern synchronous machines have stationary armatures and rotating field structures. The dc winding on the rotating field structure is connected to an external source through slip rings and brushes, or else receives brushless excitation from rotating diodes. In some respects, the stator carrying the armature windings is similar to the stator of a polyphase induction motor. In addition to the armature and field windings, a synchronous machine has *damper bars* on the rotor; these come into play during transients and start-up.

Depending upon the rotor construction, a synchronous machine may be of the *round-rotor* type [Fig. 14-23(a)] or the *salient-pole* type [Fig. 14-23(b)]. (Note that the armatures are not shown in Figs. 14-22 and 14-23.) The former type is used in high-speed machines such as turbine generators, whereas the latter type is suitable for low-speed, waterwheel generators.

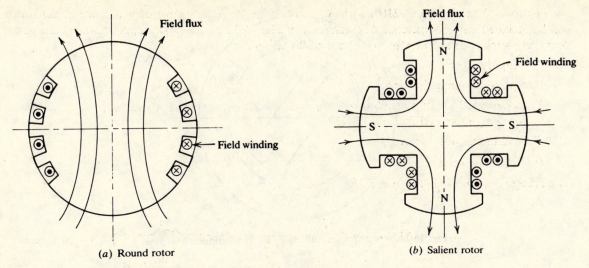

(a) Round rotor   (b) Salient rotor

Fig. 14-23

Generator and Motor Operation; the EMF Equation

To understand the generator operation, refer first to the 3-phase, round-rotor machine (Fig. 14-24), which has concentrated winding. It follows from (14.6) that the voltage induced in phase A is given by

$$v_A = V_m \sin \omega t$$

where ω is the angular velocity of the rotor. Phases B and C, being displaced from A and from each other by 120°, have induced voltages

$$v_B = V_m \sin (\omega t - 120°) \qquad v_C = V_m \sin (\omega t + 120°)$$

These voltages are sketched in Fig. 14-25. Hence, a three-phase voltage is generated, of frequency $f = \omega/2\pi$ (Hz).

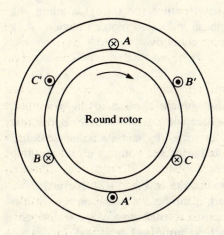

Fig. 14-24.   A 3-Phase, Round-Rotor, Synchronous Machine

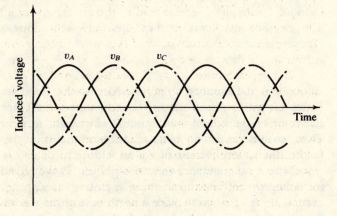

Fig. 14-25

Next, considering the salient-pole generator shown in Fig. 14-26, let the dc field winding produce the flux-density distribution

$$B(\theta) = B_m \cos \theta$$

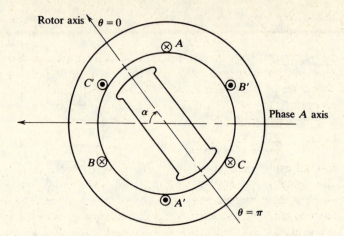

Fig. 14-26. A 3-Phase, Salient-Pole, Synchronous Machine

where θ is measured with respect to the rotor axis, as shown in Fig. 14-26. Let the N-turn armature coil corresponding to phase A have radius r and axial length ℓ. Then, when the rotor is in angular position $\alpha = \omega t$ (see Fig. 14-26), the total flux linking the coil is

$$\lambda = N \int_{(\pi/2)-\alpha}^{(3\pi/2)-\alpha} B(\theta)\ell r \, d\theta = -2NB_m\ell r \cos \alpha$$

so that, by Faraday's law, the voltage induced in the coil is

$$v_A = \frac{d\lambda}{dt} = \frac{d\lambda}{d\alpha}\frac{d\alpha}{dt} = V_m \sin \omega t \qquad (14.27)$$

where $V_m \equiv 2NB_m\ell r\omega$; similar expressions are found for phases B and C. It is seen that both round-rotor and salient-pole generators are governed by (14.27), which is known as the *emf equation* of a synchronous generator.

Turning now to the operation of a three-phase synchronous motor, and considering the salient-pole machine (Fig. 14-26), observe that the three-phase armature (or stator) winding will produce a rotating magnetic field in the air gap, as in a three-phase induction motor (Section 14.4). The speed of rotation of the field, i.e. the synchronous speed n_s, is given by

$$n_s = \frac{120f}{p} \quad \text{(rpm)} \qquad (14.28)$$

where p is the number of poles and f is the frequency of the voltage applied to the armature. However, with no short-circuited conductors on the rotor, the motor will not self-start. Suppose that the rotor of the salient-pole machine is brought to a speed close to n_s (by some auxiliary means). Then, even if there is no field excitation on the rotor, the rotor will align and rotate with the rotating field of the stator, because of the reluctance torque.

Clearly, no reluctance torque is present in the round-rotor machine of Fig. 14-24. Nevertheless, for either type of machine running at close-to-synchronous speed, if the field winding on the rotor is excited at such time as to place a north pole of the rotor field opposite a south pole of the stator field, then the two fields will lock in and the rotor will run at the synchronous speed.

In order to make the synchronous motor self-starting, it is provided with damper bars, which, like the cage of the induction motor, provide a starting torque. Once the rotor has pulled into step with the rotating stator field and runs at the synchronous speed, the damper bars go out of action. Any departure from the synchronous speed results in induced currents in the damper bars, which tend to restore the synchronous speed.

Generator No-Load, Short-Circuit, and Voltage-Regulation Characteristics

The no-load, or open-circuit, voltage characteristic of a synchronous generator is similar to that of a dc generator. Figure 14-27 shows such a characteristic, with the effect of magnetic saturation included. Now, if the terminals of the generator are short-circuited, the induced voltage is dropped internally within the generator. The short-circuit current characteristic, also shown in Fig. 14-27, is derived from the phasor relationship (on a per-phase basis):

$$\mathbf{V}_0 = \mathbf{I}_a \mathbf{Z}_s = \mathbf{I}_a(R_a + jX_s) \tag{14.29}$$

In (14.29), $\mathbf{V}_0$ is the no-load armature voltage at a certain field current, and $\mathbf{I}_a$ is the short-circuit armature current at the same value of the field current. The impedance $\mathbf{Z}_s$ is known as the *synchronous impedance*; R_a is the armature resistance and X_s is defined as the *synchronous reactance*. The synchronous reactance is readily measured for a round-rotor generator, since it is independent of the rotor position in such a machine. In salient-pole generators, however, the synchronous reactance depends on the rotor position.

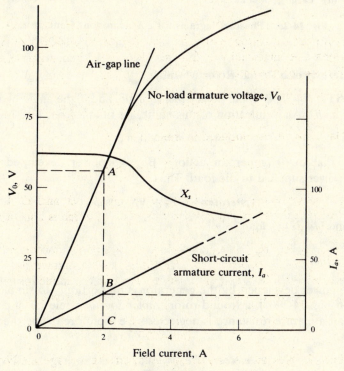

Fig. 14-27

In most synchronous machines, $R_a \ll X_s$, so that, in terms of Fig. 14-27,

$$X_s \approx Z_s = \overline{AC}/\overline{BC}$$

Thus X_s varies with field current, as indicated by the falling (because of saturation) curve in Fig. 14-27. However, for most calculations, we shall use the linear (constant) value of X_s.

As for a transformer or a dc generator, we define the *voltage regulation* of a synchronous generator at a given load by

$$\text{percent voltage regulation} \equiv \frac{V_0 - V_t}{V_t} \times 100\% \tag{14.30}$$

where V_t is the terminal voltage per phase on load and V_0 is the no-load terminal voltage per phase. Knowing X_s (for a round-rotor generator) and V_t, we can find V_0 from (14.29) and hence determine the voltage regulation.

Unlike what happens in a dc generator, the voltage regulation of a synchronous generator may become zero or even negative, depending upon the power factor and the load. Neglecting the armature resistance, we show phasor diagrams for lagging and leading power factors in Fig. 14-28.

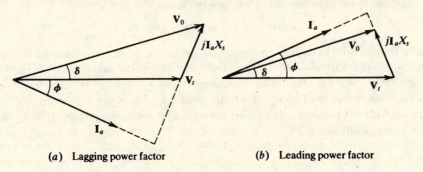

(*a*) Lagging power factor   (*b*) Leading power factor

Fig. 14-28. Phasor Diagrams for a Synchronous Generator

Power-Angle Characteristic of a Round-Rotor Machine

With reference to Fig. 14-28, ϕ is the power-factor angle, and δ, the angle by which V_0 leads V_t, is defined as the *power angle*. As justification of this name, we obtain from Fig. 14-28:

$$I_a X_s \cos \phi = V_0 \sin \delta \qquad (14.31)$$

where it is assumed that $\delta > 0$ (generator action). But the power developed (per phase) by the generator, P_d, is the power supplied to the load. Thus,

$$\text{\emph{generator}} \qquad P_d = V_t I_a \cos \phi \qquad (14.32)$$

Comparing (14.31) and (14.32) yields

$$P_d = \frac{V_0 V_t}{X_s} \sin \delta \qquad (14.33)$$

which shows that the power developed by the generator is proportional to $\sin \delta$.

As is indicated in Fig. 14-29(*a*), a (round-rotor) *motor consumes* electrical power in the amount $V_t I_a \cos \phi$ per phase, if armature resistance is neglected. We therefore define the power *developed by* the motor as

$$\text{\emph{motor}} \qquad P_d = - V_t I_a \cos \phi$$

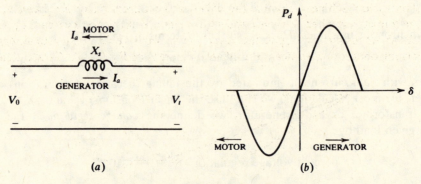

(*a*)   (*b*)

Fig. 14-29. For Round-Rotor Machines

With this understanding, (*14.33*) is also valid for a round-rotor motor, where now δ, and hence sin δ, is negative ($\mathbf{V}_t$ leads $\mathbf{V}_0$). In short, (*14.33*) is the power-angle characteristic of a round-rotor synchronous machine; a graph is given in Fig. 14-29(*b*).

For a motor, Fig. 14-29(*a*) gives

$$\mathbf{V}_t = \mathbf{V}_0 + j\mathbf{I}_a X_s \qquad (14.34)$$

If the motor operates at constant power, then (*14.31*) and (*14.33*) imply that

$$V_0 \sin \delta = I_a X_s \cos \phi = \text{constant} \qquad (14.35)$$

for a given terminal voltage V_t.

Now, V_0 depends upon the field current, I_f. Consider two cases: (1) I_f is so adjusted that $V_0 < V_t$ (the machine is *underexcited*), and (2) I_f is increased to a point where $V_0 > V_t$ (the machine is *overexcited*). The voltage-current phasor relationships for the two cases are shown in Fig. 14-30(*a*), in which single primes refer to underexcited, and double primes to overexcited, operation. At constant power, δ is less negative for $V_0 > V_t$ than for $V_0 < V_t$, as governed by (*14.35*). Notice that an underexcited motor operates at a lagging power factor ($\mathbf{I}_a$ lags $\mathbf{V}_t$), whereas an overexcited motor operates at a leading power factor. Thus, *the operating power factor of the motor is controlled by varying the field excitation* (thereby altering V_0); this is a very important property of synchronous motors. The locus of the armature current at a constant load, as given by (*14.35*), for varying field current is also shown in Fig. 14-30(*a*). From this we can obtain the variation of the armature current, I_a, with the field current, I_f (corresponding to V_0); the results for several different loads are plotted in Fig. 14-30(*b*). These curves are known as the *V-curves* of the synchronous motor. One of the applications of a synchronous motor is in power-factor correction.

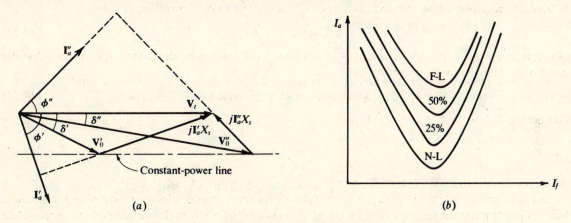

Fig. 14-30. Round-Rotor Motor Operation at Constant Power

Salient-Pole Synchronous Machines

For the salient-pole machine, we define the *direct-* and *quadrature-axis inductances*, L_d and L_q, as the values of the inductance when the rotor and stator axes are aligned and when they are antialigned (in quadrature). Analogously, we define the *d-axis* and *q-axis synchronous reactances*, X_d and X_q. Thus, for generator operation, we draw the phasor diagram of Fig. 14-31. Notice that $\mathbf{I}_a$ has been

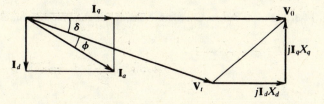

Fig. 14-31. Phasor Diagram of a Salient-Pole Generator

resolved into its d- and q-axis (fictitious) components, $\mathbf{I}_d$ and $\mathbf{I}_q$. With the help of this phasor diagram, we obtain:

$$I_d = I_a \sin(\delta + \phi) \qquad I_q = I_a \cos(\delta + \phi) \tag{14.36}$$

$$V_t \sin \delta = I_q X_q = I_a X_q \cos(\delta + \phi) \tag{14.37}$$

Expansion of (14.37) gives

$$\tan \delta = \frac{I_a X_q \cos \phi}{V_t + I_a X_q \sin \phi} \tag{14.38}$$

With δ known (in terms of ϕ), the voltage regulation may be computed from

$$V_0 = V_t \cos \delta + I_d X_d$$

$$\text{percent regulation} = \frac{V_0 - V_t}{V_t} \times 100\%$$

In fact, the phasor diagram depicts the complete performance characteristics of the machine.

Example 14.2 Figure 14-31 may be used to derive the power-angle characteristic of a salient-pole generator. If armature resistance is neglected, (14.32) applies. Now, from Fig. 14-31, the projection of $\mathbf{I}_a$ on $\mathbf{V}_t$ is

$$\frac{P_d}{V_t} = I_a \cos \phi = I_q \cos \delta + I_d \sin \delta \tag{14.39}$$

Solving

$$I_q X_q = V_t \sin \delta \qquad \text{and} \qquad I_d X_d = V_0 - V_t \cos \delta$$

for I_q and I_d, and substituting in (14.39), gives

$$P_d = \frac{V_0 V_t}{X_d} \sin \delta + \frac{V_t^2}{2}\left(\frac{1}{X_q} - \frac{1}{X_d}\right)\sin 2\delta \tag{14.40}$$

Equation (14.40) can also be established for a salient-pole motor ($\delta < 0$); the graph of (14.40) is given in Fig. 14-32. Observe that for $X_d = X_q = X_s$, (14.40) reduces to the round-rotor equation, (14.33).

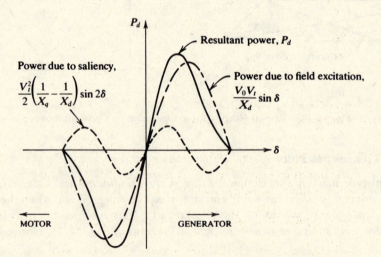

Fig. 14-32. Power-Angle Characteristics of Salient-Pole Machines

Solved Problems

INCREMENTAL-MOTION SYSTEMS

14.1 A solenoid of cylindrical geometry is shown in Fig. 14-33. (*a*) If the exciting coil carries a dc steady current *I*, derive an expression for the force on the plunger. (*b*) For the numerical values $I = 10$ A, $N = 500$ turns, $g = 5$ mm, $a = 20$ mm, $b = 2$ mm, and $\ell = 40$ mm, what is the magnitude of the force? Assume $\mu_{core} = \infty$ and neglect leakage.

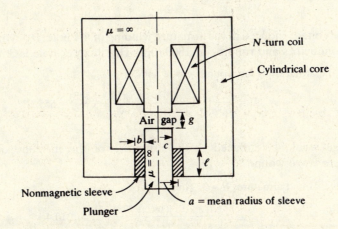

Fig. 14-33

For the magnetic circuit, the reluctance is

$$\mathcal{R} = \frac{g}{\mu_0 \pi c^2} + \frac{b}{\mu_0 2\pi a \ell} \qquad \text{where} \qquad c = a - \frac{b}{2}$$

The inductance L is then given by (*12.11*) as

$$L = \frac{N^2}{\mathcal{R}} = \frac{2\pi\mu_0 a\ell c^2 N^2}{2a\ell g + bc^2} \equiv \frac{k_1}{k_2 g + k_3}$$

where $k_1 \equiv 2\pi\mu_0 a\ell c^2 N^2$, $k_2 \equiv 2a\ell$, and $k_3 \equiv bc^2$.

(*a*) By (*14.4a*),

$$F_e = \frac{\partial}{\partial g}\left(\frac{1}{2}LI^2\right) = \frac{1}{2}I^2\frac{\partial L}{\partial g} = -\frac{I^2 k_1 k_2}{2(k_2 g + k_3)^2}$$

where the minus sign indicates that the force tends to decrease the air gap.

(*b*) Substituting the numerical values in the force expression of (*a*) yields 600 N as the magnitude of F_e.

14.2 (*a*) If the solenoid of Problem 14.1(*a*) instead carries an alternating current of 10 A (rms) at 60 Hz, what is the instantaneous force? (*b*) What is the average force, if *N*, *g*, *a*, *b*, and ℓ have the same numerical values as in Problem 14.1(*b*)?

(*a*) $$F_e = -\frac{(10\sqrt{2}\cos 120\pi t)^2 k_1 k_2}{2(k_2 g + k_3)^2} = -\frac{100 k_1 k_2}{(k_2 g + k_3)^2}\cos^2 120\pi t \quad (N)$$

(*b*) Because the $\cos^2$ has average value 1/2, the average force is the same as the force due to 10 A dc; namely, 600 N.

14.3 For a voltage-excited system, show that the electrical force can be expressed as

$$F_e = -\frac{1}{2}\phi^2\frac{\partial\mathcal{R}}{\partial x}$$

where ϕ is the core flux and $\mathcal{R}$ is the net reluctance of the magnetic circuit.

We have

$$W_m = \frac{1}{2} L i^2 = \frac{1}{2} \left(\frac{N\phi}{i} \right) i^2 = \frac{1}{2} N\phi \left(\frac{\mathcal{R}\phi}{N} \right) = \frac{1}{2} \mathcal{R}\phi^2$$

and so, by (14.4b), in which constant λ implies constant ϕ,

$$F_e = -\frac{\partial W_m}{\partial x} = -\frac{1}{2} \phi^2 \frac{\partial \mathcal{R}}{\partial x}$$

DC MACHINES

14.4 Calculate the voltage induced in the armature winding of a 4-pole, lap-wound, dc machine having 728 active conductors and running at 1800 rpm. The flux per pole is 30 mWb.

Because the armature is lap wound, $p = a$, and

$$E = \frac{\phi n z}{60} \left(\frac{p}{a} \right) = \frac{(30 \times 10^{-3})(1800)(728)}{60} = 655.2 \text{ V}$$

14.5 What voltage would be induced in the armature of the machine of Problem 14.4 if the armature were wave wound?

For a wave-wound armature, $a = 2$. Thus,

$$E = \frac{(30 \times 10^{-3})(1800)(728)}{60} \left(\frac{4}{2} \right) = 1310.4 \text{ V}$$

14.6 If the armature in Problem 14.4 is designed to carry a maximum line current of 100 A, what is the maximum electromagnetic power developed by the armature?

Because there are 4 parallel paths ($a = p = 4$) in the lap-wound armature, each path can carry a maximum current of

$$\frac{I_a}{a} = \frac{100}{4} = 25 \text{ A}$$

Nevertheless, the power developed by the armature is

$$P_d = E I_a = (655.2)(100) = 65.5 \text{ kW}$$

14.7 Calculate the electromagnetic torque developed by the armature described in Problem 14.4.

From the energy-conversion equation, $E I_a = T_e \omega_m$, and the result of Problem 14.6,

$$T_e = \frac{E I_a}{\omega_m} = \frac{65.5 \times 10^3}{2\pi(1800)/60} = 347.6 \text{ N} \cdot \text{m}$$

14.8 A 4-pole, lap-wound armature has 144 slots with two coil sides per slot, each coil having two turns. If the flux per pole is 20 mWb and the armature rotates at 720 rpm, what is the induced voltage?

Substitute $p = a = 4$, $n = 720$, $\phi = 0.020$, and $z = 144 \times 2 \times 2 = 576$ in the emf equation to obtain

$$E = \frac{(0.020)(720)(576)}{60} \left(\frac{4}{4} \right) = 138.24 \text{ V}$$

14.9 A 100-kW, 230-V, shunt generator has $R_a = 0.05 \ \Omega$ and $R_f = 57.5 \ \Omega$. If the generator operates at rated voltage, calculate the induced voltage at (a) full-load and (b) half full-load. Neglect brush-contact drop.

See Fig. 14-34; $I_f = 230/57.5 = 4$ A.

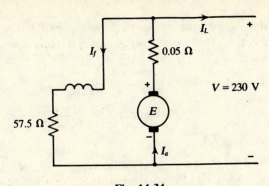

Fig. 14-34

(a)
$$I_L = \frac{100 \times 10^3}{230} = 434.8 \text{ A}$$
$$I_a = I_L + I_f = 434.8 + 4 = 438.8 \text{ A}$$
$$I_a R_a = (438.8)(0.05) = 22 \text{ V}$$
$$E = V + I_a R_a = 230 + 22 = 252 \text{ V}$$

(b)
$$I_L = 217.4 \text{ A}$$
$$I_a = 217.4 + 4 = 221.4 \text{ A}$$
$$I_a R_a = 11 \text{ V}$$
$$E = 230 + 11 = 241 \text{ V}$$

14.10 A 50-kW, 250-V, short-shunt, compound generator has the following data: $R_a = 0.06 \ \Omega$, $R_{se} = 0.04 \ \Omega$, and $R_f = 125 \ \Omega$. Calculate the induced armature voltage at rated load and terminal voltage. Take 2 V as the total brush-contact drop.

See Fig. 14-35.

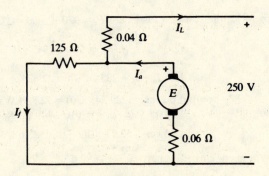

Fig. 14-35

$$I_L = \frac{50 \times 10^3}{250} = 200 \text{ A}$$
$$I_L R_{se} = (200)(0.04) = 8 \text{ V}$$
$$V_f = 250 + 8 = 258 \text{ V}$$
$$I_f = \frac{258}{125} = 2.06 \text{ A}$$
$$I_a = 200 + 2.06 = 202.06 \text{ A}$$
$$I_a R_a = (202.06)(0.06) = 12.12 \text{ V}$$
$$E = 250 + 12.12 + 8 + 2 = 272.12 \text{ V}$$

14.11 A separately-excited dc generator has a constant loss of P_c (W), and operates at a voltage V and armature current I_a. The armature resistance is R_a. At what value of I_a is the generator efficiency a maximum?

$$\text{output} = VI_a$$
$$\text{input} = VI_a + I_a^2 R_a + P_c$$
$$\text{efficiency } \eta = \frac{VI_a}{VI_a + I_a^2 R_a + P_c}$$

For η to be a maximum, $d\eta/dI_a = 0$, or

$$V(VI_a + I_a^2 R_a + P_c) - VI_a(V + 2I_a R_a) = 0 \qquad \text{or} \qquad I_a = \sqrt{P_c/R_a}$$

In other words, the efficiency is maximized when the armature loss, $I_a^2 R_a$, equals the constant loss, P_c.

14.12 The generator of Problem 14.9 has a total mechanical and core loss of 1.8 kW. Calculate (a) the generator efficiency at full-load, and (b) the horsepower output from the prime mover to drive the generator at this load.

From Problem 14.9, $I_f = 4$ A and $I_a = 438.8$ A, so that

$$I_f^2 R_f = (16)(57.5) = 0.92 \text{ kW}$$
$$I_a^2 R_a = (438.8)^2 (0.05) = 9.63 \text{ kW}$$
$$\text{total losses} = 0.92 + 9.63 + 1.8 = 12.35 \text{ kW}$$

(a)
$$\text{output} = 100 \text{ kW}$$
$$\text{input} = 100 + 12.35 = 112.35 \text{ kW}$$
$$\text{efficiency} = \frac{100}{112.35} = 89\%$$

(b)
$$\text{prime mover output} = \frac{112.35 \times 10^3 \text{ W}}{746 \text{ W/hp}} = 150.6 \text{ hp}$$

14.13 (a) At what load does the generator of Problems 14.9 and 14.12 achieve maximum efficiency? (b) What is the value of this maximum efficiency?

(a) From Problem 14.12, the constant losses are $P_c = 920 + 1800 = 2720$ W. Hence, by Problem 14.11,

$$I_a = \sqrt{\frac{2720}{0.05}} = 233.24 \text{ A}$$

and
$$I_L = I_a - I_f = 233.24 - 4 = 229.24 \text{ A}$$

(b)
$$\text{output power} = (229.24)(230) = 52.72 \text{ kW}$$
$$I_a^2 R_a = P_c = 2.72 \text{ kW} \quad \text{(by Problem 14.11)}$$
$$\text{input power} = 52.72 + 2(2.72) = 58.16 \text{ kW}$$
$$\text{maximum efficiency} = \frac{52.72}{58.16} = 90.6\%$$

14.14 A 20-hp, 250-V, shunt motor has an armature-circuit resistance (including brushes and interpoles) of 0.22 Ω and a field resistance of 170 Ω. At no-load and rated voltage, the speed is 1200 rpm and the armature current is 3.0 A. At full-load and rated voltage, the line current is 55 A, and the flux is reduced 6% (due to the effects of armature reaction) from its value at no-load. What is the full-load speed?

$$E_{\text{no-load}} = 250 - (3.0)(0.22) = 249.3 \text{ V}$$
$$I_f = \frac{250}{170} = 1.47 \text{ A}$$
$$E_{\text{full-load}} = 250 - (55 - 1.47)(0.22) = 238.2 \text{ V}$$

Since n_m is proportional to E/ϕ, we have

$$n_{m,\text{full-load}} = 1200\left(\frac{238.2}{249.3}\right)\left(\frac{1}{0.94}\right) = 1220 \text{ rpm}$$

14.15 A 10-hp, 230-V, shunt motor takes a full-load line current of 40 A. The armature and field resistances are 0.25 Ω and 230 Ω, respectively. The total brush-contact drop is 2 V and the core and friction losses are 380 W. Calculate the efficiency of the motor. Assume that stray-load loss is 1% of output.

$$\text{input} = (40)(230) \qquad\qquad\qquad\qquad = 9200 \text{ W}$$

$$\text{field-resistance loss} = \left(\frac{230}{230}\right)^2(230) \qquad = 230 \text{ W}$$

$$\text{armature-resistance loss} = (40-1)^2(0.25) = 380 \text{ W}$$

$$\text{core loss and friction loss} \qquad\qquad\quad = 380 \text{ W}$$

$$\text{brush-contact loss} = (2)(39) \qquad\qquad = 78 \text{ W}$$

$$\text{stray-load loss} = \frac{10}{100} \times 746 \qquad\qquad = \underline{75 \text{ W}}$$

$$\text{total losses} \qquad\qquad\qquad\qquad\qquad = 1143 \text{ W}$$

$$\text{power output} = 9200 - 1143 \qquad\qquad = 8057 \text{ W}$$

$$\text{efficiency} = \frac{8057}{9200} = 87.6\%$$

14.16 A 10-kW, 250-V, shunt generator, having an armature resistance of 0.1 Ω and a field resistance of 250 Ω, delivers full-load at rated voltage and 800 rpm. The machine is now run as a motor while taking 10 kW at 250 V. What is the speed of the motor? Neglect brush-contact drop.

As a generator:

$$I_f = \frac{250}{250} = 1 \text{ A} \qquad I_L = \frac{10 \times 10^3}{250} = 40 \text{ A}$$

$$I_a = 40 + 1 = 41 \text{ A} \qquad I_a R_a = (41)(0.1) = 4.1 \text{ V}$$

$$E_g = 250 + 4.1 = 254.1 \text{ V}$$

As a motor:

$$I_L = \frac{10 \times 10^3}{250} = 40 \text{ A} \qquad I_f = \frac{250}{250} = 1 \text{ A}$$

$$I_a = 40 - 1 = 39 \text{ A} \qquad I_a R_a = (39)(0.1) = 3.9 \text{ V}$$

$$E_m = 250 - 3.9 = 246.1 \text{ V}$$

Now,

$$\frac{n_m}{n_g} = \frac{E_m}{E_g} \qquad \text{or} \qquad n_m = \frac{E_m}{E_g} n_g = \frac{246.1}{254.1}(800) = 774.8 \text{ rpm}$$

INDUCTION MOTORS

14.17 A 4-pole, 3-phase induction motor is energized from a 60-Hz supply, and is running at a load condition for which the slip is 0.03. Determine: (a) rotor speed, in rpm; (b) rotor current frequency, in Hz; (c) speed of the rotor's rotating magnetic field with respect to the stator frame, in rpm; (d) speed of the rotor's rotating magnetic field with respect to the stator's rotating magnetic field, in rpm.

$$n_s = \frac{120 f_1}{p} = \frac{120(60)}{4} = 1800 \text{ rpm}$$

(a)
$$n = (1-s)n_s = (1-0.03)(1800) = 1746 \text{ rpm}$$

(b)
$$f_2 = sf_1 = (0.03)(60) = 1.8 \text{ Hz}$$

(c) The p poles on the stator induce an equal number of poles on the rotor. Now, the same argument that led to (14.20) can be applied to the rotor. Thus, the rotor produces a rotating magnetic field whose speed *relative to the rotor* is

$$n_r = \frac{120 f_2}{p} = \frac{120 s f_1}{p} = s n_s$$

But the speed of the rotor relative to the stator is $n = (1 - s)n_s$. Therefore, the speed of the rotor field with respect to the stator is $n_s' = n_r + n = n_s$, or 1800 rpm.

(d)   Zero.

14.18 A 60-Hz induction motor has 2 poles and runs at 3510 rpm. Calculate (a) the synchronous speed and (b) the percent slip.

(a)
$$n_s = \frac{120 f_1}{s} = \frac{120(60)}{2} = 3600 \text{ rpm}$$

(b)
$$s = \frac{n_s - n}{n_s} = \frac{3600 - 3510}{3600} = 0.025 = 2.5\%$$

14.19 Using the rotor equivalent circuit of Fig. 14-19(b), show that an induction motor will have a maximum starting torque when its rotor resistance (regarded as variable) is equal to its leakage reactance. All quantities are on a per-phase basis.

From Fig. 14-19(b), the developed power, P_d, is given by

$$P_d = I_2^2 \frac{R_2}{s} - I_2^2 R_2 = T_e \omega_m \tag{1}$$

and the rotor current, I_2, is such that

$$I_2^2 = \frac{E_2^2}{(R_2/s)^2 + X_2^2} \tag{2}$$

Also, the mechanical angular velocity is

$$\omega_m = (1 - s)\omega_s \tag{3}$$

where ω_s is the synchronous angular velocity. These three equations give:

$$T_e = \frac{E_2^2 s}{\omega_s} \frac{R_2}{R_2^2 + s^2 X_2^2} = \frac{E_2^2}{2 X_2 \omega_s} \left[1 - \frac{(R_2 - s X_2)^2}{R_2^2 + s^2 X_2^2} \right] \tag{4}$$

By inspection of (4), T_e is a maximum for $R_2 = s X_2$, or, at starting ($s = 1$), for $R_2 = X_2$.

14.20 The rotor of a 3-phase, 60-Hz, 4-pole induction motor takes 120 kW at 3 Hz. Determine (a) the rotor speed and (b) the rotor copper losses.

$$s = \frac{f_2}{f_1} = \frac{3}{60} = 0.05 \qquad n_s = \frac{120 f_1}{p} = \frac{120(60)}{4} = 1800 \text{ rpm}$$

(a)
$$n = (1 - s)n_s = (1 - 0.05)(1800) = 1710 \text{ rpm}$$

(b)   From Fig. 14-22(b),

$$\text{rotor copper losses} = I_2^2 R_2' = s \times (\text{rotor input}) = (0.05)(120) = 6 \text{ kW}$$

14.21 The motor of Problem 14.20 has a stator copper loss of 3 kW, a mechanical loss of 2 kW, and a stator core loss of 1.7 kW. Calculate (a) the motor output at the shaft, and (b) the efficiency. Neglect rotor core loss.

From Problem 14.20, the rotor input is 120 kW and the rotor copper loss is 6 kW.

(a)
$$\text{motor output} = 120 - 6 - 2 = 112 \text{ kW}$$

(b)
$$\text{motor input} = 120 + 3 + 1.7 = 124.7 \text{ kW}$$

$$\text{efficiency} = \frac{\text{output}}{\text{input}} = \frac{112}{124.7} = 89.7\%$$

14.22 A 6-pole, 3-phase, 60-Hz, induction motor takes 48 kW in power at 1140 rpm. The stator copper loss is 1.4 kW, stator core loss is 1.6 kW, and rotor mechanical losses are 1 kW. Find the motor efficiency.

$$n_s = \frac{120f_1}{p} = \frac{120(60)}{6} = 1200 \text{ rpm} \qquad s = \frac{n_s - n}{n_s} = \frac{1200 - 1140}{1200} = 0.05$$

By Fig. 14-22(b):

rotor input = stator output = (stator input) − (stator losses) = 48 − (1.4 + 1.6) = 45 kW

rotor output = (1 − s) × (rotor input) = (1 − 0.05)(45) = 42.75 kW

motor output = (rotor output) − (rotational losses) = 42.75 − 1 = 41.75 kW

$$\text{motor efficiency} = \frac{41.75}{48} = 87\%$$

14.23 The synchronous speed of an induction motor is 900 rpm. Under a blocked-rotor condition, the input power to the motor is 45 kW at 193.6 A. The stator resistance per phase is 0.2 Ω and the transformation ratio is $a = 2$. Calculate (a) the ohmic value of the rotor resistance per phase, and (b) the motor starting torque. The stator and rotor are wye-connected.

(a) Approximately,

$$R_1 + a^2 R_2 = \frac{P_s}{I_s^2} \qquad \text{or} \qquad 0.2 + 4R_2 = \frac{(45 \times 10^3)/3}{(193.6)^2}$$

whence $R_2 = 0.05$ Ω.

(b) Referred to the stator, the rotor resistance per phase is $R_2' = a^2 R_2 = 0.2$ Ω. Then

$$\text{starting torque} = \frac{3I_s^2 R_2'}{\omega_s} = \frac{3(193.6)^2(0.2)}{2\pi(900)/60} = 238.6 \text{ N} \cdot \text{m}$$

14.24 The per-phase parameters of the equivalent circuit, Fig. 14-22(a), for a 400-V, 60-Hz, 3-phase, wye-connected, 4-pole induction motor are:

$$R_1 = 2R_2' = 0.2 \ \Omega \qquad X_1 = 0.5 \ \Omega \qquad X_2' = 0.2 \ \Omega \qquad X_m = 20 \ \Omega$$

If the total mechanical and iron losses at 1755 rpm are 800 W, compute (a) input current, (b) input power, (c) output power, (d) output torque, and (e) efficiency (all at 1755 rpm).

$$n_s = \frac{120(60)}{4} = 1800 \text{ rpm} \qquad s = \frac{1800 - 1755}{1800} = \frac{1}{40}$$

From the given circuit, the equivalent impedance per phase is

$$\mathbf{Z}_e = (0.2 + j0.5) + \frac{(j20)(4 + j0.2)}{4 + j(20 + 0.2)}$$

$$= (0.2 + j0.5) + (3.77 + j0.944) = 4.223\underline{/20°} \ \Omega$$

and the phase voltage is $400/\sqrt{3} = 231$ V.

(a) $$\text{input current} = \frac{231}{4.223} = 54.65 \text{ A}$$

(b) $$\text{total input power} = \sqrt{3}(400)(54.65)(\cos 20°) = 35.58 \text{ kW}$$

(c) The total power crossing the air gap, P_g, is the power in the three 3.77 Ω resistances (see the expression for $\mathbf{Z}_e$ above). Thus,

$$P_g = 3(54.65)^2(3.77) = 33.789 \text{ kW}$$

[Or, by subtraction of the stator losses, $P_g = 35\,580 - 3(54.65)^2(0.2) = 33.788$ kW.] The total developed power is then

$$P_d = (1 - s)P_g = (0.975)(33.79) = 32.94 \text{ kW}$$

and the total output power is $P_o = P_d - (800 \text{ W}) = 32.14$ kW.

(d)

$$\text{output torque} = \frac{P_o}{\omega_m} = \frac{32\,140}{2\pi(1755)/60} = 174.9 \text{ N} \cdot \text{m}$$

(e)

$$\text{efficiency} = \frac{32.14}{35.58} = 90.3\%$$

SYNCHRONOUS MACHINES

14.25 A 4-pole induction motor, running with 5% slip, is supplied by a 60-Hz synchronous generator. (a) Calculate the speed of the motor. (b) What is the generator speed if it has six poles?

(a)

$$n = (1-s)n_s = (1-0.05)\frac{120(60)}{4} = 1710 \text{ rpm}$$

(b)

$$n_s = \frac{120f}{p} = \frac{120(60)}{6} = 1200 \text{ rpm}$$

14.26 For a 60-Hz synchronous generator, list six possible combinations of number of poles and speed.

From $f = pn_s/120$ (Hz), we must have $pn_s = 120(60) = 7200$ rpm. Hence Table 14-1.

Table 14-1

No. of Poles	Speed, rpm
2	3600
4	1800
6	1200
8	900
10	720
12	600

14.27 A 3-phase, wye-connected, round-rotor synchronous generator, rated at 10 kVA and 230 V, has a synchronous reactance of 1.2 Ω per phase and an armature resistance of 0.5 Ω per phase. Calculate the percent voltage regulation at full-load with 0.8 lagging power factor.

The phasor diagram is shown in Fig. 14-36, from which (ϕ being negative)

$$V_0 = \sqrt{(V_t \cos\phi + I_a R_a)^2 + (|V_t \sin\phi| + I_a X_s)^2}$$

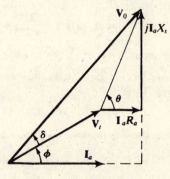

Fig. 14-36

Substituting $\qquad V_t = \dfrac{230}{\sqrt{3}} = 132.8$ V $\qquad I_a = \dfrac{(10 \times 10^3)/3}{230/\sqrt{3}} = 25.1$ A

and the other data yields:

$$V_0 = \sqrt{(106.24 + 12.55)^2 + (79.68 + 30.12)^2} = 161.76 \text{ V}$$

Then $\qquad \dfrac{V_0 - V_t}{V_t} \times 100\% = \dfrac{161.76 - 132.8}{132.8} \times 100\% = 21.8\%$

14.28 Repeat Problem 14.27 for the case of 0.8 leading power factor, other data remaining unchanged.

Let $\mathbf{V}_t = 132.8 + j0$ V be the reference phasor. Then,

$$\mathbf{I}_a = (25.1)(0.8 + j0.6) \quad \text{A}$$
$$\mathbf{Z}_s = 0.5 + j1.2 \quad \Omega$$
$$\mathbf{I}_a\mathbf{Z}_s = -8 + j31.6 \quad \text{V}$$
$$\mathbf{V}_0 = \mathbf{V}_t + \mathbf{I}_a\mathbf{Z}_s = 124.8 + j31.6 \quad \text{V}$$

or $V_0 = 128.7$ V. Hence,

$$\text{percent regulation} = \dfrac{128.7 - 132.8}{132.8} \times 100\% = -3.1\%$$

Notice that this problem has been solved without using the phasor diagram, and that the voltage regulation at full-load with 0.8 leading power factor is negative.

14.29 For the generator of Problem 14.27, determine the power factor such that the voltage regulation is zero on full-load.

Let ϕ be the required power-factor angle, so that $\mathbf{I}_a = 25.1\underline{/\phi}$ A. Then,

$$\mathbf{Z}_s = 0.5 + j1.2 = 1.3\underline{/67.38°} \quad \Omega$$
$$\mathbf{I}_a\mathbf{Z}_s = 32.63\underline{/\phi + 67.38°} = 32.63 \cos(\phi + 67.38°) + j32.63 \sin(\phi + 67.38°) \quad \text{V}$$
$$\mathbf{V}_0 = 132.8 + j0 + 32.63 \cos(\phi + 67.38°) + j32.63 \sin(\phi + 67.38°) \quad \text{V}$$

For zero voltage regulation, $V_0 = V_t = 132.8$ V; i.e.,

$$(132.8)^2 = [132.8 + 32.63 \cos(\phi + 67.38°)]^2 + [32.63 \sin(\phi + 67.38°)]^2$$

which gives $\qquad \cos(\phi + 67.38°) = \dfrac{-32.63}{2(132.8)} \qquad$ or $\qquad \phi = +29.67°$

Thus, $\cos \phi = 0.869$ leading.

14.30 A 20-kVA, 220-V, wye-connected, 3-phase, salient-pole synchronous generator supplies rated load at 0.707 lagging power factor. The reactances per phase are $X_d = 2X_q = 4$ Ω. Neglecting the armature resistance, determine (a) the power angle and (b) the percent voltage regulation.

(a) From (14.38) and the phasor diagram of Fig. 14-31, with

$$V_t = \dfrac{220}{\sqrt{3}} = 127 \text{ V} \qquad I_a = \dfrac{(20 \times 10^3)/3}{220/\sqrt{3}} = 52.5 \text{ A} \qquad \phi = \cos^{-1} 0.707 = 45°$$

we get $\qquad \tan \delta = \dfrac{I_aX_q \cos \phi}{V_t + I_aX_q \sin \phi} = \dfrac{(52.5)(2)(0.707)}{127 + (52.5)(2)(0.707)} = 0.369$

or $\delta = 20.25°$.

(b) $\qquad V_0 = V_t \cos \delta + I_aX_d = V_t \cos \delta + I_aX_d \sin(\delta + \phi)$
$$= 127 \cos 20.25° + (52.5)(4) \sin(20.25° + 45°) = 309.8 \text{ V}$$

$$\text{percent voltage regulation} = \dfrac{309.8 - 127}{127} \times 100\% = 144\%$$

14.31 (*a*) Determine the power developed by the generator of Problem 14.30 and verify that it is equal to the power supplied to the load. (*b*) How much power is developed due to saliency?

(*a*) The power developed per phase is given by (*14.40*) as

$$P_d = \frac{(309.8)(127)}{4}\sin 20.25° + \frac{(127)^2}{2}\left(\frac{1}{2}-\frac{1}{4}\right)\sin 40.50° = 3404.46 + 1309.37 = 4713.8 \text{ W}$$

The power supplied to the load per phase is $(20 \times 10^3)(0.707)/3 = 4713.3$ W.

(*b*) From (*a*), the power due to saliency is $3(1309.37) = 3928$ W.

14.32 A 3-phase, wye-connected load takes 50 A in current at 0.707 lagging power factor, with 220 V between the lines. A 3-phase, wye-connected, round-rotor synchronous motor, having a synchronous reactance of 1.27 Ω per phase, is connected in parallel with the load. The power developed by the motor is 33 kW at a power angle of 30°. Neglecting the armature resistance, calculate (*a*) the reactive power (in kvar) of the motor, and (*b*) the overall power factor of the motor and the load.

(*a*) The circuit and phasor diagrams, on a per-phase basis, are shown in Fig. 14-37. From (*14.33*),

$$P_d = \frac{33 \times 10^3}{3} = \frac{220}{\sqrt{3}}\frac{V_0}{1.27}\sin 30° \qquad \text{or} \qquad V_0 = 220 \text{ V}$$

By the parallel connection, Fig. 14-37(*a*), $I_a X_s = V_t = (220/\sqrt{3})$ V. Then, from the isosceles triangle in Fig. 14-37(*b*),

$$2\delta + 90° + \phi_a = 180° \qquad \text{or} \qquad \phi_a = 90 - 2\delta = 30°$$

and

$$\text{motor reactive power} = 3V_t I_a \sin\phi_a = 3\frac{V_t^2}{X_s}\sin\phi_a$$

$$= 3\frac{(220/\sqrt{3})^2}{1.27}\left(\frac{1}{2}\right) = 19\,000 \text{ var} = 19 \text{ kvar}$$

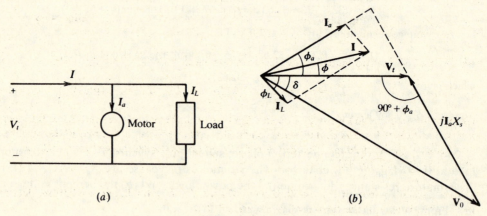

(*a*)   (*b*)

Fig. 14-37

(*b*) From Fig. 14-37(*b*), the projection of **I** on $\mathbf{V}_t$ is

$$I\cos\phi = I_a\cos\phi_a + I_L\cos\phi_L$$

and its projection perpendicular to $\mathbf{V}_t$ is (taking account of the fact that ϕ_L is negative)

$$I\sin\phi = I_a\sin\phi_a + I_L\sin\phi_L$$

Hence $$\tan\phi = \frac{I_a\sin\phi_a + I_L\sin\phi_L}{I_a\cos\phi_a + I_L\cos\phi_L}$$

Substituting the values

$$I_a = \frac{220/\sqrt{3}}{1.27} = 100 \text{ A} \qquad I_L = 50 \text{ A} \qquad \phi_a = 30° \qquad \phi_L = -45°$$

we obtain $\tan \phi = 0.120$, or $\cos \phi = 0.993$ leading.

Supplementary Problems

INCREMENTAL-MOTION SYSTEMS

14.33   An electromagnet, shown in Fig. 14-38, is required to exert a 500-N force on the iron at an air gap of 1 mm, while the exciting coil is carrying 25 A dc. The core cross section at the air gap is 600 mm² in area. Calculate the required number of turns of the exciting coil.   *Ans.*   65 turns

14.34   (a) How many turns must the exciting coil of the electromagnet of Fig. 14-38 have in order to produce a 500-N (average) force if the coil is excited by a 60-Hz alternating current having a maximum value of 35.35 A? (b) Is the average force frequency-dependent?   *Ans.*   (a) 65 turns; (b) no

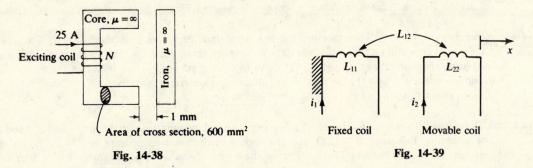

Fig. 14-38   Fig. 14-39

14.35   Figure 14-39 shows two mutually coupled coils, for which

$$L_{11} = L_{22} = 3 + \frac{2}{3x} \quad \text{(mH)} \qquad L_{12} = L_{21} = \frac{1}{3x} \quad \text{(mH)}$$

where x is in meters. (a) If $i_1 = 5$ A dc and $i_2 = 0$, what is the mutual electrical force between the coils at $x = 0.01$ m? (b) If $i_1 = 5$ A dc and the second coil is open-circuited and moved in the positive x-direction at a constant speed of 20 m/s, determine the voltage across the second coil at $x = 0.01$ m.
Ans.   (a) 83.33 N; (b) 333.3 V

14.36   For the two-coil system of Fig. 14-39, if $i_1 = 7.07 \sin 377t$ (A), $i_2 = 0$, and $x = 0.01$ m, determine (a) the instantaneous and (b) the time-average electrical force.
Ans.   (a) $166.67 \sin^2 377t$ (N); (b) 83.33 N

14.37   (a) The two coils of Fig. 14-39 are connected in series, with 5 A dc flowing in them. Determine the electrical force between the coils at $x = 0.01$ m. Does the force tend to increase or decrease x? (b) Next, the coils are connected in parallel across a 194-V, 60-Hz source. Compute the average electrical force at $x = 0.01$ m, neglecting the coil resistances.   *Ans.*   (a) 250 N, decrease x; (b) 250 N

DC MACHINES

14.38   The armature of a 6-pole, lap-wound, dc generator has 720 active conductors. The generator is designed to generate 420 V at 1720 rpm. Determine the flux per pole.   *Ans.*   20.35 mWb

14.39 The armature of Problem 14.38 is reconnected as wave-wound. At what speed must the generator operate to induce 630 V in the armature?   *Ans.*   860 rpm

14.40 At what speed, in rpm, must the armature of a dc machine run to develop 572 kW at a torque of 4605 N · m?   *Ans.*   1187 rpm

14.41 The armature of a dc machine running at 1200 rpm carries 45 A in current. If the induced armature voltage is 130 V, what is the torque developed by the armature?   *Ans.*   46.5 N · m

14.42 A self-excited shunt generator supplies a load of 12.5 kW at 125 V. The field resistance is 25 Ω and the armature resistance is 0.1 Ω. The total voltage drop because of brush contact and armature reaction at this load is 3.5 V. Calculate the induced armature voltage.   *Ans.*   139 V

14.43 A 6-pole, lap-wound armature, having 720 conductors, rotates in a flux of 20.35 mWb per pole. (*a*) If the armature current is 78 A, what is the torque developed by the armature? (*b*) If the induced armature voltage is 420 V, what is the motor speed?   *Ans.*   (*a*) 181.9 N · m; (*b*) 1720 rpm

14.44 A separately-excited motor runs at 1045 rpm, with a constant field current, while taking an armature current of 50 A at 120 V. The armature resistance is 0.1 Ω. If the load on the motor changes such that it now takes 95 A at 120 V, determine the motor speed at the new load.   *Ans.*   1004 rpm

14.45 A long-shunt compound generator supplies 50 kW at 230 V. The total field- and armature-circuit resistances are 46 Ω and 0.03 Ω, respectively. The brush-contact drop is 2 V. Determine the percent voltage regulation. Neglect armature reaction.   *Ans.*   3.77%

14.46 A separately-excited dc generator has the following data: armature resistance, 0.04 Ω; field resistance, 110 Ω; total core and mechanical losses, 960 W; voltage across the field, 230 V. The generator supplies a load at a terminal voltage of 230 V. Calculate (*a*) the armature current at which the generator has a maximum efficiency, and (*b*) the maximum value of the generator efficiency.
Ans.   (*a*) 189.8 A; (*b*) 93.8%

14.47 A shunt motor operates at a flux of 25 mWb per pole, is lap-wound, and has 2 poles and 360 conductors. The armature resistance is 0.12 Ω and the motor is designed to operate at 115 V, taking 60 A in armature current at full-load. (*a*) Determine the value of the external resistance to be inserted in the armature circuit so that the armature current shall not exceed twice its full-load value at starting. (*b*) When the motor has reached a speed of 400 rpm, the external resistance is cut by 50%. What is the armature current then, at this speed? (*c*) The external resistance is completely cut out when the motor reaches its final speed; the armature current is then at its full-load value. Calculate the motor speed.
Ans.   (*a*) 0.838 Ω; (*b*) 102 A; (*c*) 718.6 rpm

14.48 A 230-V shunt motor, having an armature resistance of 0.05 Ω and a field resistance of 75 Ω, draws a line current of 7 A while running light at 1120 rpm. For a load at which the line current is 46 A, determine (*a*) the motor speed, (*b*) motor efficiency, and (*c*) total core and mechanical losses.
Ans.   (*a*) 1110.5 rpm; (*b*) 83.9%; (*c*) 903.9 W

INDUCTION MOTORS

14.49 A 3-phase, 60-Hz induction motor has 8 poles and operates with a slip of 0.05 for a certain load. Compute (in rpm) the (*a*) speed of the rotor with respect to the stator, (*b*) speed of the rotor with respect to the stator magnetic field, (*c*) speed of the rotor magnetic field with respect to the rotor, (*d*) speed of the rotor magnetic field with respect to the stator, (*e*) speed of the rotor field with respect to the stator field.   *Ans.*   (*a*) 855 rpm; (*b*) 45 rpm; (*c*) 45 rpm; (*d*) 900 rpm; (*e*) 0

14.50 A 3-phase, 6-pole, 60-Hz induction motor runs (*a*) on no-load at 1160 rpm and (*b*) on full-load at 1092 rpm. Determine the slip and frequency of rotor currents on no-load and on full-load.
Ans.   (*a*) 0.034, 2 Hz; (*b*) 0.09, 5.4 Hz

14.51   A 3-phase, 60-Hz, 4-pole induction motor has a rotor leakage reactance of 0.8 Ω per phase and a rotor resistance of 0.1 Ω per phase. How much additional resistance must be inserted in the rotor circuit so that the motor shall have the maximum starting torque? Use the rotor circuit of Fig. 14-22(a) for your calculations.   *Ans.*   0.7 Ω

14.52   A 20-hp, 3-phase, 400-V, 60-Hz, 4-pole induction motor delivers full-load at 5% slip. The mechanical rotational losses are 400 W. Calculate (a) the electromagnetic torque, (b) the shaft torque, and (c) the rotor copper loss.   *Ans.*   (a) 85.5 N · m; (b) 83.3 N · m; (c) 806.3 W

14.53   A 3-phase, 6-pole induction motor is rated 400 Hz, 150 V, 10 hp, 3% slip at rated power output. The windage and friction loss is 200 W at rated speed. With the motor operating at rated voltage, frequency, slip, and power output, determine (a) rotor speed, (b) frequency of rotor current, (c) rotor copper loss, (d) power crossing the air gap, (e) output torque.
Ans.   (a) 7760 rpm; (b) 12 Hz; (c) 237 W; (d) 7897 W; (e) 9.2 N · m

14.54   A 3-phase, wye-connected, 12-pole induction motor is rated 500 hp, 2200 V, 60 Hz. The stator resistance per phase is 0.4 Ω, the rotor resistance per phase in stator terms is 0.2 Ω, and the total rotor and stator reactance per phase in stator terms is 2 Ω. With rated voltage and frequency applied, the motor slip is 0.02. For this condition, find, on a per-phase basis, (a) the stator current (neglect magnetizing current), (b) the developed torque, (c) the rotor power input, (d) the rotor copper loss.
Ans.   (a) 120 A; (b) 2292 N · m; (c) 144 kW; (d) 2880 W

SYNCHRONOUS MACHINES

14.55   What is the maximum speed at which (a) a 60-Hz, (b) a 50-Hz, synchronous machine can be operated?
Ans.   (a) 3600 rpm; (b) 3000 rpm

14.56   A 25-kVA, 3-phase, wye-connected, 400-V synchronous generator has a synchronous impedance of 0.05 $+j1.6$ Ω per phase. Determine the full-load voltage regulation at (a) 0.8 power factor lagging, (b) unity power factor, (c) 0.8 power factor leading.   *Ans.*   (a) 22%; (b) 10.6%; (c) −5.5%

14.57   Determine the power angles for the three cases in Problem 14.56.   *Ans.*   (a) 7.2°; (b) 13°; (c) 15°

14.58   The generator of Problem 14.56 is to have zero voltage regulation at half full-load. Neglecting the armature resistance, find the operating power factor and the developed power.
Ans.   0.997 leading; 12.5 kW

14.59   A 500-kVA, 6-pole, 500-V, 3-phase, wye-connected synchronous generator has a synchronous impedance of $0.1 + j1.5$ Ω per phase. If the generator is driven at 1000 rpm, what is the frequency of the generated voltage? Determine the excitation voltage and the power angle on full-load and 0.8 lagging power factor.
Ans.   50 Hz; 1098 V; 37.6°

14.60   A 100-kVA, 400-V, wye-connected, salient-pole synchronous generator runs at full-load and 0.8 leading power factor. If $X_d = 2X_q = 1.1$ Ω per phase and R_a is negligible, calculate (a) the voltage regulation, (b) the power angle, and (c) the developed power.   *Ans.*   (a) −28.7%; (b) 15.5°; (c) 80 kW

14.61   A 30-kVA, 3-phase, 230-V, wye-connected synchronous generator has a synchronous reactance of 0.8 Ω per phase. The armature resistance is negligible. Calculate the percent voltage regulation on (a) full-load at 0.8 leading power factor, (b) 50% full-load at unity power factor, (c) 25% full-load at 0.8 lagging power factor.   *Ans.*   (a) −18.7%; (b) 2.5%; (c) 7.2%

14.62   A 400-V, 3-phase, wye-connected, round-rotor synchronous motor operates at unity power factor while developing a power of 60 kW. If the synchronous reactance is 1.0 Ω per phase and the armature resistance is negligible, calculate (a) the induced voltage per phase and (b) the power angle.
Ans.   (a) 246.6 V; (b) −20.5°

14.63 An overexcited 2300-V, 3-phase, wye-connected synchronous motor runs at a power angle of $-21°$. The per-phase synchronous impedance is $0.1 + j2$ Ω. If the motor takes a line current of 350 A, determine the power factor.   *Ans.*   0.87 leading

14.64 What are (*a*) the power factor, and (*b*) the line current, of the motor of Problem 14.63, if the internal induced voltage is the same as the line voltage and the power angle is $-20°$? (*c*) Find the developed power.   *Ans.*   (*a*) 0.99 lagging; (*b*) 242 A; (*c*) 938 kW

14.65 A 400-V, 3-phase, round-rotor synchronous motor has an efficiency of 92% while delivering 18 hp (at the shaft). The per-phase synchronous impedance is $0.5 + j1.5$ Ω. If the motor operates at 0.9 lagging power factor, determine (*a*) the power angle and (*b*) the field current. The motor saturation characteristic is shown in Fig. 14-40.   *Ans.*   (*a*) $-7.4°$; (*b*) 4.5 A

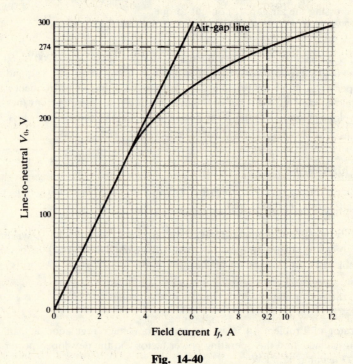

Fig. 14-40

14.66 An overexcited synchronous motor is connected across a 100-kVA inductive load having a 0.8 lagging power factor. The motor takes 10 kW in power while idling (on no-load). Calculate the kVA-rating of the motor if it is desired to bring the overall power factor to unity. The motor is not used to carry any load.   *Ans.*   60.8 kVA

Chapter 15

Concepts of Control.
Transfer Functions

A *control system* is an interrelated collection of elements that directs or regulates (*controls*) energy or exerts restraint, from a source point to a usage point. To be acceptable, it must have physical properties and performance characteristics that satisfy all concerned observers. Physical properties can be peculiar to an application, and they are usually subject to economic constraints presenting their own challenge to the designer. However, different physical systems have performance similarities that logically lead to a unified control-system theory.

15.1 DEFINITIONS AND TERMINOLOGY

1. The *plant* is the original or basic system over which control is desired; it is also called the *controlled system*.

2. The *control elements*, or *controller*, is the agency that applies a *control signal* to the plant, which signal represents a change in a *manipulated variable* of the system.

3. The *controlled variable* is that characteristic of the plant which is being regulated. Usually, plant output is the controlled variable.

4. The *reference command* or *reference input* is an external signal applied to a control system in order to bring about a specific action of the plant (i.e., a specific value of the controlled variable).

5. *Disturbance inputs* are undesired stimuli of the plant which affect the value of the controlled variable.

6. A *comparator* is a control system element that compares (by addition or subtraction) two or more signals.

7. *Feedback* is that property of a control system whereby the output (controlled variable) is compared to the reference input, and an *actuating signal* or *control force*, depending on both the output and the input, is generated. This signal actuates the controller. In control systems without feedback, the reference input serves as the actuating signal.

8. A *feedback element* converts the controlled variable into a form (the *feedback signal*) which may be directly compared with the reference command.

15.2 SYSTEM CLASSIFICATIONS

Control systems are first of all classified according to their requirement of external intervention. In a *manual control system*, an intelligent being forms one of the interdependent elements. A control system for which all interdependent elements are nonintelligent components is called an *automatic control system*; it is this latter type on which most control system theory centers.

Example 15.1 (*a*) A man flipping a light switch to illuminate a darkened room constitutes a manual control system. The switch and light bulb are the plant, with luminous flux (measured in lumens) as controlled variable or output. The movement of the switch handle is the actuating signal or control force. (*b*) An automobile driver adjusting the degree of accelerator depression to maintain a desired speed is a manual control system. Movement of the accelerator pedal, linkage, and carburetor butterfly valve is the actuating signal. The engine and drive train can be considered the plant. (c) A light bulb switched on by a relay activated by circuitry that senses degree of darkness through use of a photocell is an automatic control system. The photocell and switching circuitry are considered the controller supplying the control force (relay coil voltage) to the plant, which consists of the relay and light bulb. (d) A mechanical alarm clock forms an automatic control system, in which the bell (plant) generates an audible sound (controlled variable). The control force is retraction of the bell hammer stop by the clock's time-indicating movement and linkage.

Another classification of control systems is according to their use of feedback. An *open-loop control system* has an actuating signal which is independent of any other variable of the system. A control system for which the control force is somehow dependent on the output or another indirectly controlled variable (using feedback) is called a *closed-loop control system*. Closed-loop control systems are subdivided into two categories. A *regulator system* has as its primary function to maintain the controlled variable within a specified range, regardless of the presence of disturbance inputs; its reference command is changed infrequently. A *followup system* is one that is required to maintain a close correspondence between a frequently changed reference command and the controlled variable.

Example 15.2 (a) A fireplace can be considered an open-loop control system. The system output is heat, with the rate at which the logs burn being independent of room temperature. (b) A thermostat-operated home heating unit forms a closed-loop control system. Room temperature is sensed by a bimetal strip within the thermostat, the movement due to expansion of which (feedback signal) acts to tip a mercury switch.

Example 15.3 (a) The home heating unit of Example 15.2(b) is a regulator system. The bimetal strip is oriented by a precalibrated rotary mechanism (reference command) so that the mercury switch is "off" when the amount of expansion of the strip corresponds to a desired room temperature. Once a comfortable room temperature has been achieved, the reference command is not often changed. (b) The control system of an air-to-air heat-seeking missile is a followup system that constantly adjusts the airfoils of the missile to maintain a course toward the exhaust heat of its target aircraft.

Control systems are further classified in terms that describe their variables. In a *continuous (-variable) system*, all variables are continuous, differentiable functions of time; behavior of the system is therefore described by differential equations. A *discrete (-variable) system* has at least one variable known only at particular instants in time, necessitating description by difference equations.

Example 15.4 (a) The flyball-governed internal combustion engine exemplifies the continuous-variable control system. The engine speed is monitored for all time, and the fuel flow rate can be adjusted at any instant. (b) A tachometer utilizes a microprocessor to count the number of teeth on a 60-tooth gear that pass a reference point in a second; it then divides the count by 60 to give speed in revolutions per second. If this tachometer functions as a speed feedback element in a control system, then that control system is a discrete-variable system, since the value of speed is known only once each second.

15.3 FUNCTIONAL BLOCK DIAGRAMS

The workings of a control system, as outlined in Section 15.1, may be represented in a flow diagram such as Fig. 15-1, called the *functional block diagram* of the system. It is to be understood that any or all of $r(t)$, $q(t)$, and $c(t)$ may represent multiple signals.

Ordinarily the functional block diagram is drawn as a preliminary to establishing the formal mathematical relationship connecting $c(t)$ with $r(t)$ and $q(t)$.

Example 15.5 Make a functional block diagram of an open-loop room heating system that consists of a gas-fired space heater, with a broken window pane constituting a disturbance.

See Fig. 15-2. Air heated by the flame and cold air through the broken window pane are forcing functions in the heat flow equation that describes the room heating process and determines the room temperature as the system output.

Example 15.6 Draw a functional block diagram of the flyball-governed engine of Example 15.4(a).

See Fig. 15-3, where the restraining spring acts as the reference command and engine speed is the controlled variable.

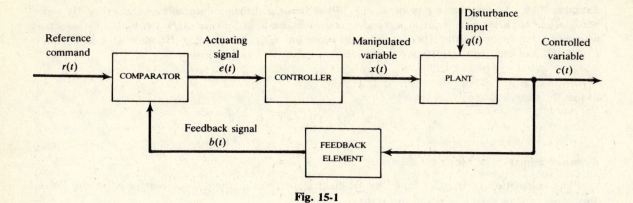

Fig. 15-1

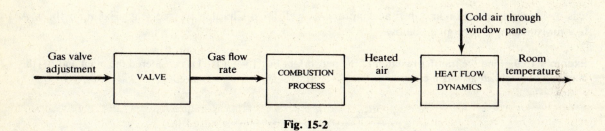

Fig. 15-2

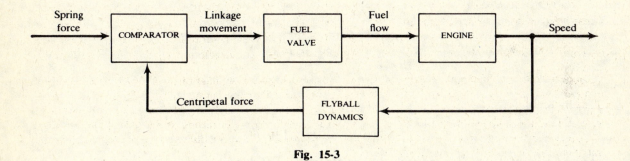

Fig. 15-3

15.4 TRANSFER FUNCTIONS

The first requirement for analysis of a (linear) control system is that one must be able to describe each physical component of the system by a mathematical function. These mathematical models are then combined to produce a composite mathematical model of the system, which usually takes the form of a differential equation with time as the independent variable. The Laplace transform methods of Section 6.3 provide the easiest way of solving such an equation.

The *transfer function* of a transmission path through a system component is defined as the ratio of the Laplace transform of the component's output signal to the Laplace transform of the component's input signal, with all initial conditions taken to be zero (i.e., the component is initially "relaxed," or unenergized). This assumption represents no real restriction, since any nonzero initial conditions can be introduced later as input. (See Problem 16.4.)

Example 15.7 If $q(t) \equiv 0$ (no disturbance input) in Fig. 15-1, the transfer function of the plant is $G_1(s) \equiv C(s)/X(s)$.

Example 15.8 Assume that the plant of Fig. 15-1 is linear and that a disturbance input exists. By super-position, $c(t)$ has a component due to $x(t)$ and a component due to $q(t)$; hence, a transfer function is associated with either input-output path. These transfer functions are $G_1(s)$, as found in Example 15.7, and $G_2(s) \equiv C(s)/Q(s)$, computed with $x(t) \equiv 0$. We then have

$$\mathscr{L}\{c(t)\} = G_1(s)X(s) + G_2(s)Q(s)$$

so that the controlled variable is given by

$$c(t) = \mathscr{L}^{-1}\{G_1(s)X(s) + G_2(s)Q(s)\}$$

Transfer Functions of Electrical Elements

The calculation of transfer functions of electrical circuit elements was illustrated in the Solved Problems of Chapter 6 (although the ratio

$$\frac{\text{output transform}}{\text{input transform}}$$

was not given a special name in that chapter), as well as in Chapter 10 (see, e.g., Example 10.7). A few more examples are given below.

Example 15.9 Find the transfer function of the network of Fig. 15-4, if $v_i(t)$ is considered the input and $i(t)$ the output. Terminals ab are open-circuit.

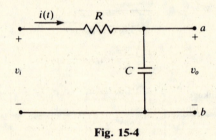

Fig. 15-4

By KVL,

$$v_i(t) = Ri(t) + \frac{1}{C}\int i(t)\,dt \tag{1}$$

Assume zero initial conditions and apply the Laplace transform (Table 6-1) to (1) to obtain

$$V_i(s) = RI(s) + \frac{1}{sC}I(s) \tag{2}$$

Solving (2) for the ratio of output and input transforms gives the transfer function as

$$F(s) \equiv \frac{I(s)}{V_i(s)} = \frac{sC}{sRC + 1}$$

Example 15.10 Find the transfer function of the RC filter network of Fig. 15-4 when $v_i(t)$ is considered the input and $v_o(t)$ is considered the output. Terminals ab are open-circuit.

Voltage division, applied to the s-domain version of Fig. 15-4, gives at once

$$F(s) \equiv \frac{V_o(s)}{V_i(s)} = \frac{1/sC}{R + (1/sC)} = \frac{1}{RCs + 1}$$

Example 15.11 Find the transfer function of the dc servomotor of Fig. 15-5, in which the field current (i_f) is constant. Shaft angular position, $\theta(t)$, is considered the output and terminal voltage, $v_T(t)$, the input. The parameter J (kg·m²) is the combined polar moment of inertia of the motor armature and connected load; β (N·m·s/rad) is a viscous friction coefficient that accounts for load rotational losses.

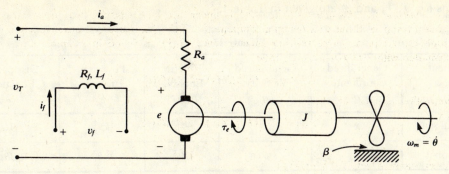

Fig. 15-5

Under the assumption of constant field current, (14.11) gives the counter emf as $e(t) = K\dot{\theta}(t)$. Armature current is given by KVL as

$$i_a(t) = \frac{v_T(t) - e(t)}{R_a} = \frac{v_T(t)}{R_a} - \frac{K\dot{\theta}(t)}{R_a}$$

and (14.13) now gives the shaft torque as

$$\tau_e(t) = K i_a(t) = \frac{K}{R_a} v_T(t) - \frac{K^2}{R_a} \dot{\theta}(t)$$

But, by Newton's second law for rotational motion, the net torque on the shaft-plus-armature must equal its time rate of change of angular momentum:

$$\tau_e(t) - \beta\dot{\theta}(t) = J\ddot{\theta}(t)$$

$$J\ddot{\theta}(t) + \left(\beta + \frac{K^2}{R_a}\right)\dot{\theta}(t) = \frac{K}{R_a} v_T(t)$$

Application of the Laplace transform to this differential equation, under zero initial conditions, yields

$$F(s) \equiv \frac{\Theta(s)}{V_T(s)} = \frac{K/R_a}{s[Js + \beta + (K^2/R_a)]}$$

Transfer Functions of Mechanical Elements

Mechanical systems can be classified as translational or rotational. The elements used to describe a translational system are mass (stores kinetic energy), linear spring (stores potential energy), and linear viscous damper (creates frictional force). The analogous elements of a rotational system are polar moment of inertia, torsional spring, and torsional viscous damper.

The number of *degrees of freedom* (N) of a mechanical system is the number of nodes (points of independent motion) in the system at any instant. Let q_i ($i = 1, 2, \ldots, N$) be the generalized coordinate associated with the ith node of a mechanical system. Let $\mathcal{F}_i$ ($i = 1, 2, \ldots, N$) denote the generalized applied force (or torque, for a rotational system) associated with the generalized coordinate q_i; $\mathcal{F}_i$ is positive when it tends to increase q_i. Further, define three energy functions, as follows:

$\mathcal{T} \equiv$ sum of all kinetic energies associated with elements in the system, expressed in terms of the generalized coordinates and their time derivatives

$\mathcal{V} \equiv$ sum of all potential energies associated with elements in the system, expressed in terms of the generalized coordinates

$\mathcal{D} =$ sum of all dissipation functions associated with elements in the system, expressed in terms of the time derivatives of the generalized coordinates

Then, the *Lagrangian equations of motion* of the mechanical system are

$$\frac{d}{dt}\left(\frac{\partial\mathcal{T}}{\partial\dot{q}_i}\right) - \frac{\partial\mathcal{T}}{\partial q_i} + \frac{\partial\mathcal{V}}{\partial q_i} + \frac{\partial\mathcal{D}}{\partial\dot{q}_1} = \mathcal{F}_i \qquad (i = 1, 2, \ldots, N) \tag{15.1}$$

Formulations for $\mathscr{T}$, $\mathscr{V}$, and $\mathscr{D}$ are given in Table 15-1.

Table 15-1

Element Type	Generalized Coordinate, q	Generalized Velocity, $\dot{q}$	Kinetic Energy, $\mathscr{T}$	Potential Energy, $\mathscr{V}$	Dissipation Function, $\mathscr{D}$
translational	x	$\dot{x}$	$\frac{1}{2}M\dot{x}^2$	$\frac{1}{2}kx^2$	$\frac{1}{2}\beta\dot{x}^2$
rotational	θ	$\dot{\theta}$	$\frac{1}{2}J\dot{\theta}^2$	$\frac{1}{2}k\theta^2$	$\frac{1}{2}\beta\dot{\theta}^2$

Example 15.12 Find the transfer function of the mass-spring-dashpot system of Fig. 15-6, where $p(t)$ is the applied external force.

The Lagrangian energy functions are $\mathscr{T} = \frac{1}{2}M\dot{x}^2$, $\mathscr{V} = \frac{1}{2}kx^2$, $\mathscr{D} = \frac{1}{2}\beta\dot{x}^2$. Applying (15.1) and noting that $p(t)$ acts to increase $x(t)$, we have

$$M\ddot{x} + \beta\dot{x} + kx = p(t) \tag{1}$$

Using the Laplace transform on (1) and solving for the transfer function gives

$$F(s) \equiv \frac{X(s)}{P(s)} = \frac{1}{s^2M + s\beta + k}$$

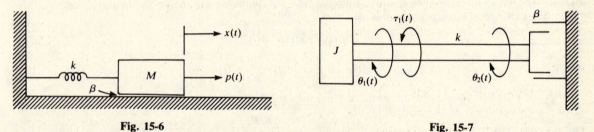

Fig. 15-6   **Fig. 15-7**

Example 15.13 For the rotational system of Fig. 15-7, with applied torque $T_1(t)$, find the transfer functions $\Theta_1(s)/T_1(s)$ and $\Theta_2(s)/T_1(s)$.

The Lagrangian energy functions are

$$\mathscr{T} = \frac{1}{2}J\dot{\theta}_1^2 \qquad \mathscr{V} = \frac{1}{2}k(\theta_1 - \theta_2)^2 \qquad \mathscr{D} = \frac{1}{2}\beta\dot{\theta}_2^2$$

Noting that $N = 2$ and that torque $\tau_1(t)$ acts to decrease $\theta_1(t)$, we have from (15.1):

$$J\ddot{\theta}_1 + k(\theta_1 - \theta_2) = -\tau_1(t) \tag{1}$$
$$\beta\dot{\theta}_2 - k(\theta_1 - \theta_2) = 0 \tag{2}$$

Application of the Laplace transform to (1) and (2) leads to

$$(s^2J + k)\Theta_1(s) - k\Theta_2(s) = -T_1(s) \tag{3}$$
$$-k\Theta_1(s) + (s\beta + k)\Theta_2(s) = 0 \tag{4}$$

Simultaneous solution of (3) and (4) yields

$$\frac{\Theta_1(s)}{T_1(s)} = \frac{-(s\beta + k)}{s(s^2J\beta + sJk + k\beta)}$$
$$\frac{\Theta_2(s)}{T_1(s)} = \frac{-k}{s(s^2J\beta + sJk + k\beta)}$$

15.5 THERMAL AND FLUID-FLOW RELATIONSHIPS

Thermal systems tend to be nonlinear, but first-order approximations can be utilized to gain understanding of a system prior to cumbersome nonlinear analysis. Elementary analysis of heat flow can be made after introduction of two concepts, analogs of electrical resistance and capacitance, to be used along with the conservation of energy.

Consider the interface between a thermally conductive body and a constant-temperature reservoir. Fourier's law gives, to the first order,

$$Q = -hA \, \Delta T = -hA(T - T_s) \qquad (15.2)$$

where

$Q \equiv$ heat flow into the body, W

$h \equiv$ heat transfer coefficient, W/m$^2 \cdot$ K or W/m$^2 \cdot$ °C

$A \equiv$ area of interface, m^2

$T \equiv$ temperature of body, K or °C

$T_s \equiv$ temperature of reservoir, K or °C

(If T exceeds T_s, the heat flow is from the body to the reservoir.) Comparing (15.2) with Ohm's law,

$$i = \frac{-\Delta v}{R}$$

where Δv is the voltage rise across the resistor R, we define

$$thermal \; resistance \equiv R_t = \frac{1}{hA} \qquad (15.3)$$

Furthermore, the first law of thermodynamics requires that the heat flow into the body equal the rate of increase in its internal energy, assuming the body performs no work.

$$Q = \frac{d}{dt}(mcT) = mc \frac{dT}{dt} \qquad (15.4)$$

where

$m \equiv$ mass of the body, kg

$c \equiv$ specific heat of the body, J/kg $\cdot$ K or J/kg $\cdot$ °C

The analogy of (15.4) to the electrical equation for a capacitor,

$$i = C \frac{dv}{dt}$$

leads to the definition

$$thermal \; capacitance \equiv C_t = mc \qquad (15.5)$$

Example 15.14 Find the dynamic equation describing the temperature change of the room of Example 15.5, if the broken pane has been repaired and the only escape for heat to the outside is through the walls. Assume that the air in the room is of uniform temperature at any moment. The walls have thermal resistance R_{tw}, the room air has thermal capacitance C_{ta}, and the outside temperature is T_o (a constant).

Assume a constant output, Q_h (W), for the space heater. The room loses heat to the outside at the rate

$$Q_o = +\frac{1}{R_{tw}}(T - T_o) \equiv \frac{1}{R_{tw}} T_\delta$$

where T_δ is the room temperature with respect to the outside. The net heat influx to the room air is therefore $Q_h - Q_o$, and this must equal the rate of increase in the room air's internal energy:

$$Q_h - \frac{1}{R_{tw}} T_\delta = C_{ta} \frac{dT}{dt} = C_{ta} \frac{dT_\delta}{dt}$$

or

$$\frac{dT_\delta}{dt} + \frac{1}{R_{tw}C_{ta}} T_\delta = \frac{Q_h}{C_{ta}}$$

The steady-state solution of the above dynamic equation is obviously $T_\delta = R_{tw}Q_h$.

Linear, incompressible fluid-flow problems to be discussed are of two kinds: (1) liquid levels in containers and (2) fluid power systems utilizing valves. Both types must obey the conservation of energy and the conservation of mass, which latter takes the form

$$q = Av = \text{constant} \qquad (15.6)$$

where

$q \equiv$ volumetric flow rate, m^3/s

$A \equiv$ local cross-sectional area, m^2

$v \equiv$ local flow speed, m/s

It is convenient to define a *hydraulic resistance* (R_h) for modeling pressure changes (P) across valves or other constrictions:

$$R_h \equiv \frac{P}{q} \qquad (15.7)$$

Commonly, a constant pressure drop across a valve is assumed, leading to a flow rate through the valve that is proportional to its opening (x), or

$$q = k_v x \qquad (15.8)$$

Example 15.15 A pilot-valve-controlled hydraulic actuator is shown in Fig. 15-8. Mechanical feedback is present, so that $y(t)$ is controlled by $x(t)$. Find the transfer function $Y(s)/X(s)$.

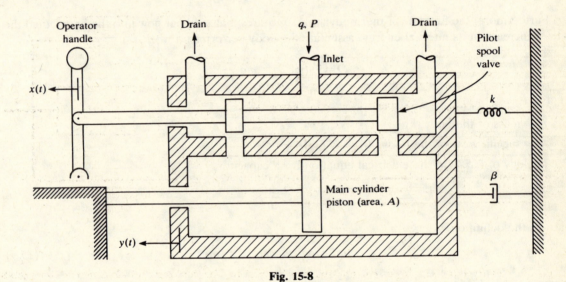

Fig. 15-8

Fluid flow through the pilot spool valve is determined by $x' \equiv x - y$, the displacement of the valve with respect to the actuator housing. In fact, if $x > y$, (15.8) gives for the flow into the left chamber of the main cylinder

$$q_{\text{left}} = k_v x' = k_v(x - y) \qquad (1)$$

To accommodate this influx the left chamber must expand at a rate $A\dot{y}$, where

$$A\dot{y} = k_v(x - y) \qquad \text{or} \qquad A\dot{y} + k_v y = k_v x \qquad (2)$$

The right chamber will contract at this same rate, and, by incompressibility, the flow out the right drain will be equal to q. It is easy to verify that (2) also holds for $x < y$.

Taking the Laplace transform of (2), one finds:

$$F(s) \equiv \frac{Y(s)}{X(s)} = \frac{k_v}{As + k_v}$$

This expression for the transfer function is only approximate, since (1) ceases to be valid when $|x'|$ becomes too large.

Solved Problems

15.1 A constant-armature-current, dc servomotor is shown in a positioning control system in Fig. 15-9(*a*). Draw a functional block diagram of the system.

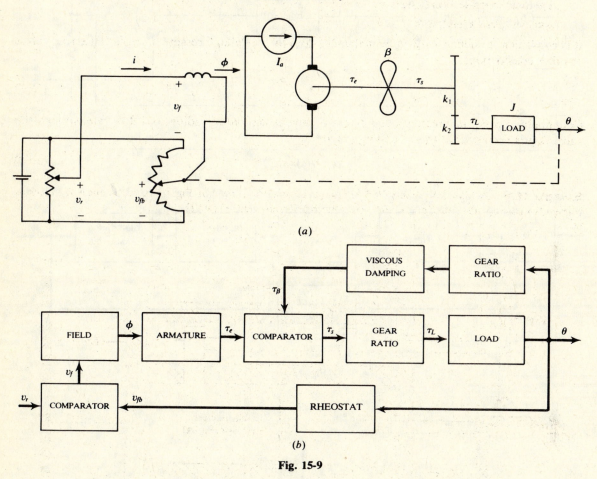

(*a*)

(*b*)

Fig. 15-9

An operator sets a desired position by changing v_r. The functional block diagram with signals labeled is shown in Fig. 15-9(*b*). Any motor armature inertia is lumped with load inertia.

15.2 The constant-speed engine of Fig. 15-10(*a*) drives both a dc generator and an exciter generator. Voltage feedback from the dc generator output is used to form a voltage regulator. Draw a functional block diagram of the system.

Voltage v_r is the reference command. The controlled variable is dc generator terminal voltage v_t. A functional block diagram is given in Fig. 15-10(*b*).

15.3 Draw a functional block diagram of the system of Example 15.2(*b*).

Reference command would be setting the thermostat to a desired room temperature. The thermostat also contains a bimetal strip utilized as a basic part of the comparator. The method of actuation of the furnace fuel valve is on/off control. A functional block diagram is depicted in Fig. 15-11.

15.4 Draw a functional block diagram of the photocell-controlled light of Example 15.1(*c*).

Reference command to this open-loop system is absence of outdoor light. The controlled variable is indoor or room light. A functional block diagram is presented in Fig. 15-12.

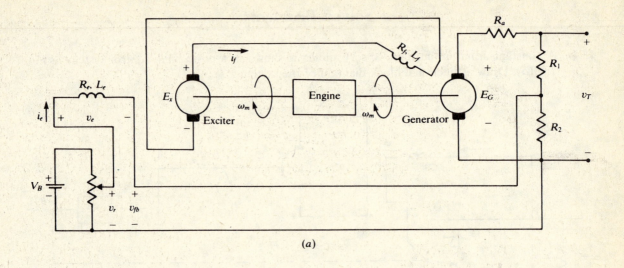

(a)

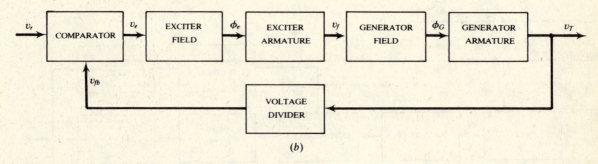

(b)

Fig. 15-10

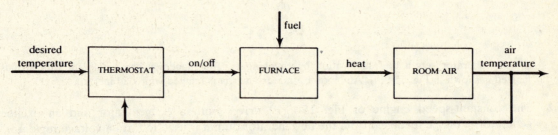

Fig. 15-11

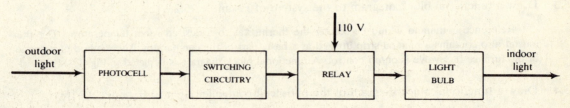

Fig. 15-12

15.5 Find the transfer function of a system if its output $(t \geq 0)$ due to an impulse $2\delta(t)$ is given by (a) $12e^{-3t}$, (b) $5e^{-7t} + 6e^{-2t}$, (c) $6t^2$, (d) $e^{-2t} \cos \omega t$.

Because the Laplace transform of the input is 2, the transfer function is one-half the Laplace transform of the output.

(a)
$$F(s) = \frac{1}{2} \mathscr{L}\{12e^{-3t}\} = \frac{1}{2}\left(\frac{12}{s+3}\right) = \frac{6}{s+3}$$

(b)
$$F(s) = \frac{1}{2} \mathscr{L}\{5e^{-7t} + 6e^{-2t}\} = \frac{5/2}{s+7} + \frac{3}{s+2} = \frac{(11/2)s + 26}{s^2 + 9s + 14}$$

(c)
$$F(s) = \frac{1}{2} \mathscr{L}\{6t^2\} = \frac{6}{s^3}$$

(d)
$$F(s) = \frac{1}{2} \mathscr{L}\{e^{-2t} \cos \omega t\} = \frac{1}{2}\left[\frac{s+7}{(s+7)^2 + \omega^2}\right] = \frac{s+7}{2(s^2 + 14s + 49 + \omega^2)}$$

15.6 The dc servomotor of Fig. 15-5 is field-controlled, with armature current (i_a) constant. Find the transfer function relating output speed (ω_m) to input field voltage (v_f). Assume magnetic linearity.

Apply KVL to the field circuit, to get

$$v_f(t) = R_f i_f(t) + L_f \frac{di_f(t)}{dt} \qquad (1)$$

Since armature current is constant and magnetic linearity prevails, developed motor torque is proportional to i_f, or $\tau_e(t) = K' i_f(t)$. By Newton's law,

$$J\frac{d\omega_m}{dt} + \beta\omega_m = \tau_e(t) = K' i_f(t) \qquad (2)$$

Take the Laplace transforms of (1) and (2), and simultaneously solve the two equations for the transfer function:

$$\frac{\Omega_m(s)}{V_f(s)} = \frac{K'}{(sJ + \beta)(sL_f + R_f)}$$

15.7 An automobile suspension system is modeled in Fig. 15-13, where the dashpot represents a shock absorber. If the road profile is considered the input, and the displacement of the car body is the output, find the transfer function.

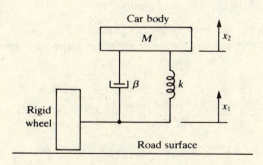

Fig. 15-13

The Lagrangian energy functions are

$$\mathscr{T} = \tfrac{1}{2}m\dot{x}_2^2 \qquad \mathscr{V} = \tfrac{1}{2}k(x_1 - x_2)^2 \qquad \mathscr{D} = \tfrac{1}{2}\beta(\dot{x}_1 - \dot{x}_2)^2$$

The Lagrangian equation of motion for node 2 is

$$M\ddot{x}_2 + \beta(\dot{x}_2 - \dot{x}_1) + k(x_2 - x_1) = 0$$

and the Laplace transform of this equation yields the transfer function as

$$\frac{X_2(s)}{X_1(s)} = \frac{s\beta + k}{s^2 M + s\beta + k}$$

15.8   For the torsional system of Fig. 15-14, find the transfer functions $\Theta_1(s)/T_2(s)$ and $\Theta_2(s)/T_2(s)$.

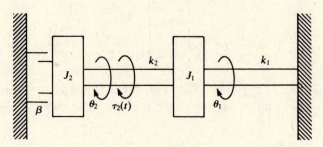

Fig. 15-14

The Lagrangian energy functions are

$$\mathcal{T} = \tfrac{1}{2}J_1\dot{\theta}_1^2 + \tfrac{1}{2}J_2\dot{\theta}_2^2 \qquad \mathcal{V} = \tfrac{1}{2}k_1\theta_1^2 + \tfrac{1}{2}k_2(\theta_2 - \theta_1)^2 \qquad \mathcal{D} = \tfrac{1}{2}\beta\dot{\theta}_2^2$$

Then, by (15.1),

$$J_1\ddot{\theta}_1 + (k_1 + k_2)\theta_1 - k_2\theta_2 = 0$$
$$J_2\ddot{\theta}_2 + \beta\dot{\theta}_2 + k_2\theta_2 - k_2\theta_1 = \tau_2(t)$$

Application of the Laplace transform to this system and simultaneous solution for $\Theta_1(s)$ and $\Theta_2(s)$ leads to

$$\frac{\Theta_1(s)}{T_2(s)} = \frac{k_2}{\Delta} \qquad \frac{\Theta_2(s)}{T_2(s)} = \frac{J_1 s^2 + k_1 + k_2}{\Delta}$$

where $\Delta \equiv J_1 J_2 s^4 + \beta J_1 s^3 + (J_2 k_2 + J_2 k_1 + J_1 k_2)s^2 + \beta(k_1 + k_2)s + k_1 k_2$.

15.9   Rework Example 15.14 by electrical analogy.

Under the analogy, heat corresponds to electric current and temperature differences to potential differences. The equivalent circuit for the room/heater system is given in Fig. 15-15; KCL yields

$$Q_h = C_{ta}\frac{dT_\delta}{dt} + \frac{T_\delta}{R_{tw}}$$

which is the dynamic equation found in Example 15.14.

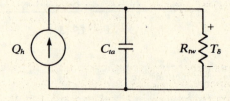

Fig. 15-15

15.10 The two cascaded fluid tanks of Fig. 15-16 form a system that receives an input fluid of weight density γ (N/m^3) at a volumetric flow rate $q_i(t)$. Considering $y_1(t)$ to be the output, find the transfer function $Y_1(s)/Q_i(s)$.

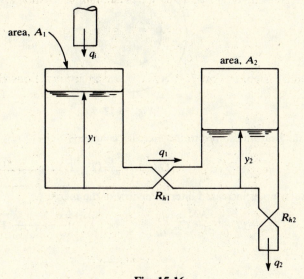

Fig. 15-16

The pressure drop (P_1) across hydraulic resistance R_{h1} is $\gamma(y_1 - y_2)$. Using (15.7),

$$q_1 = \frac{P_1}{R_{h1}} = \frac{\gamma(y_1 - y_2)}{R_{h1}} \qquad (1)$$

Further, incompressibility requires that

$$q_i - q_1 = A_1 \frac{dy_1}{dt} \qquad (2)$$

Substitute (1) into (2) and rearrange to give

$$A_1 \frac{dy_1}{dt} + \frac{\gamma}{R_{h1}}(y_1 - y_2) = q_i \qquad (3)$$

Since the second tank exhausts to atmospheric pressure, the pressure drop across R_{h2} is $P_2 = \gamma y_2$. Therefore,

$$A \frac{dy_2}{dt} = q_1 - q_2 = q_1 - \frac{\gamma y_2}{R_{h2}} \qquad (4)$$

Substitute (1) into (4) and get

$$A_2 \frac{dy_2}{dt} + \left(\frac{\gamma}{R_{h1}} + \frac{\gamma}{R_{h2}}\right)y_2 - \frac{\gamma}{R_{h1}}y_1 = 0 \qquad (5)$$

Now apply the Laplace transform to (3) and (5) and solve the two resulting simultaneous equations for $Y_1(s)$; this gives

$$F(s) \equiv \frac{Y_1(s)}{Q_i(s)} = \frac{sA_2 + \left(\dfrac{\gamma}{R_{h1}} + \dfrac{\gamma}{R_{h2}}\right)}{s^2 A_1 A_2 + s\left[\dfrac{\gamma}{R_{h1}}(A_1 + A_2) + \dfrac{\gamma}{R_{h2}} A_1\right] + \dfrac{\gamma^2}{R_{h1}R_{h2}}}$$

15.11 Figure 15-17 represents a mercury-filled thermometer. If the thermometer is introduced into a constant ambient temperature, T_a, find the differential equation relating the temperature (T) of the mercury column to the ambient temperature. Let R_{tg} be the thermal resistance of the glass wall and C_{tm} be the thermal capacitance of mercury.

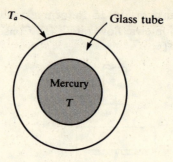

Fig. 15-17

By (*15.2*), the heat transferred to the mercury column is

$$Q_i(t) = -\frac{1}{R_{tg}}(T - T_a) \tag{1}$$

Here Q_i must show up as increased internal energy of the mercury:

$$-\frac{1}{R_{tg}}(T - T_a) = C_{tm}\frac{dT}{dt} \qquad \text{or} \qquad \frac{dT}{dt} + \frac{1}{R_{tg}C_{tm}}T = \frac{T_a}{R_{tg}C_{tm}}$$

Supplementary Problems

15.12 If the system response to a unit step function at $t = 0$ is $c(t) = 10 + 5\sin 2t$ $\quad$ ($t \geq 0$), find the transfer function of the system.

$\quad$ *Ans.* $\quad 10 + \dfrac{10s}{s^2 + 4}$

15.13 After interchanging R and C in Fig. 15-4, find the transfer function $V_o(s)/V_i(s)$.
$\quad$ *Ans.* $\quad sRC/(sRC + 1)$

15.14 Write the transfer function of the general parallel RLC circuit, if applied voltage is considered the input and total circuit current the output.

$\quad$ *Ans.* $\quad Y(s) = \dfrac{1}{R} + \dfrac{1}{sL} + sC$

15.15 Find the transfer function of the dc servomotor of Example 15.11 relating speed (ω_m) to terminal voltage. [*Hint:* $\Omega_m(s) = s\Theta(s)$.]

$\quad$ *Ans.* $\quad \dfrac{K/R_a}{Js + \beta + (K^2/R_a)}$

15.16 Find the transfer function for the dc servomotor of Example 15.11 if armature current (i_a) is the input and angular position of the shaft (θ) is the output. $\quad$ *Ans.* $\quad \Theta(s)/I_a(s) = K/[s(Js + \beta)]$

15.17 Show that the dc servomotor of Example 15.11 becomes an integrator as $R_a \to 0$.

$\quad$ *Ans.* $\quad \dfrac{\Theta(s)}{V_T(s)} \to \dfrac{1}{Ks}$

15.18 Show that the actuator of Example 15.15 becomes an integrator if feedback is removed.

$\quad$ *Ans.* $\quad$ If $x' \equiv x$, then $A\dot{y} = k_v x$, or $y(t) = (k_v/A)\displaystyle\int x(t)\, dt$.

15.19 The float-valve of Fig. 15-18 obeys $q_i = k_v y$ over the range of concern. (a) Find the differential equation describing the behavior of $y(t)$. (b) What is the condition for the tank to empty? (c) Under what condition does the fluid level in the tank remain unchanged?

Ans.   (a) $A\dfrac{dy}{dt} + \left(\dfrac{\gamma}{R_{ho}} - k_v\right)y = 0$; (b) $\dfrac{\gamma}{R_{ho}} > k_v$; (c) $\dfrac{\gamma}{R_{ho}} = k_v$

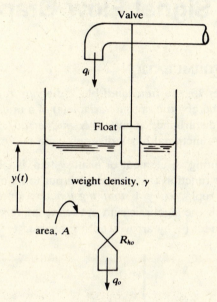

Fig. 15-18

15.20 The heat exchanger of Fig. 15-19 must remove Q_0 (J/m³) of thermal energy from a liquid with volumetric flow rate q_h. The inlet cooling water is at fixed temperature T_1. The cooling water thoroughly mixes upon entering the exchanger to a uniform temperature T_2. No heat escapes through the exchanger walls. The thermal capacity of the cooling water is C_{tw}, the specific heat of water is c_w, and the density of water is ρ_w (kg/m³). Find the equation governing the temperature (T_2) of the outlet cooling water.

Ans. $C_{tw}\dfrac{dT_2}{dt} + \rho_w q_c c_w T_2 = Q_0 q_h + \rho_w q_c c_w T_1$

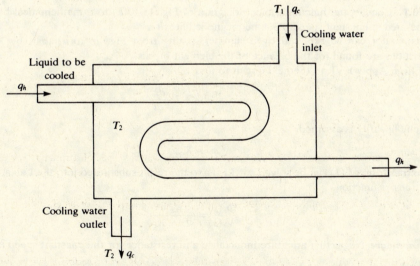

Fig. 15-19

Chapter 16

Block Diagrams and
Signal Flow Graphs

16.1 BLOCK DIAGRAM FORMULATION

As introduced in Chapter 15, the *functional* block diagram represents in black-box fashion a set of simultaneous equations whereby the output variable(s) of a linear control system is (are) related to the reference input(s) and disturbance input(s). A *mathematical block diagram* (or simply *block diagram*) is formed just as a functional block diagram, with two exceptions:

1. Each block, representing an element or a subsystem, is labeled not with a descriptive name, but with the transfer function that relates its input to its output.
2. Each comparator is replaced by a *summing junction*, or *summer*, as illustrated in Fig. 16-1. The outgoing signal is the algebraic sum of the incoming signals, where each incoming signal has been summed direct (+ by arrowhead) or inverted (− by arrowhead).

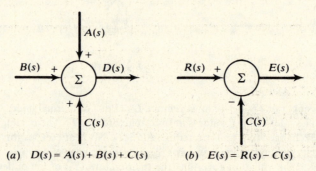

(a) $D(s) = A(s) + B(s) + C(s)$ (b) $E(s) = R(s) - C(s)$

Fig. 16-1

Example 16.1 Convert the functional block diagram of Fig. 15-10(*b*) to a mathematical block diagram. The engine is operated at constant speed. Assume magnetic linearity.

The comparator can be replaced by a summer, noting from Fig. 15-10(*a*) that, by KVL, $v_e = v_r - v_{fb}$. Transfer functions are found for the balance of the element blocks.

Exciter field. Apply KVL to the field circuit to find

$$v_e = R_e i_e + L_e \frac{di_e}{dt} \tag{1}$$

Since magnetic linearity is assumed,

$$\phi_e = K_{\phi e} i_e \tag{2}$$

The Laplace transform of (*1*) can be solved for $I_e(s)$ and the result substituted into the Laplace transform of (*2*) to give the transfer function.

$$\frac{\Phi_e(s)}{V_e(s)} = \frac{K_{\phi e}}{sL_e + R_e} \tag{3}$$

Exciter armature. Neglecting armature inductance and resistance for the constant-speed exciter,

$$v_f = K_e \phi_e \qquad \text{or} \qquad \frac{V_f(s)}{\Phi_e(s)} = K_e \tag{4}$$

256

Generator field. Proceeding as with the exciter field,

$$\frac{\Phi_G(s)}{V_f(s)} = \frac{K_{\phi G}}{sL_f + R_f} \tag{5}$$

Generator armature. So long as the generator is operated open-circuit, $E_G = v_T$, and the transfer function between ϕ_G and v_T is simply a constant:

$$\frac{V_T(s)}{\Phi_G(s)} = K_G \tag{6}$$

Voltage divider. If the series combination of the exciter field resistance and the reference-command potentiometer resistance is much larger than R_2, then voltage division is valid and the transfer function relating v_T to v_{fb} is

$$\frac{V_{fb}(s)}{V_T(s)} = \frac{R_2}{R_1 + R_2} \tag{7}$$

The above transfer functions and summer signal have been used to form the (mathematical) block diagram of Fig. 16-2.

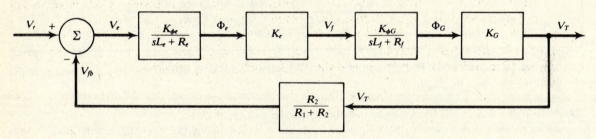

Fig. 16-2

16.2 BLOCK DIAGRAM ALGEBRA

For a control system of any complexity, the block diagram will contain many transfer functions in series and parallel arrangements. It is expedient to reduce the block diagram to more tractable form by applying the rules of *block diagram algebra*, as summarized in Table 16-1.

Table 16-1

Rule	Original System	Reduced System
1. Cascaded blocks	$R \rightarrow [G_1] \xrightarrow{A} [G_2] \rightarrow C$	$R \rightarrow [G_1 G_2] \rightarrow C$
2. Parallel paths	$R \rightarrow [G_1] \xrightarrow{+} \Sigma \rightarrow C$, $[G_2] \xrightarrow{\pm}$	$R \rightarrow [G_1 \pm G_2] \rightarrow C$
3. Moving a pickoff point	$R \rightarrow [G] \rightarrow C$, $B \leftarrow$	$R \rightarrow [G] \rightarrow C$, $B \leftarrow [1/G] \leftarrow$

Table 16-1   *(cont.)*

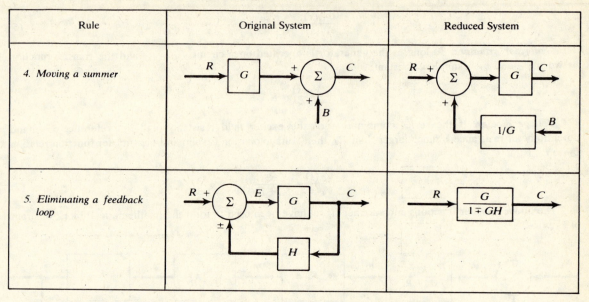

Rule	Original System	Reduced System
4. Moving a summer		
5. Eliminating a feedback loop		

The original system of Rule 5 exhibits the *canonical form* into which any control system with feedback can be transformed.

Example 16.2   Reduce the block diagram of Fig. 16-3(*a*) to a single block.

Move the pickup point for feedback signal $W(s)$ from the left to the right of block $C(s)$, using Rule 3. The result is shown in Fig. 16-3(*b*).

Combine cascaded (series-connected) blocks in the forward and feedback paths by Rule 1, giving the reduced block diagram of Fig. 16-3(*c*).

Finally, the simple negative feedback loop is eliminated by Rule 5, to yield the diagram of Fig. 16-3(*d*).

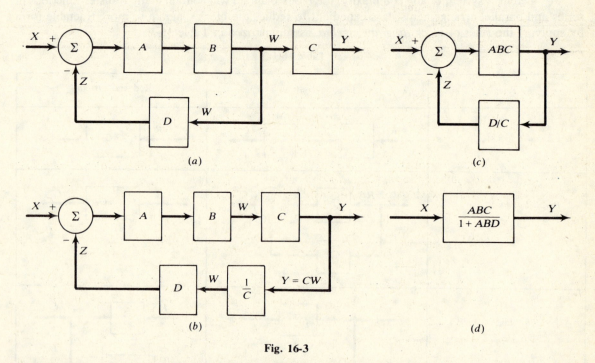

Fig. 16-3

Example 16.3   For the block diagram of Fig. 16-4(*a*), find the output, $C(s)$, due to the reference command, $R(s)$, and disturbance, $Q(s)$, jointly acting on the system.

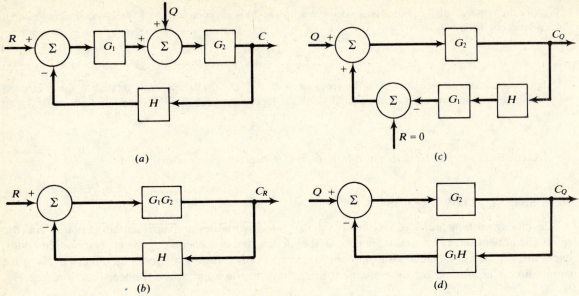

Fig. 16-4

Superposition is used, as in Examples 15.7 and 15.8. With $Q \equiv 0$, the intermediate summer is of no consequence and can be removed. Cascaded blocks G_1 and G_2 can be combined, reducing the system to the canonical form of Fig. 16-4(b). Then, by Rule 5, the component of the controlled variable due to R alone is

$$C_R = \frac{G_1 G_2 R}{1 + G_1 G_2 H}$$

If $R \equiv 0$, the block diagram can be redrawn as shown in Fig. 16-4(c). The only function of the summer to the left of G_1 is to provide negative feedback by changing the sign of the signal leaving G_1; this summer can be eliminated by proper sign adjustment as the signal enters the "input" summer, as illustrated in Fig. 16-4(d). Combining the cascaded blocks G_1 and H and using Rule 5, we find as the component of response due to Q alone

$$C_Q = \frac{G_2 Q}{1 + G_1 G_2 H}$$

The complete response, C, is the sum of C_R and C_Q:

$$C = C_R + C_Q = \frac{G_2(G_1 R + Q)}{1 + G_1 G_2 H}$$

Example 16.4 For the control system diagramed in Fig. 16-5(a), find the response of (a) the controlled variable C_1, and (b) the indirectly controlled variable C_2.

Again by superposition, each output may be treated as if the sole output.

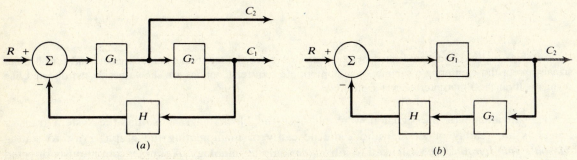

Fig. 16-5

(a) With the presence of C_2 ignored, cascaded blocks G_1 and G_2 are combined by Rule 1, and the response C_1
     is found by Rule 5 as

$$C_1 = \frac{G_1 G_2 R}{1 + G_1 G_2 H}$$

(b) Ignoring C_1, the block diagram can be redrawn as in Fig. 16-5(b), where C_2 is treated as the controlled
     variable. Combine cascaded blocks H and G_2 in the feedback path and apply Rule 5 to find

$$C_2 = \frac{G_1 R}{1 + G_1 G_2 H}$$

[Alternatively, C_2 may be determined from $G_2 \equiv C_1/C_2$ and the result of (a)].

16.3   SIGNAL FLOW GRAPHS

In the *signal flow graph* of a control system, system variables are represented as *nodes*, and the
relationship between two variables is indicated by a directed *branch* joining the corresponding nodes,
each branch being labeled with the appropriate transfer function. See Fig. 16-6(a). Negative
summation is accounted for by prefixing a minus sign to the appropriate transfer function; see Fig.
16-6(b).

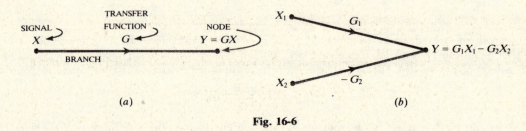

Fig. 16-6

With experience, signal flow graphs can be drawn directly from the system equations; but
construction by reference to the block diagram is safe and systematic.

Example 16.5   Construct the signal flow graph of the system whose block diagram is given in Fig. 16-7(a).

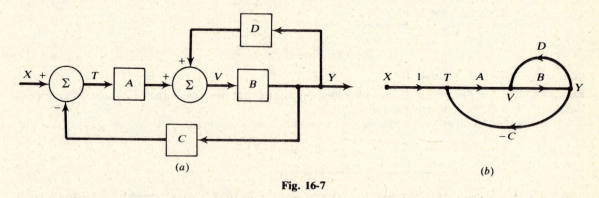

Fig. 16-7

See Fig. 16-7(b). Nodes (indicated by dots) are drawn for the input and output signals, and for all other
signals except those entering a summer. The appropriate transfer functions are shown beside the directed line
segments (branches) that interconnect signals.

The advantage of the signal flow graph over the block diagram is that the system transfer
functions are readily obtained without the necessity of manipulating the graph. Prior to stating
Mason's gain formula, we shall need to introduce some terminology. A simple, continuously directed
path leading from a node P of a signal flow graph to a node Q is called a *forward path* from P to Q.

If P and Q coincide, the forward path is termed a *loop*. The *gain* of a forward path (loop) is the product of the transfer functions over all branches comprising the forward path (loop). A set of loops (or arbitrary subgraphs) of a signal flow graph is *nontouching* if no two of the loops have a node in common.

$$\textbf{\textit{Mason's gain formula}} \qquad F(s) \equiv \frac{C(s)}{R(s)} = \frac{1}{\Delta} \sum_{k=1}^{n} M_k \Delta_k$$

where

$M_k \equiv$ gain of the kth forward path $(k = 1, 2, \ldots, n)$ from input node $R(s)$ to output node $C(s)$

$\Delta \equiv 1 -$ (sum of all loop gains) $+$ (sum of products of loop gains over all sets of *two* nontouching loops) $-$ (sum of products of loop gains over all sets of *three* nontouching loops) $+ \cdots$

$\Delta_k \equiv$ value of Δ if all loops that touch the kth forward path from R to C are excepted

The function $\Delta(s)$ is known as the *determinant* of the signal flow graph.

Example 16.6 Use Mason's gain formula to find the transfer function $Y(s)/X(s)$ for the signal flow graph of Fig. 16-7(b).

There is only one forward path from X to Y, with gain

$$M_1 = (1)(A)(B) = AB$$

There are just two loops in the graph, and they are touching. Thus,

$$\Delta = 1 - [(A)(B)(-C) + (B)(D)] = 1 + ABC - BD$$

Furthermore, both loops touch the single forward path, so that $\Delta_1 = 1$. Then, by Mason's gain formula,

$$F(s) = \frac{M_1}{\Delta} = \frac{AB}{1 + B(AC - D)}$$

Solved Problems

16.1 Verify Rule 1 of Table 16-1.

$$C = G_2 A = G_2(G_1 R) = (G_1 G_2)R$$

16.2 Verify Rule 5 of Table 16-1.

$$C = GE = G(R \pm HC)$$

Solve for C, obtaining

$$C = \left(\frac{G}{1 \mp GH}\right) R$$

16.3 The block diagram of Fig. 16-7(a) represents a control system. (a) Reduce the block diagram to a single loop with positive feedback. (b) Has the system been rendered unstable?

(a) The negative feedback path through C and the positive feedback path through D can be moved by Rule 4 of Table 16-1, to give the equivalent block diagram of Fig. 16-8(a). Next, the negative feedback loop comprised of B and AC in Fig. 16-8(a) can be reduced by Rule 5, yielding the single positive feedback loop of Fig. 16-8(b). Cascaded blocks A and $B/(1 + ABC)$ can be combined by Rule 1 if desired.

(b) Manipulation using block diagram algebra preserves the system transfer function: if the system is stable before manipulation, it will be stable afterwards.

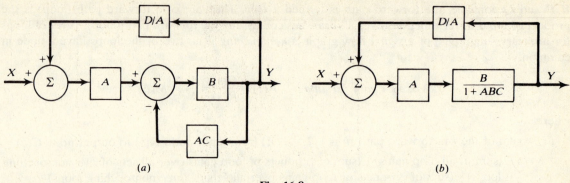

(a)   (b)

Fig. 16-8

16.4 A system is described by the differential equation $\ddot{y} + 5\dot{y} + 6y = 3x(t)$. Draw a block diagram of the system, if the initial conditions are: (a) $y(0) = \dot{y}(0) = 0$; (b) $y(0)$ and $\dot{y}(0)$ arbitrary.

The Laplace transform of the differential equation is

$$s^2 Y(s) - sy(0) - \dot{y}(0) + 5sY(s) - 5y(0) + 6Y(s) = 3X(s) \tag{1}$$

(a) For zero initial conditions, (1) may be solved to give

$$Y(s) \doteq \frac{3}{s^2 + 5s + 6} X(s)$$

Figure 16-9(a) is the corresponding block diagram.

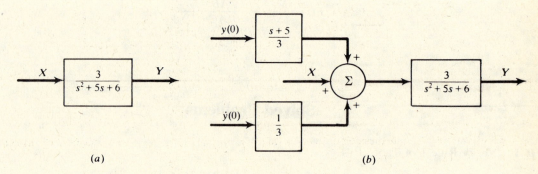

(a)   (b)

Fig. 16-9

(b) For nonzero initial conditions, (1) yields

$$Y(s) = \frac{3}{s^2 + 5s + 6} X(s) + \frac{s + 5}{s^2 + 5s + 6} y(0) + \frac{1}{s^2 + 5s + 6} \dot{y}(0)$$

An appropriate block diagram, Fig. 16-9(b), is constructed by adding two inputs to the diagram of Fig. 16-9(a).

16.5 Rework Problem 16.4 if only integrator blocks $(1/s)$ and constant gains are permitted in the block diagrams. (Such a simulation is called an *analog computer implementation.*)

Solve (1) of Problem 16.4 for the term of highest order in s, making its coefficient unity:

$$s^2 Y(s) = -5sY(s) - 6Y(s) + 3X(s) + (s + 5)y(0) + \dot{y}(0) \tag{1}$$

(a) $$s^2 Y(s) = -5sY(s) - 6Y(s) + 3X(s)$$

Observe that $sY(s)$ can be generated by one integration of $s^2 Y(s)$; also, $Y(s)$ follows from one integration of $sY(s)$. Two cascaded integrator blocks are thus suggested to form $Y(s)$ from $s^2 Y(s)$, the latter being formed from a summation involving two negative feedback terms. See Fig. 16-10(a).

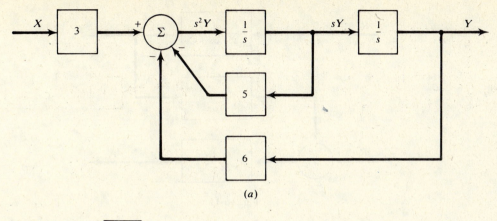

(a)

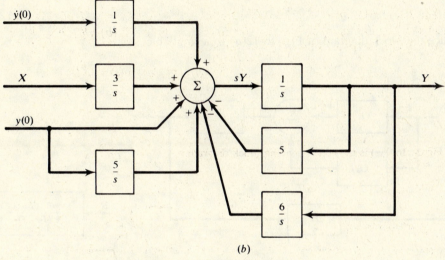

(b)

Fig. 16-10

(b)   In order to handle the coefficient of $y(0)$ in (1) without use of a differentiator, divide each term of (1) by s to give

$$sY(s) = -5Y(s) - \frac{6}{s}Y(s) + \frac{3}{s}X(s) + \left(1 + \frac{5}{s}\right)y(0) + \frac{1}{s}\dot{y}(0) \qquad (2)$$

$Y(s)$ results from a single integration of $sY(s)$, which is formed using the summation suggested by (2); Fig. 16-10(b) is an appropriate block diagram.

16.6   A system with two inputs (r_1, r_2) and two outputs (x_1, x_2) is described by the coupled differential equations

$$\ddot{x}_1 + 3\dot{x}_1 + 4x_1 = r_1(t)$$

$$\dot{x}_2 - \dot{x}_1 + 7x_2 = r_2(t)$$

Draw a block diagram for the system if all initial conditions are zero.

Note that x_2 depends on x_1, but not vice versa. Taking Laplace transforms and solving,

$$X_1 = \frac{1}{s^2 + 3s + 4}R_1 \qquad X_2 = \frac{1}{s+7}(R_2 + sX_1) \qquad (1)$$

whence the block diagram of Fig. 16-11.

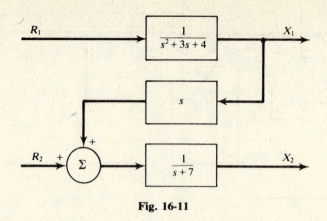

Fig. 16-11

16.7   Reduce the block diagram of Fig. 16-12(a) to a single block.

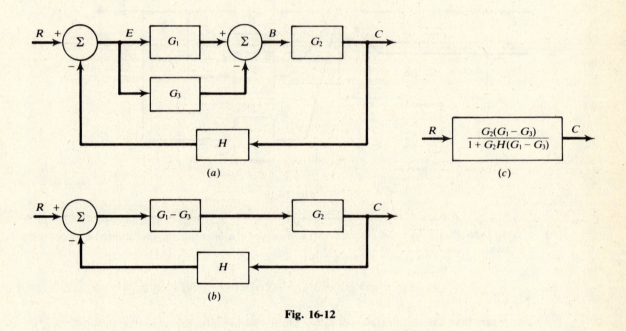

Fig. 16-12

The parallel branches, which give two forward paths from R to C, can be reduced by Rule 2 of Table 16-1, resulting in the block diagram of Fig. 16-12(b). Now the cascaded blocks $G_1 - G_3$ and G_2 are combined by Rule 1 and the resulting diagram is reduced by Rule 5, yielding the final single block of Fig. 16-12(c).

16.8   Reduce the block diagram of Fig. 16-13(a) to a single block.

The pickoff point and the summer of the unity feedback loop can be moved using Rules 3 and 4 of Table 16-1, to give the block diagram of Fig. 16-13(b).

The feedback loops through H_1 and H_2 can be removed by Rule 5, yielding the block diagram of Fig. 16-13(c).

Finally, the cascaded blocks of Fig. 16-13(c) are combined by Rule 1 and the resulting canonical feedback system reduced by Rule 5, to give the single block of Fig. 16-13(d).

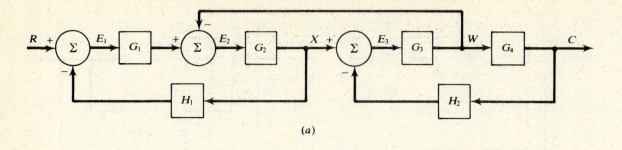

(a)

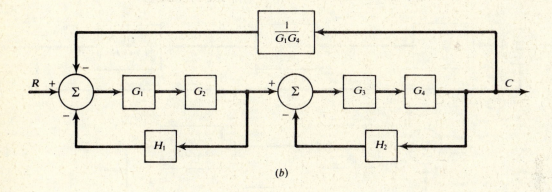

(b)

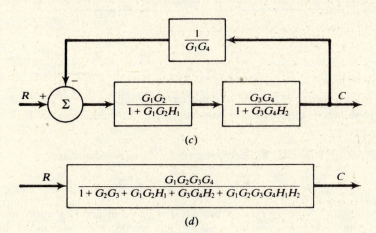

(c)

$$\begin{array}{c} R \longrightarrow \boxed{\dfrac{G_1G_2G_3G_4}{1+G_2G_3+G_1G_2H_1+G_3G_4H_2+G_1G_2G_3G_4H_1H_2}} \longrightarrow C \end{array}$$

(d)

Fig. 16-13

16.9 The system of Fig. 16-14(a) has a disturbance input modeled by $Q(s)$. Find the transfer functions (a) $C(s)/R(s)$ if $Q(s) \equiv 0$, and (b) $C(s)/Q(s)$ if $R(s) \equiv 0$.

(a) The inner feedback loop can be reduced to a single block by Rule 5 of Table 16-1. With $Q(s) \equiv 0$, the block diagram can be redrawn as in Fig. 16-14(b). If the cascaded blocks in the forward path are combined by Rule 1, then Rule 5 can be used to give the transfer function:

$$\frac{C}{R} = \frac{G_1G_2G_3}{1+G_3H+G_1G_2G_3}$$

(b) With $R(s) \equiv 0$, the block diagram is arranged as in Fig. 16-14(c). Application of Rule 1 and Rule 5 leads to the transfer function

$$\frac{C}{Q} = \frac{G_2G_3}{1+G_3H+G_1G_2G_3}$$

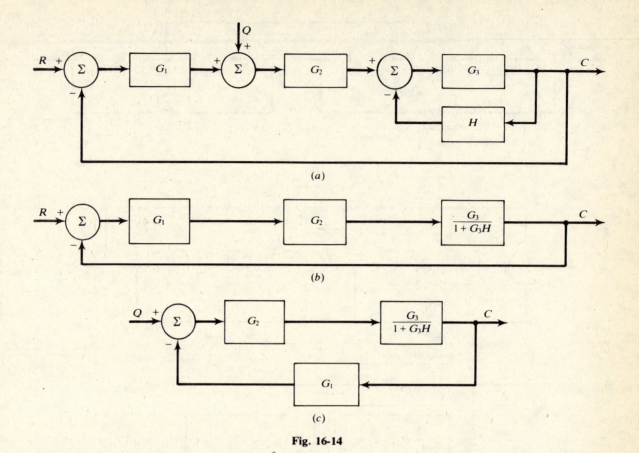

Fig. 16-14

16.10 For the *RC* filter network of Fig. 15-4, consider terminals *ab* open-circuit and directly draw a signal flow graph, where current $i(t)$ is considered an intermediate variable. The input is v_i and the output is v_o.

The input voltage $V_i(s)$ must be operated upon by the input admittance

$$Y_i(s) = \frac{1}{R + (1/sC)} = \frac{sC}{sRC + 1}$$

to give the manipulated variable $I(s)$. In turn, $I(s)$ is operated upon by impedance $Z_L(s) = 1/sC$ to produce $V_o(s)$. The signal flow graph suggested by these two operations is depicted in Fig. 16-15.

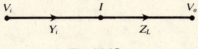

Fig. 16-15

16.11 Draw a signal flow graph for the dc servomotor of Fig. 15-5 without use of a block diagram. The motor is operated with constant field current. Voltage $v_T(t)$ is the input and speed $\omega_m(t)$ is the output.

Application of KVL and solution for armature current in the *s*-domain gives

$$I_a(s) = \frac{1}{R_a}[V_T(s) - E(s)] \tag{1}$$

which suggests negative feedback (inherent feedback) of the motor counter-emf (E) for comparison with input (V_T), and then operating on their difference by $1/R_a$ to yield I_a.

With constant field current, motor torque is proportional to armature current, and so

$$T_e(s) = KI_a(s) \tag{2}$$

The differential equation of rotation of the motor (see Example 15.11) has the Laplace transform

$$sJ\Omega_m(s) + \beta\Omega_m(s) = T_e(s) \tag{3}$$

Finally, output speed is operated upon by a feedback gain to produce counter-emf E:

$$E(s) = K\Omega_m(s) \tag{4}$$

where K is the same constant as in (2).

A signal flow graph that preserves (1) through (4) is displayed in Fig. 16-16.

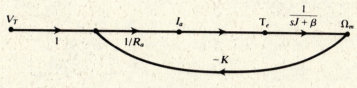

Fig. 16-16

16.12 The system represented by the block diagram of Fig. 16-12(a) is to be analyzed by signal flow graph theory. (a) Draw the signal flow graph. (b) Apply Mason's gain formula to the signal flow graph of (a) to find the transfer function $C(s)/R(s)$.

(a)   See Fig. 16-17.

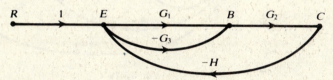

Fig. 16-17

(b)   There are two forward paths from R to C, having gains $M_1 = G_1G_2$ and $M_2 = -G_2G_3$. The graph has two loops, with gains $-G_1G_2H$ and G_2G_3H. As these loops touch,

$$\Delta = 1 - (-G_1G_2H + G_2G_3H) = 1 + G_1G_2H - G_2G_3H$$

Both loops touch both forward paths; hence, $\Delta_1 = \Delta_2 = 1$. Application of Mason's gain formula yields

$$\frac{C}{R} = \frac{M_1\Delta_1}{\Delta} + \frac{M_2\Delta_2}{\Delta} = \frac{G_2(G_1 - G_3)}{1 + G_2H(G_1 - G_3)}$$

16.13 Analyze the system represented by Fig. 16-13(a) by signal flow graph theory. (a) Draw the signal flow graph from the given block diagram. (b) Find $C(s)/R(s)$ by Mason's gain formula.

(a)   The signal flow graph is drawn in Fig. 16-18.

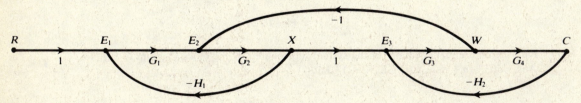

Fig. 16-18

(b)   There is one forward path from R to C, with gain $M_1 = G_1G_2G_3G_4$. The three loops have gains $-G_1G_2H_1$, $-G_3G_4H_2$, and $-G_2G_3$; the first two of these loops compose a nontouching set. The graph determinant is therefore

$$\Delta = 1 - (-G_1G_2H_1 - G_3G_4H_2 - G_2G_3) + [(-G_1G_2H_1)(-G_3G_4H_2)]$$
$$= 1 + G_1G_2H_1 + G_3G_4H_2 + G_2G_3 + G_1G_2G_3G_4H_1H_2$$

As all loops touch the forward path, $\Delta_1 = 1$, and Mason's gain formula gives

$$\frac{C}{R} = \frac{M_1 \Delta_1}{\Delta} = \frac{G_1 G_2 G_3 G_4}{1 + G_2 G_3 + G_1 G_2 H_1 + G_3 G_4 H_2 + G_1 G_2 G_3 G_4 H_1 H_2}$$

16.14 A system is described by the differential equation $a_3 \dddot{x} + a_2 \ddot{x} + a_1 \dot{x} + a_0 x = r(t)$. Decompose the differential equation into a set of first-order differential equations and draw the signal flow graph directly from this set of equations.

Let $x_1 \equiv x$, $x_2 \equiv \dot{x}$, and $x_3 \equiv \ddot{x}$; then

$$\dot{x}_1 = x_2 \tag{1}$$
$$\dot{x}_2 = x_3 \tag{2}$$
$$\dot{x}_3 = -\frac{a_2}{a_3} x_3 - \frac{a_1}{a_3} x_2 - \frac{a_0}{a_3} x_1 + \frac{1}{a_3} r(t) \tag{3}$$

Inspection of (1) shows that x_1 is formed by integration of x_2; likewise, (2) indicates that x_2 is given by integration of x_3. Equation (3) suggests a linear combination of x_1, x_2, x_3 and $r(t)$. The signal flow graph of Fig. 16-19 preserves (1), (2), and (3).

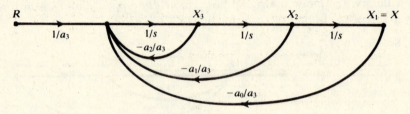

Fig. 16-19

Supplementary Problems

16.15 Verify Rules 2, 3, and 4 of Table 16-1.

16.16 Draw a signal flow graph for the system diagramed in Fig. 16-10(a), and use it to verify the transfer function found in Problem 16.4(a).   *Ans.*   See Fig. 16-20.

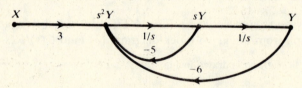

Fig. 16-20

16.17 Find the differential equation for the system represented by the block diagram of Fig. 16-21.
Ans. $\dot{c}(t) + (a + K_1 K_2) c(t) = K_1 K_2 r(t) + K_2 q(t)$

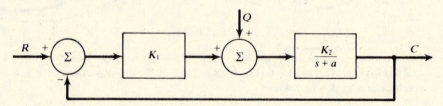

Fig. 16-21

16.18   Using block diagram algebra, find the transfer function $C(s)/R(s)$ for the system represented by Fig. 16-22.   *Ans.* $G_1(G_2 + G_3)/(1 + G_1H)$

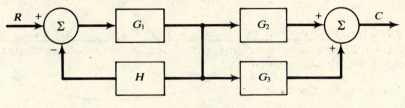

Fig. 16-22

16.19   For the system with block diagram Fig. 16-23, find the transfer function $C(s)/R(s)$.   *Ans.* $G_1(1 + G_2)/(1 + G_1)$

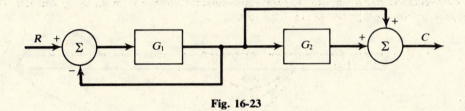

Fig. 16-23

16.20   Find the transfer function $C(s)/R(s)$ for the block diagram of Fig. 16-24.   *Ans.* $G_1G_2(A + BD)$

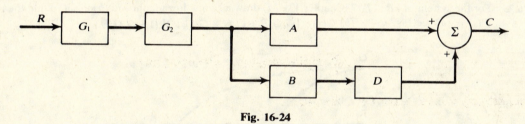

Fig. 16-24

16.21   From (*3*) and (*4*) of Example 15.13, draw the signal flow graph for the rotational system of Fig. 15-7.   *Ans.*   See Fig. 16-25.

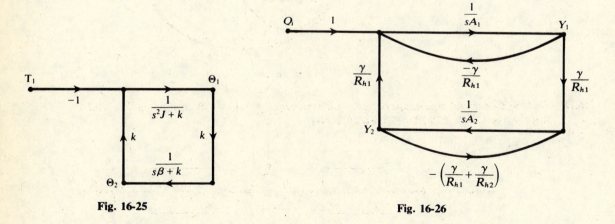

Fig. 16-25

Fig. 16-26

16.22   The two cascaded fluid tanks of Fig. 15-16 are described by (*3*) and (*5*) of Problem 15.10. Use these differential equations to draw a signal flow graph of the system.   *Ans.*   See Fig. 16-26.

16.23 For the signal flow graph of Fig. 16-27, find the transfer function $C(s)/R(s)$ by Mason's gain formula.
Ans. $G_1(G_2 + G_3)/(1 + G_1H_1 + G_1G_2H_2 + G_1G_3H_2)$

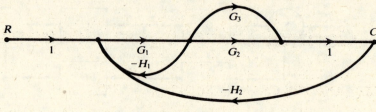

Fig. 16-27

16.24 In Fig. 8-7(a), connect a load resistor R_L from C to E. (a) Draw a signal flow graph relating input v_{be} to output v_{ce}, and (b) use it to find the voltage gain $V_{ce}(s)/V_{be}(s)$.
Ans. (a) see Fig. 16-28; (b) $-h_{fe}R_L/(h_{ie} + h_{ie}h_{oe}R_L - h_{fe}h_{re}R_L)$

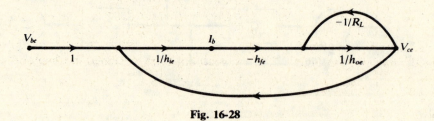

Fig. 16-28

16.25 For the system of Fig. 17-17 (Chapter 17), (a) draw a signal flow graph, and (b) use it to find the response $C(s)$. (a) see Fig. 16-29; (b) $C(s) = [G_cG_pR + G_p(1 - G_dG_c)Q]/(1 + G_cG_pH)$

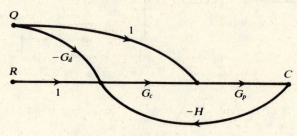

Fig. 16-29

Control Criteria and Response

17.1 STABILITY AND THE CHARACTERISTIC EQUATION

A linear system is *stable* if and only if *every bounded input results in a bounded output.* While this criterion is conceptually simple, the test it implies cannot be carried out: a more practical criterion is needed.

Our discussion will be limited to systems describable by a linear differential equation with constant real coefficients in a single output variable (controlled variable) $c(t)$ and a single input variable (reference command) $r(t)$:

$$\sum_{k=0}^{n} a_{n-k} \frac{d^{n-k}c(t)}{dt^{n-k}} = \sum_{k=0}^{m} b_{m-k} \frac{d^{m-k}r(t)}{dt^{m-k}} \tag{17.1}$$

with $a_n > 0$. In many cases the right-hand side of (17.1) will reduce to $r(t)$, and $r(t)$ will often be one of the functions graphed in Fig. 17-1. These functions were examined in Chapter 6.

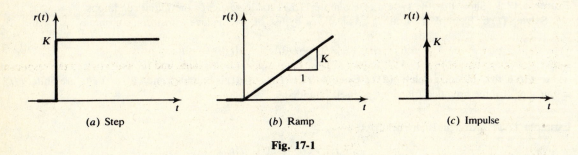

(*a*) Step   (*b*) Ramp   (*c*) Impulse

Fig. 17-1

Under zero initial conditions, the Laplace transform of (17.1) gives the transfer function of the system as

$$F(s) \equiv \frac{C(s)}{R(s)} = \frac{N(s)}{D(s)} \tag{17.2}$$

in which

$$N(s) \equiv \sum_{k=0}^{m} b_{m-k}s^{m-k} \qquad \text{and} \qquad D(s) \equiv \sum_{k=0}^{n} a_{n-k}s^{n-k}$$

are polynomials in s; $F(s)$ is thus a rational function of s. $D(s)$ is called the *characteristic polynomial* of the system; it may be rewritten in factored form as

$$D(s) \equiv a_n \prod_{i=1}^{n} (s - s_i) \tag{17.3}$$

where the s_i are the (not necessarily distinct) roots of the *characteristic equation* $D(s) = 0$. As the following examples will show, it is the characteristic roots that essentially determine the stability or instability of the system.

Example 17.1 A system is characterized by

$$\ddot{c}(t) + 4\dot{c}(t) + 3c(t) = r(t) \tag{1}$$

for $t \geq 0$. If initial conditions are zero, show that the system is stable.

Here, $D(s) = (s + 1)(s + 3)$, and (17.2) gives

$$F(s) = \frac{1}{(s + 1)(s + 3)} = \frac{0.5}{s + 1} - \frac{0.5}{s + 3}$$

whence

$$f(t) = \mathcal{L}^{-1}\{F(s)\} = 0.5\, e^{-t} - 0.5\, e^{-3t}$$

The convolution theorem (Table 6-1, line 14) now gives:

$$c(t) = \int_0^t r(\tau)f(t - \tau)\, d\tau = 0.5 \int_0^t r(\tau)e^{-(t-\tau)}\, d\tau - 0.5 \int_0^t r(\tau)e^{-3(t-\tau)}\, d\tau \qquad (2)$$

If the input $r(t)$ is bounded—say, $|r(t)| < M$ for $t > 0$—then we have from (2):

$$|c(t)| \leq 0.5 \int_0^t M e^{-(t-\tau)}\, d\tau + 0.5 \int_0^t M e^{-3(t-\tau)}\, d\tau$$

$$= 0.5 M(1 - e^{-t}) + 0.5 M \left(\frac{1 - e^{-3t}}{3} \right)$$

$$< 0.5 M + 0.5 M \left(\frac{1}{3} \right) = \frac{2}{3} M$$

i.e., $c(t)$ is also bounded.

Essential to the proof was the fact that the characteristic roots, $s = -1$ and $s = -3$, were *real and negative*. This made $f(t)$ a sum of *decreasing* exponentials, which in turn produced an exponentially decreasing $c(t)$. The argument would go through unchanged for complex characteristic roots with *negative real parts*.

Example 17.2   Show that the system of Example 17.1 is stable whatever the initial conditions.

Solving (1) of Example 17.1 either classically or by use of the Laplace transform, we obtain

$$c(t) = c_f(t) + \frac{1}{2}[3c(0) + \dot{c}(0)]e^{-t} - \frac{1}{2}[c(0) + \dot{c}(0)]e^{-3t}$$

where $c_f(t)$ is the function given by (2) of Example 17.1. Thus, if $r(t)$ is bounded and $c(0)$ and $\dot{c}(0)$ are finite, $c(t)$, as the sum of three bounded functions, is itself bounded.

Example 17.3   A system is characterized by

$$\ddot{c}(t) + 4\dot{c}(t) - 5c(t) = r(t)$$

for $t \geq 0$. Show that the system is unstable.

One may as well assume that $c(0) = \dot{c}(0) = 0$; for instability under zero initial conditions implies instability under any initial conditions. The characteristic roots are now $s = +1$ and $s = -5$. Proceeding as in Example 17.1, one finds

$$c(t) = \frac{1}{6} e^t \int_0^t r(\tau)e^{-\tau}\, d\tau - \frac{1}{6} e^{-5t} \int_0^t r(\tau)e^{5\tau}\, d\tau$$

In particular, for the bounded input $r(t) \equiv 1$,

$$c(t) = \frac{1}{6}(e^t - 1) - \frac{1}{30}(1 - e^{-5t})$$

an unbounded function because of the term in e^t.

The system was unstable because one of the characteristic roots was *real and positive*; it also would have been so had a characteristic root been complex with *positive real part*.

The examples just given, plus an examination of the case of purely imaginary or zero characteristic roots, render plausible the following categorization:

1. A system is *absolutely stable* (or just *stable*) if all characteristic roots lie in the left-half s-plane (LHSP).

2. A system is *unstable* if any characteristic root lies in the right-half s-plane (RHSP) or if repeated characteristic roots lie on the imaginary axis (in particular, at the origin).

3. A system is *marginally stable*, or *limitedly stable*, if no characteristic root lies in the RHSP, but nonrepeated characteristic roots lie on the imaginary axis (in particular, at the origin).

Marginally stable systems are not stable, although they do have bounded responses to "many" bounded inputs. Because such systems are not important practically, we shall henceforth consider a system either as *stable* (Case 1) or *unstable* (Cases 2 and 3).

17.2 ROUTH-HURWITZ STABILITY CRITERION

The Routh-Hurwitz procedure determines the stability or unstability of a system from its unfactored characteristic equation,

$$D(s) \equiv a_n s^n + a_{n-1} s^{n-1} + \cdots + a_1 s + a_0 = 0 \qquad (17.4)$$

where $a_n > 0$. A *necessary, but not sufficient,* condition for absolute stability is that all coefficients in (17.4) be positive. If any coefficient a_{n-1} through a_0 is negative or zero, there is at least one characteristic root in the right-half s-plane (RHSP) or on the imaginary axis; the system is unstable. Whether or not the necessary condition is met, a *Routhian array* may be constructed as follows:

$$
\begin{array}{c|ccccc}
s^n   & a_n   & a_{n-2} & a_{n-4} & \ldots & 0 \\
s^{n-1} & a_{n-1} & a_{n-3} & a_{n-5} & \ldots & 0 \\
s^{n-2} & b_1   & b_2   & b_3   & \ldots & 0 \\
s^{n-3} & c_1   & c_2   & \ldots & 0   &   \\
\cdots & \multicolumn{5}{c}{\cdots\cdots\cdots\cdots\cdots\cdots\cdots} \\
s^1   & i_1   & 0   &   &   &   \\
s^0   & j_1   & 0   &   &   &
\end{array}
$$

The third row of the array is formed with determinants:

$$b_1 = \frac{-\begin{vmatrix} a_n & a_{n-2} \\ a_{n-1} & a_{n-3} \end{vmatrix}}{a_{n-1}} \qquad b_2 = \frac{-\begin{vmatrix} a_n & a_{n-4} \\ a_{n-1} & a_{n-5} \end{vmatrix}}{a_{n-1}} \qquad \cdots$$

until b_m becomes and remains zero. Likewise, the fourth row is formed as

$$c_1 = \frac{-\begin{vmatrix} a_{n-1} & a_{n-3} \\ b_1 & b_2 \end{vmatrix}}{b_1} \qquad c_2 = \frac{-\begin{vmatrix} a_{n-1} & a_{n-5} \\ b_1 & b_3 \end{vmatrix}}{b_1} \qquad \cdots$$

until c_m becomes and remains zero. The pattern is continued until an array of $n+1$ rows is constructed.

Stability Theorem: The number of characteristic roots (counted by multiplicity) in the RHSP is equal to the number of sign changes in the first column of the Routhian array.

Example 17.4 If a system has as its characteristic equation

$$D(s) = s^5 + s^4 + 3s^3 + 9s^2 + 16s + 10 = 0$$

find the number of RHSP roots.

The final Routhian array is

$$
\begin{array}{c|cccl}
s^5 & 1 & 3 & 16 & 0 \\
s^4 & 1 & 9 & 10 & 0 \\
s^3 & -1 & 1 & 0 & \textbf{\textit{b}-row} \\
s^2 & 1 & 1 & 0 & \textbf{\textit{c}-row} \\
s^1 & 2 & 0 &   & \textbf{\textit{d}-row} \\
s^0 & 1 & 0 &   & \textbf{\textit{e}-row}
\end{array}
$$

where, to simplify the arithmetic (with no effect on the final result), b-row entries were multiplied by 1/6 and c-row entries by 1/10 prior to entering the array. In the first column of the array, the sign changes to negative for the third row and back to positive for the fourth row. Therefore, the characteristic equation has two simple roots or one double root in the RHSP, and the system is unstable.

Two exceptions or special cases may arise when constructing the Routhian array: (I) occurrence of a zero in the first column of a row, followed by at least one nonzero entry; (II) occurrence of an entire row of zeros (this includes the occurrence of a zero in the first column of either of the last two rows).

Special Case I. Upon occurrence of a first-column zero followed by nonzero entries, immediately replace the zero by ϵ. Complete the array and then examine the first column for sign changes as $\epsilon \to 0$ through positive values.

Special Case II. A zero row arises when the characteristic equation has a pair of roots whose sum is zero (the two roots are symmetric about the s-plane origin); it is handled as follows:

(i)   From the last nonvanishing row of the Routhian array form an auxiliary polynomial in s using the coefficients of that row and alternate powers of s, beginning with the power of s that labels the nonvanishing row.

(ii)   Differentiate the auxiliary polynomial with respect to s and substitute the coefficients of the derived polynomial into the vanishing row.

(iii)   Complete the array and apply the Stability Theorem to determine the number of RHSP roots.

It is to be remarked that the symmetrical pairs of characteristic roots are the roots of the auxiliary polynomial equations generated in the applications of step (i).

The Routh-Hurwitz criterion may now be stated as follows: *A system is stable if and only if its Routhian array shows no sign changes in the first column and there occurred no zero row in the course of the computation.*

17.3   RESPONSE OF FIRST-ORDER SYSTEMS

The *order* of a system is defined as the degree of its characteristic polynomial. If the transfer function has the form $N(s)/D(s)$, with $D(s)$ a polynomial of degree n and $N(s)$ a polynomial of degree zero, the system is called a *simple nth-order system*.

The typical simple first-order system is given in canonical form by Fig. 17-2(a), which can be reduced to the form of Fig. 17-2(b). The transfer function is

$$F(s) \equiv \frac{C(s)}{R(s)} = \frac{N(s)}{D(s)} \equiv \frac{1}{\tau s + 1} \tag{17.5}$$

It is seen that the system is stable if and only if $\tau > 0$.

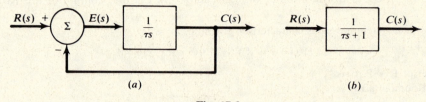

(a)   (b)

Fig. 17-2

Example 17.5 If a step command, Fig. 17-1(a), is applied to a stable, simple first-order system, (a) find the response for $t \ge 0$, identifying the transient and steady-state components; (b) find the *settling time*.

(a)   By (*17.5*),

$$C(s) = \frac{1}{\tau s + 1} R(s) = \frac{1}{\tau s + 1} \frac{K}{s}$$

or, inverting to the time domain,

$$c(t) = K - Ke^{-t/\tau} \qquad (t \ge 0) \tag{1}$$

The first term of (1) is the steady-state response, the second term is the transient. The smaller the value of τ (the *time constant*), the faster the system response along the exponential curve of Fig. 17-3. In the steady state, the value of $c(t)$ approaches that of $r(t)$, or the system tracks without error.

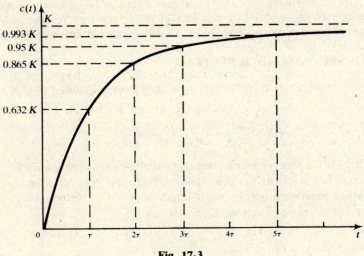

Fig. 17-3

(b)　Settling time (response maintained to within 5% of final value) is illustrated in Fig. 17-3, where $t_s \approx 3\tau$.

Example 17.6　A ramp command, Fig. 17-1(b), is applied to a stable, simple first-order system. (a) Find an expression for the system response. (b) Identify and discuss the system error.

(a)
$$C(s) = \frac{1}{\tau s + 1}\, R(s) = \frac{1}{\tau s + 1}\frac{K}{s^2} = \frac{K}{s^2} - \frac{K\tau}{s} + \frac{K\tau}{s + (1/\tau)}$$

and inverse Laplace transformation yields

$$c(t) = K(t - \tau) + K\tau e^{-t/\tau} \qquad (t \geq 0)$$

Note that $c(t)$ is unbounded, even though the system is stable. The reason is that $r(t)$ is unbounded.

(b)　A sketch of $r(t)$ and the resulting response $c(t)$ is displayed in Fig. 17-4. The error,

$$e(t) \equiv r(t) - c(t)$$

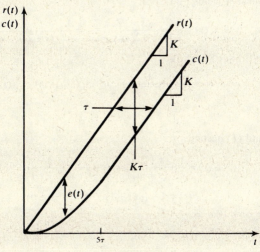

Fig. 17-4

is the vertical separation of the two curves. The error increases until $t \approx 5\tau$, at which time it has essentially attained the constant value $K\tau$. Under steady-state conditions, from the time that $r(t)$ reaches a particular value until $c(t)$ reaches that same value there is an elapsed interval of τ (s), called the *steady-state lag*.

It is important to note the generalized sense of the word "steady-state" in the above applications. The forced solution, $K(t - \tau)$, and along with it $c(t)$, becomes infinite as $t \to \infty$; strictly speaking, no steady state exists in this case. In the balance of this chapter we shall continue to use "steady-state" in reference to the nondecaying component of a response, whether or not that component has a finite limit as $t \to \infty$.

17.4 RESPONSE OF SECOND-ORDER SYSTEMS

The transfer function of physically realizable, simple second-order systems can be written in parametric form as

$$F(s) \equiv \frac{C(s)}{R(s)} = \frac{N(s)}{D(s)} \equiv \frac{\omega_n^2}{s^2 + 2\zeta\omega_n s + \omega_n^2} \tag{17.6}$$

where ζ (dimensionless) is the *damping ratio* and ω_n (rad/s) is the *natural frequency*. Both parameters are real numbers, and ω_n is positive. The value ω_n^2 chosen for $N(s)$ is simply a convenient normalization; any other constant might be substituted.

The roots of the characteristic equation, $D(s) = 0$, are

$$s_{1,2} = -\zeta\omega_n \pm \omega_n\sqrt{\zeta^2 - 1} \tag{17.7}$$

Case 1.   If $\zeta > 1$, the roots are real, negative, and unequal.

Case 2.   If $\zeta = 1$, the roots are real, negative, and equal.

Case 3.   If $0 < \zeta < 1$, the roots are complex conjugates having negative real parts.

Case 4.   If $\zeta = 0$, the roots are pure imaginaries (complex conjugates).

Case 5.   If $\zeta < 0$, the roots are complex conjugates having positive real parts.

Cases 1, 2, and 3 (i.e. $\zeta > 0$) correspond to stable systems. For further study of case 3 it is usual to rewrite (17.7) as

$$s_{1,2} = -\zeta\omega_n \pm j\omega_n\sqrt{1 - \zeta^2} \equiv -\zeta\omega_n \pm j\omega_d \tag{17.8}$$

where $\omega_d \equiv \omega_n\sqrt{1 - \zeta^2}$ is called the *damped (natural) frequency*.

Example 17.7   A unit step command is applied to a simple second-order system. Find the response for (a) $0 < \zeta < 1$, (b) $\zeta = 1$, and (c) $\zeta > 1$.

(a)   Solving (17.6) for $C(s)$, applying a unit step input [Fig. 17-1(a), with $K = 1$], for which $R(s) = 1/s$, and using a partial-fractions expansion yields

$$C(s) = \frac{1}{s} - \frac{s + 2\zeta\omega_n}{s^2 + 2\zeta\omega_n s + \omega_n^2} = \frac{1}{s} - \frac{s + 2\zeta\omega_n}{(s + \zeta\omega_n)^2 + \omega_d^2} \tag{1}$$

Inverting (see Table 6-1, lines 8 and 9),

$$c(t) = 1 - \frac{1}{\sqrt{1 - \zeta^2}} e^{-\zeta\omega_n t} \sin(\omega_d t + \phi) \qquad (t \geq 0) \tag{2}$$

where $\phi = \cos^{-1}\zeta$.

(b)   If $\zeta = 1$, then $\omega_d = 0$ and (1) becomes

$$C(s) = \frac{1}{s} - \frac{s + 2\omega_n}{(s + \omega_n)^2} = \frac{1}{s} - \frac{1}{s + \omega_n} - \frac{\omega_n}{(s + \omega_n)^2} \tag{3}$$

$$c(t) = 1 - (\omega_n t + 1)e^{-\omega_n t} \qquad (t \geq 0) \tag{4}$$

(c)   If $\zeta > 1$, (17.7) may be rewritten as

$$s_1 = -(\zeta - \sqrt{\zeta^2 - 1})\omega_n \equiv -\lambda_1\omega_n \qquad\qquad s_2 = -(\zeta + \sqrt{\zeta^2 - 1})\omega_n \equiv -\lambda_2\omega_n$$

where λ_1 and λ_2 are real and positive. Then,

$$C(s) = \frac{1}{s} - \frac{s + 2\zeta\omega_n}{(s + \lambda_1\omega_n)(s + \lambda_2\omega_n)} = \frac{1}{s} + \frac{\lambda_1/2\sqrt{\zeta^2 - 1}}{s + \lambda_1\omega_n} - \frac{\lambda_2/2\sqrt{\zeta^2 - 1}}{s + \lambda_2\omega_n} \qquad (5)$$

and the inverse Laplace transform gives

$$c(t) = 1 - \frac{1}{2\sqrt{\zeta^2 - 1}}(\lambda_1 e^{-\lambda_1\omega_n t} - \lambda_2 e^{-\lambda_2\omega_n t}) \qquad (t \geq 0) \qquad (6)$$

Figure 17-5 is a graph of (2) of Example 17.7 for $\zeta = 0.1, 0.2, 0.5, 0.7$; of (4); and of (6) for $\zeta = 2$. The independent variable is the dimensionless time, $\omega_n t$. When $\zeta = 0$, the system is *undamped* and the response is purely oscillatory. For $0 < \zeta < 1$, the system is *underdamped* and overshoots prior to settling down to a steady-state value. If $\zeta = 1$, the system is *critically damped*; this is the minimum value of ζ for which there is no overshoot of the steady-state value. For $\zeta \geq 1$, there is no overshoot, and the response is similar to that of a first-order system.

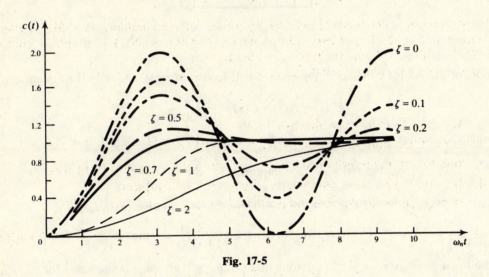

Fig. 17-5

17.5 STEADY-STATE ERROR

By adding and subtracting a unity feedback path, the canonical-form block diagram of Fig. 17-6(a) can be manipulated to the equivalent block diagram of Fig. 17-6(b). Reduction of the inner loop then gives

$$G_u(s) = \frac{G(s)}{1 + G(s)H(s) - G(s)} \qquad (17.9)$$

and the system can be modeled in the unity feedback form of Fig. 17-6(c). In general, $G_u(s)$ is rational, of the form

$$G_u(s) = \frac{K_u P(s)}{s^n Q(s)} \qquad (17.10)$$

where $P(s)$ and $Q(s)$ are polynomials such that $P(0) = Q(0) = 1$. K_u is called the *forward gain* of the unity feedback system. The nonnegative integer n is the *type number* of the system; there are effectively n integrations in the feedback loop.

Conventionally, control systems are classified according to their ability to follow unit step commands in position, velocity, and acceleration; in each case steady-state error is a measure of goodness. With position as the controlled variable $c(t)$, the commands to be investigated are therefore

step in position $r(t) = u(t) \leftrightarrow R(s) = 1/s$ *(17.11)*

step in velocity $r(t) = tu(t) \leftrightarrow R(s) = 1/s^2$ *(17.12)*

step in acceleration $r(t) = \frac{1}{2}t^2 u(t) \leftrightarrow R(s) = 1/s^3$ *(17.13)*

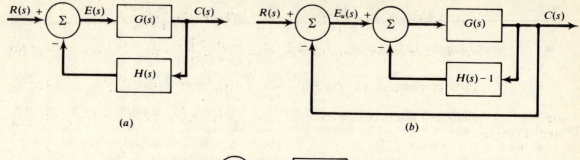

(a)   (b)

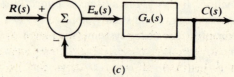

(c)

Fig. 17-6

In the model of Fig. 17-6(c), the actuating error signal (control force) is exactly the system error:

$$E_u(s) = R(s) - C(s) = \frac{R(s)}{1 + G_u(s)}$$

According to the final-value theorem of the Laplace transform (Table 6-1, line 15),

$$\lim_{t \to \infty} f(t) = a \qquad \text{implies} \qquad \lim_{s \to 0} sF(s) = a$$

Therefore, *for a stable system*, we define the *steady-state error* (e_{ss}) as

$$e_{ss} \equiv \lim_{s \to 0} sE_u(s) = \lim_{s \to 0} \left[\frac{sR(s)}{1 + G_u(s)} \right] \tag{17.14}$$

Even if stability is guaranteed, it is still possible that (*17.14*) will give an infinite value for e_{ss}, owing to an unbounded input $r(t)$; some instances will be found in Table 17-1 below.

Table 17-1

System Type	Position Step		Velocity Step		Acceleration Step	
	K_p	e_{pss}	K_v	e_{vss}	K_a	e_{ass}
0	K_u	$1/(1 + K_u)$	0	∞	0	∞
1	∞	0	K_u	$1/K_u$	0	∞
2	∞	0	∞	0	K_u	$1/K_u$

Use of the reference commands (*17.11*), (*17.12*), (*17.13*) in (*17.14*) leads to general expressions for steady-state error in terms of a set of error coefficients (K_p, K_v, K_a):

$$\textit{steady-state position error} \equiv e_{pss} = \lim_{s \to 0} \left[\frac{s(1/s)}{1 + G_u(s)} \right] = \frac{1}{1 + K_p} \tag{17.15}$$

$$\textit{static position error coefficient} \equiv K_p = \lim_{s \to 0} [G_u(s)] = \lim_{s \to 0} \left[\frac{K_u}{s^n} \right] \tag{17.16}$$

$$\textit{steady-state velocity error} \equiv e_{vss} = \lim_{s \to 0} \left[\frac{s(1/s^2)}{1 + G_u(s)} \right] = \frac{1}{K_v} \tag{17.17}$$

$$\text{static velocity error coefficient} \equiv K_v = \lim_{s \to 0} [sG_u(s)] = \lim_{s \to 0} \left[\frac{K_u}{s^{n-1}} \right] \tag{17.18}$$

$$\text{steady-state acceleration error} \equiv e_{ass} = \lim_{s \to 0} \left[\frac{s(1/s^3)}{1 + G_u(s)} \right] = \frac{1}{K_a} \tag{17.19}$$

$$\text{static acceleration error coefficient} \equiv K_a = \lim_{s \to 0} [s^2 G_u(s)] = \lim_{s \to 0} \left[\frac{K_u}{s^{n-2}} \right] \tag{17.20}$$

Static error coefficients and associated steady-state errors are summarized in Table 17-1 for type-0, -1, and -2 systems.

Error Due to Disturbance

Figure 17-7(a) is a block diagram for the case of disturbance at the plant input; $C(s)$ is the controlled response if $Q(s)$ were not present and $C_d(s)$ is the response due to $Q(s)$ alone. Since the system is linear, superposition is justified, allowing solution for $c_d(t)$ and $c(t)$ with $r(t) \equiv 0$ and $q(t) \equiv 0$, respectively.

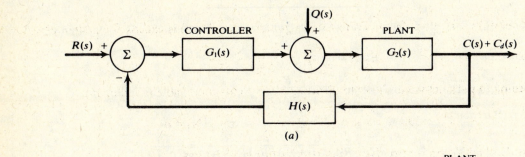

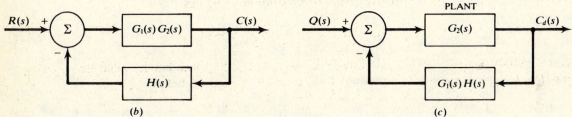

(a)

(b) (c)

Fig. 17-7

Example 17.8 Using the final-value theorem, find a general expression for the steady-state error, e_{ss}, of the system of Fig. 17-7(a) (presumed stable).

Figure 17-7(b) is a reduction of original network for $Q(s) \equiv 0$. Likewise, with $R(s) \equiv 0$, Fig. 17-7(c) is the reduced system representation. If $R(s) \equiv 0$, the response is given by

$$C_d(s) = \frac{G_2(s)Q(s)}{1 + G_1(s)G_2(s)H(s)} \tag{1}$$

Similarly, with $Q(s) \equiv 0$,

$$C(s) = \frac{G_1(s)G_2(s)R(s)}{1 + G_1(s)G_2(s)H(s)} \tag{2}$$

Any response due to $Q(s)$ being undesired, $C_d(s)$ is additional error beyond that of the undisturbed system; thus, the s-domain expression for the error is

$$E(s) = R(s) - C(s) + C_d(s) = \frac{R(s)\{1 + G_1(s)G_2(s)[H(s) - 1]\} + G_2(s)Q(s)}{1 + G_1(s)G_2(s)H(s)} \tag{3}$$

The final-value theorem gives for the value of the error as $t \to \infty$

$$e_{ss} = \lim_{s \to 0} sE(s) \tag{4}$$

17.6 SENSITIVITY ANALYSIS

Sensitivity is a quantitative measure of the effect upon a variable of interest wrought by changes from nominal in some parameter of concern. Denoting by S_B^A the sensitivity of A to changes in B, we have:

$$S_B^A \equiv \frac{\text{normalized change in } A}{\text{normalized change in } B} = \frac{\Delta A/A}{\Delta B/B} \approx \frac{dA/A}{dB/B} = \frac{B \, dA}{A \, dB} \tag{17.21}$$

Example 17.9 The voltage divider of Fig. 17-8 has nominal resistances $R_1 = 2 \, \Omega$ and $R_2 = 3 \, \Omega$. Find the percentage change in v_2 for a $+5\%$ change in R_1 caused by heating effects.

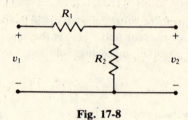

Fig. 17-8

By voltage division,

$$v_2 = \frac{R_2}{R_1 + R_2} v_1$$

The sensitivity function is

$$S_{R_1}^{v_2} = \frac{R_1}{v_2} \frac{dv_2}{dR_1} = \frac{R_1}{R_2 v_1/(R_1 + R_2)} \frac{d}{dR_1} \left(\frac{R_2}{R_1 + R_2} \right) = \frac{-R_1}{R_1 + R_2}$$

and from (17.21),

$$\frac{\Delta v_2}{v_2} = S_{R_1}^{v_2} \frac{\Delta R_1}{R_1} = \left(\frac{-R_1}{R_1 + R_2} \right)(+5\%) = \left(\frac{-2}{2+3} \right)(5\%) = -2\%$$

That is, v_2 is reduced by 2% if R_1 increases by 5%.

If $A(s)$ and $B(s)$ are quantities in the s-domain, S_B^A will also be a function of s. The *static sensitivity* is found by setting $s = 0$; it measures the sensitivity of $a(\infty)$ to changes in the parameter of concern when a unit step is applied (dc excitation). If s is replaced by $j\omega$, where ω is a frequency of interest in the spectrum of $r(t)$, then the sensitivity is called the *dynamic sensitivity*.

Solved Problems

17.1 For a system in the canonical form of Fig. 17-6(a), evaluate $G(s)H(s)$ at a root of the characteristic equation.

By Rule 5 of Table 16-1, the system transfer function is

$$F(s) \equiv \frac{C(s)}{R(s)} = \frac{G}{1 + GH} \tag{1}$$

where the dependence of G and H on s is understood. For any system described by a differential equation of the form (17.1), G and H will be rational functions of s; i.e. $G = N_G/D_G$ and $H = N_H/D_H$, where the N's and D's are polynomials in s. Thus, (1) can be written as

$$F(s) = \frac{D_H N_G}{D_G D_H + N_G N_H} \equiv \frac{N}{D} \tag{2}$$

The poles of the rational function $F(s)$ are given either by the roots of $D = 0$ or by the roots of $1 + GH = 0$; the two sets of roots must therefore be identical. Hence,

$$GH = -1$$

at any root of $D = 0$.

17.2   A system has the transfer function

$$F(s) = \frac{2}{(s+2)(s+4)}$$

A reference command is applied for which $R(s) = 1 + (4/s)$. (a) Find the system response for $t \geq 0$ if initial conditions are zero. (b) Why is the choice of this reference command an example of *mode suppression control*?

(a) System response in the Laplace domain is given by

$$C(s) = F(s)R(s) = \frac{2}{(s+2)(s+4)}\left(\frac{s+4}{s}\right) = \frac{2}{s(s+2)} = \frac{1}{s} - \frac{1}{s+2} \tag{1}$$

Inverting,

$$c(t) = 1 - e^{-2t} \qquad (t \geq 0) \tag{2}$$

(b) Mode suppression control forces zero coefficients for undesired transient response modes. For the case at hand, the cancellation of $(s+4)$ in (1) prevented the appearance of a term in e^{-4t} in (2).

17.3   Write the form of expected transient response and discuss stability for each of the characteristic equations below:

(a) $D(s) = (s+1)(s+100)(s+4) = 0$ (c) $D(s) = s^2(s-2)(s+2) = 0$

(b) $D(s) = s(s^2 + 2s + 5) = 0$ (d) $D(s) = (s+3)(s^2 - 6s + 18) = 0$

(a) The characteristic equation has three roots all in LHSP; the system thus is absolutely stable.

$$c_{tr}(t) = A_1 e^{-t} + A_2 e^{-100t} + A_3 e^{-4t} \qquad (t \geq 0)$$

(b) Factoring the quadratic term of $D(s)$ gives roots $s_{1,2} = -1 \pm j2$, which lie in LHSP. However, one root ($s_3 = 0$) is located at the origin of the s-plane, giving instability (marginal stability).

$$c_{tr}(t) = A_1 u(t) + B_1 e^{-t} \cos(2t + \phi) \qquad (t \geq 0)$$

(c) The characteristic equation has a double root at the origin, which renders the system unstable. Further, a root is located at $s = +2$, which also results in an unstable system.

$$c_{tr}(t) = A_1 u(t) + A_2 t + A_3 e^{+2t} + A_4 e^{-2t} \qquad (t \geq 0)$$

(d) The quadratic term factors to give roots $s_{1,2} = +3 \pm j3$, which lie in the RHSP; the system is unstable.

$$c_{tr}(t) = A_1 e^{-3t} + A_2 e^{+3t} \cos(3t + \phi) \qquad (t \geq 0)$$

17.4   A control system in the canonical form of Fig. 17-6(a) has $G(s) = K/(s+1)$ and $H(s) = 0.05$. An element in the forward path is an inverting amplifier, so that $K = -10$. (a) Is this system stable? (b) Interpret the sign of $F(s)$.

(a)

$$F(s) = \frac{G(s)}{1 + G(s)H(s)} = \frac{-10/(s+1)}{1 + \left(\frac{-10}{s+1}\right)(0.05)} = \frac{-10}{s+0.5} \equiv \frac{N(s)}{D(s)}$$

The sole root of $D(s) = 0$ lies in the LHSP; the system is stable. For $K < -20$ the system would be unstable.

(b) The negative sign indicates a reversal in sense or direction between the command $r(t)$ and the response $c(t)$.

17.5   The s-domain response of a stable, simple first-order system is given by

$$C(s) = \frac{4A}{s^2} - \frac{A}{s} + \frac{A}{s+4}$$

What is the form of the system command in the time domain?

By finding the lowest common denominator, the response can be rewritten as

$$C(s) = F(s)R(s) = \frac{16A}{s^2(s+4)} \tag{1}$$

Since the system is stable, simple, and first-order, the transfer function must be of the form

$$F(s) = \frac{K}{s + \tau^{-1}}$$

Inspection of (1) now shows that $R(s)$ must be of the form B/s^2, or, in the time domain, $r(t)$ is a ramp function:

$$r(t) = Bt \qquad (t \geq 0)$$

17.6   Apply the Routh-Hurwitz criterion to the following characteristic equation to determine stability of system:

$$D(s) = 3s^5 + 12s^4 + 15s^2 + 21s + 33 = 0$$

Since the coefficient of s^3 in $D(s)$ is zero, $D(s)$ fails the necessity test for absolute stability: there will be at least one characteristic root with nonnegative real part. Construct the first two rows of the Routhian array after multiplying all coefficients by 1/3. Likewise, b-, c-, d-, and e-entries are multiplied by positive constants prior to placement in the array.

s^5	1	0	7	
s^4	4	5	11	
s^3	-5	17		b-row
s^2	93	55		c-row
s^1	1			d-row
s^0	1			e-row

$$b_1 = \frac{-\begin{vmatrix} 1 & 0 \\ 4 & 5 \end{vmatrix}}{1} = -\frac{5}{4}$$

$$b_2 = \frac{-\begin{vmatrix} 1 & 7 \\ 4 & 11 \end{vmatrix}}{4} = \frac{17}{4}$$

$$c_1 = \frac{-\begin{vmatrix} 4 & 5 \\ -5 & 17 \end{vmatrix}}{-5} = \frac{93}{5} \qquad c_2 = \frac{-\begin{vmatrix} 4 & 11 \\ -5 & 0 \end{vmatrix}}{-5} = \frac{55}{5}$$

$$d_1 = \frac{-\begin{vmatrix} -5 & 17 \\ 93 & 55 \end{vmatrix}}{93} = \frac{1856}{93} \qquad e_1 = \frac{-\begin{vmatrix} 93 & 55 \\ 1 & 0 \end{vmatrix}}{1} = 55$$

Two sign changes occur in the first column of the Routhian array; therefore, the characteristic equation has two RHSP roots. (By factorization, the five roots are -4.222, $-0.7235 \pm j0.8364$, $+0.8346 \pm j1.197$.)

17.7   Test the following characteristic equation for RHSP roots by the Routh-Hurwitz method:

$$D(s) = s^5 + s^4 + s^3 + s^2 + 4s + 1 = 0$$

The necessary condition for stability is met. Set up the Routhian array and calculate row entries.

s^5	1	1	4
s^4	1	1	1
s^3	ϵ	3	b-row
s^2	$\dfrac{\epsilon - 3}{\epsilon}$	1	c-row
s^1	$3 - \dfrac{\epsilon^2}{\epsilon - 3}$		d-row
s^0	1		e-row

$$b_1 = \frac{-\begin{vmatrix} 1 & 1 \\ 1 & 1 \end{vmatrix}}{1} = 0$$

$$b_2 = \frac{-\begin{vmatrix} 1 & 4 \\ 1 & 1 \end{vmatrix}}{1} = 3$$

Upon calculation of the b-row entries, $b_1 = 0$ but the remaining b's are not all zeros. This is Special Case I. Set $b_1 = \epsilon$ and continue construction of the array.

$$c_1 = \frac{-\begin{vmatrix} 1 & 1 \\ \epsilon & 3 \end{vmatrix}}{\epsilon} = \frac{\epsilon - 3}{\epsilon} \qquad c_2 = \frac{-\begin{vmatrix} 1 & 1 \\ \epsilon & 0 \end{vmatrix}}{\epsilon} = 1$$

$$d_1 = \frac{-\begin{vmatrix} \epsilon & 3 \\ \frac{\epsilon-3}{\epsilon} & 1 \end{vmatrix}}{\frac{\epsilon-3}{\epsilon}} = 3 - \frac{\epsilon^2}{\epsilon - 3} \qquad e_1 = \frac{-\begin{vmatrix} \frac{\epsilon-3}{\epsilon} & 1 \\ 3 - \frac{\epsilon^2}{\epsilon - 3} & 0 \end{vmatrix}}{3 - \frac{\epsilon^2}{\epsilon-3}} = 1$$

After completing the array, let $\epsilon \to 0+$, making $b_1 > 0$, $c_1 < 0$, $d_1 > 0$. It is seen that two sign changes occur; $D(s) = 0$ has two RHSP roots.

17.8 Apply the Routh-Hurwitz procedure to $D(s) = s^5 + s^4 + 4s + 4$ to determine the number of RHSP roots.

The system is necessarily unstable. The final Routhian array is

s^5	1	0	4
s^4	1	0	4
s^3	4	0	b-row
s^2	ϵ	4	c-row
s^1	$-16/\epsilon$		d-row
s^0	4		e-row

In attempting to calculate the b-row, it is found that all entries are zero, signaling Special Case II. The characteristic equation has roots symmetric about the s-plane origin. Backing up to the s^4-row, construct the auxiliary polynomial in alternate powers of s, starting with s^4:

$$A(s) = 1s^4 + 0s^2 + 4s^0 \qquad \frac{dA(s)}{ds} = 4s^3 + 0s^1$$

Place the coefficients of dA/ds in the b-row and continue as normal.

$$c_1 = \frac{-\begin{vmatrix} 1 & 0 \\ 4 & 0 \end{vmatrix}}{4} = 0 \qquad c_2 = \frac{-\begin{vmatrix} 1 & 4 \\ 4 & 0 \end{vmatrix}}{4} = 4$$

Replace c_1 with ϵ (Special Case I).

$$d_1 = \frac{-\begin{vmatrix} 4 & 0 \\ \epsilon & 4 \end{vmatrix}}{\epsilon} = -\frac{16}{\epsilon} \qquad e_1 = \frac{-\begin{vmatrix} \epsilon & 4 \\ -16/\epsilon & 0 \end{vmatrix}}{-16/\epsilon} = 4$$

Let $\epsilon \to 0+$ and observe that two sign changes occur in the first column of the array, signifying two RHSP roots.

For this particular case, the auxiliary equation, $A(s) = 0$, is of fourth order, implying two pairs of roots symmetric about the s-plane origin. If $A(s)$ is factored, the roots are found to be $-1 \pm j1$ and $+1 \pm j1$; thus, either symmetric pair determines a diagonal of a square.

17.9 The characteristic equation of a control system contains a parameter K:

$$D(s) = 2s^3 + 6s^2 + 6s + (K + 1) = 0$$

Determine the range of K for which the system is stable.

Set up the Routhian array, multiplying the coefficients of the s^3-row by $\frac{1}{2}$ prior to insertion.

$$
\begin{array}{c|cc}
s^3 & 1 & 3 \\
s^2 & 6 & K+1 \\
s^1 & 17-K & \textbf{b-row} \\
s^0 & K+1 & \textbf{c-row}
\end{array}
\qquad
\begin{aligned}
b_1 &= \dfrac{-\begin{vmatrix} 1 & 3 \\ 6 & K+1 \end{vmatrix}}{6} = 17-K \\[2ex]
c_1 &= \dfrac{-\begin{vmatrix} 6 & K+1 \\ 17-K & 0 \end{vmatrix}}{17-K} = K+1
\end{aligned}
$$

There will be no sign change in the first column, hence the system will be stable, if and only if $-1 < K < 17$.

17.10 A control system has as its transfer function

$$
F(s) = \frac{N(s)}{As^3 + Bs^2 + Cs + D} \qquad (A > 0)
$$

Determine the minimum number of conditions on the coefficients in the characteristic equation that will ensure only LHSP roots.

Form the Routhian array:

$$
\begin{array}{c|cc}
s^3 & A & C \\
s^2 & B & D \\
s^1 & \dfrac{BC - AD}{B} & \\
s^0 & D &
\end{array}
$$

wherein b_1 and c_1 were computed in the usual way. With A given as positive, there will be no sign changes in the first column if and only if $B > 0$, $D > 0$, and

$$
BC - AD > 0 \qquad \text{or} \qquad C > AD/B
$$

Any further constraints besides these three would be redundant.

17.11 Reverse the sign of the unity feedback path in the block diagram of Fig. 17-2(a). (As it stands, the system is stable, provided τ is positive.) (a) If a minute burst of noise occurs—say, $r(t) = 10^{-10}\delta(t)$—find the value of $c(t)$ as $t \to \infty$. (b) Relate part (a) to physical control systems. (c) If the block diagram of Fig. 17-2(a) had represented a series RC filter network, with $r(t)$ the input voltage and $c(t)$ the capacitor output voltage, could the sign reversal on the feedback loop be physically accomplished?

(a) Implementing the reversal of sign,

$$
C(s) = \frac{1}{\tau s - 1} R(s) = \frac{1}{\tau s - 1} 10^{-10}
$$

or
$$
c(t) = 10^{-10} e^{+t/\tau} \qquad (t \geq 0)
$$

For this system, rendered unstable by positive feedback, $c(t) \to \infty$ as $t \to \infty$, although the command was negligibly small and has long since vanished.

(b) Many times, upon connection of systems with positive feedback, violent oscillations or extreme limiting values of the controlled variable are observed. Even though the command may not have been "formally" applied, noise is nearly always present to initiate the unstable response.

(c) If the system were the described RC filter network, the negative sign would imply a negative value of R or C. Negative values of passive elements do not exist; thus the sign change could not be physically accomplished. However, for a simple active integrator, the positive sign could be synthesized.

17.12 For a simple second-order system, (a) show that the characteristic roots lie on quarter-circles of radius ω_n in the second and third quadrants of the complex plane, if $0 \leq \zeta \leq 1$. (b) What are the root loci as ζ increases beyond 1?

(a)   From $D(s)$, as given in (17.6),

$$s_1 + s_2 = -2\zeta\omega_n \qquad s_1 s_2 = \omega_n^2$$

which, since s_1 and s_2 are complex conjugates, imply

$$\text{Re}\,(s_1) = \text{Re}\,(s_2) = -\zeta\omega_n \leq 0 \qquad |s_1| = |s_2| = \omega_n$$

These relations define the stated quarter-circles. See Fig. 17-9.

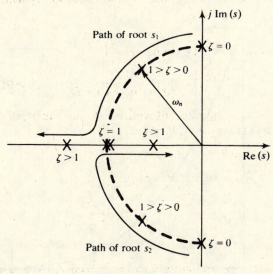

Fig. 17-9

(b)   For $\zeta = 1$, it is seen from (a) that both characteristic roots are located at $s = -\omega_n$ (i.e., a double root). As ζ increases, the roots remain real, one moving to the right toward the origin and the other to the left toward $-\infty$. See Fig. 17-9.

17.13 From Example 17.7 and Fig. 17-5, it is noted that if $0 < \zeta < 1$, the response of a simple second-order system to a unit step input overshoots the final or steady-state value. Find expressions for (a) the time to peak overshoot (t_p), and (b) $c(t_p) \equiv C_p$, the maximum of $c(t)$ during the transient period.

(a)   We wish to locate the first maximum of $c(t)$. Differentiating (2) of Example 17.7,

$$\frac{dc}{dt} = -\frac{e^{-\zeta\omega_n t}}{\sqrt{1 - \zeta^2}}\left[-\zeta\omega_n \sin(\omega_d t + \phi) + \omega_d \cos(\omega_d t + \phi)\right] = 0$$

or, since $\omega_d = \omega_n\sqrt{1 - \zeta^2}$ and $\zeta = \cos\phi$,

$$\tan(\omega_d t + \phi) = \tan\phi \qquad\qquad (1)$$

The tangent function has period π; therefore, the smallest positive root of (1) is

$$\omega_d t_p = \pi \qquad \text{or} \qquad t_p = \frac{\pi}{\omega_d}$$

(b)

$$C_p = c(t_p) = 1 - \frac{1}{\sqrt{1 - \zeta^2}}\, e^{-\zeta\pi\omega_n/\omega_d} \sin(\pi + \phi)$$

$$= 1 - \frac{1}{\sin\phi}\, e^{-\zeta\pi/\sqrt{1-\zeta^2}}(-\sin\phi)$$

$$= 1 + e^{-\zeta\pi/\sqrt{1-\zeta^2}}$$

17.14 Using the expression for C_p found in Problem 17.13(b), (a) derive an expression for the *normalized peak overshoot* (NPO), and (b) plot it for $0 < \zeta < 1$.

(a) By (2) of Example 17.7, the steady-state value of $c(t)$ is $c_{ss} = 1$. Then,

$$\text{NPO} \equiv \frac{C_p - c_{ss}}{c_{ss}} = e^{-\zeta\pi/\sqrt{1-\zeta^2}} \tag{1}$$

(b) See Fig. 17-10. Note that for $0.85 < \zeta < 1$, the overshoot is less than 1% and consequently negligible.

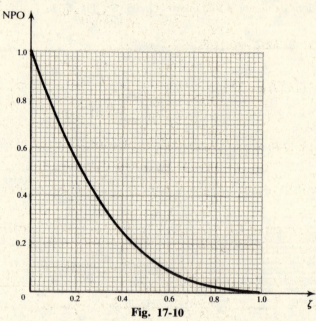

Fig. 17-10

17.15 *Rate feedback* is sometimes used to reduce the overshoot (increase the damping ratio) of a second-order control system. The system of Fig. 17-11 is a speed-control system, with tachometer feedback through K_t. The derivative of output (controlled speed) is scaled and fed back by sK_r. (a) Find the characteristic equation and derive a formula for the percentage increase in damping ratio due to addition of rate feedback. (b) If $K_1 = 9$, $\tau = 0.8333$, and $K_t = 1$, calculate the percent overshoot in response for $K_r = 0$ and $K_r = 0.4$.

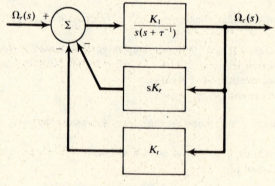

Fig. 17-11

(a) By block diagram algebra,

$$\frac{\Omega_c(s)}{\Omega_r(s)} = \frac{K_1}{s^2 + (\tau^{-1} + K_r K_1)s + K_1 K_t} \tag{1}$$

Compare (1) with (17.6) to see that $\omega_n = \sqrt{K_1 K_t}$ and that

$$\zeta = \frac{\tau^{-1} + K_r K_1}{2\omega_n} = \frac{\tau^{-1} + K_r K_1}{2\sqrt{K_1 K_t}} \tag{2}$$

The desired formula is

$$\% \text{ increase in } \zeta = \frac{(\tau^{-1} + K_r K_1) - \tau^{-1}}{\tau^{-1}} \times 100\% = 100\tau K_r K_1 \tag{3}$$

(b) From the data, $\omega_n = \sqrt{(9)(1)} = 3$ rad/s. If $K_r = 0$, (2) gives

$$\zeta = \frac{1}{2\tau\omega_n} = \frac{1}{2(0.8333)(3)} = 0.2$$

From Fig. 17-10, for $\zeta = 0.2$, NPO = 0.53, and

$$\% \text{ overshoot} = \text{NPO} \times 100\% = 53\%$$

If $K_r = 0.4$, (2) gives

$$\zeta = \frac{(0.8333)^{-1} + (0.4)(9)}{2(3)} = 0.8$$

Entering Fig. 17-10 at $\zeta = 0.8$ gives NPO = 0.015, and so

$$\% \text{ overshoot} = \text{NPO} \times 100\% = 1.5\%$$

The effectiveness of rate feedback in reduction of overshoot is clearly evident.

17.16 For the control system of Fig. 17-12, find the error at $t = 2$ s if $q(t) \equiv 0$ and $r(t) = 10\,u(t) - 5t$.

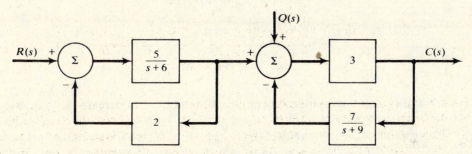

Fig. 17-12

The Laplace response is found by block diagram algebra to be

$$C(s) = \frac{15(s+9)}{(s+16)(s+30)} R(s) = \frac{15(s+9)}{(s+16)(s+30)} \left(\frac{10s-5}{s^2} \right) = \frac{75(s+9)(2s-1)}{s^2(s+16)(s+30)} \tag{1}$$

The roots of $D(s) = 0$ are -16 and -30, so the system is stable; however, $r(t)$ is unbounded. Because the time constants of the two transient modes ($\frac{1}{16}$ s and $\frac{1}{30}$ s) are small compared with 2 s, it is sufficient to find the steady-state error and evaluate it at $t = 2$ s.

Partial-fractions expansion of (1) gives

$$C(s) = \frac{A}{s^2} + \frac{B}{s} + \text{(transient terms)} \tag{2}$$

and, by the usual methods, one finds $A = -1.4063$ and $B = 2.791$. The steady-state response results from inverse transformation of (2):

$$c_{ss}(t) = 2.791 - 1.4063\,t$$

The steady-state error is then

$$e_{ss}(t) = r(t) - c_{ss}(t) = (10 - 5t) - (2.791 - 1.4063\,t) = 7.209 - 3.594\,t$$

and $e_{ss}(2) = 0.0216$.

17.17 A stable control system is described by the block diagram of Fig. 17-7(b) with $G_1(s) = 1/(s+3)$, $G_2(s) = 1/s$, and $H(s) = 1$. (a) What is the system type number? Predict the steady-state error for (b) a step input, $Ku(t)$; (c) a ramp input, Kt; (d) a parabolic input, Kt^2.

(a) Putting the system into unity feedback form, Fig. 17-6(c),

$$G_u(s) = G_1(s)G_2(s) = \frac{\frac{1}{3}}{s(\frac{1}{3}s + 1)} \qquad (1)$$

Comparing (1) with (17.10), it is apparent that $n = 1$, or the system is type-1.

(b) Since $K_u = \frac{1}{3}$, (17.16) gives

$$K_p = \lim_{s \to 0} \left[\frac{\frac{1}{3}}{s} \right] = \infty$$

Since $R(s) = K/s$, the steady-state error is found by scaling (17.15):

$$e_{pss} = \frac{K}{1 + K_p} = \frac{K}{1 + \infty} = 0$$

(c) By (17.18),

$$K_v = \lim_{s \to 0} \left[\frac{\frac{1}{3}}{s^0} \right] = \frac{1}{3}$$

With $R(s) = K/s^2$, (17.17) gives, after scaling,

$$e_{vss} = \frac{K}{K_v} = \frac{K}{\frac{1}{3}} = 3K$$

(d) From (17.20),

$$K_a = \lim_{s \to 0} \left[\frac{\frac{1}{3}}{s^{-1}} \right] = 0$$

With $R(s) = 2K/s^3$, (17.19) yields, after scaling,

$$e_{ass} = \frac{2K}{K_a} = \frac{2K}{0} = \infty$$

17.18 For a control system in canonical form, Fig. 17-6(a),

$$G(s) = \frac{10}{s(s + 10)} \qquad H(s) = 1$$

Use Table 17-1 to predict the steady-state error to $r(t) = 6u(t) + 4t$.

It is easy to verify that this system is stable. Since $H(s) = 1$,

$$G_u(s) = G_1(s) = \frac{10}{s(s + 10)} = \frac{1}{s(0.1s + 1)} \qquad (1)$$

Compare (1) with (17.10) to see that the system is type-1 with $K_u = 1$. It must be noted that Table 17-1 was derived for commands as described by (17.11), (17.12), and (17.13), so that tabulated errors must be appropriately scaled. Also, for a linear system, superposition allows the total error to be found as the sum of the errors due to each component of $r(t)$ acting individually. Thus,

$$e_{ss} = 6e_{pss} + 4e_{vss} = (6)(0) + 4(1/K_u) = 4$$

where e_{pss} and e_{vss} are respectively the errors to step and ramp inputs to a type-1 system, as given in Table 17-1.

17.19 A control system can be modeled by Fig. 17-12, with $r(t) = R_0 t$ and $q(t) = Q_0 u(t)$. Find an expression for $e_{ss}(t)$, the steady-state error.

The system was shown to be stable in Problem 17.16. By block diagram reduction, the Laplace response due to $r(t)$ only is

$$C(s) = \frac{15(s + 9)}{(s + 16)(s + 30)} R(s) = \frac{15(s + 9)}{(s + 16)(s + 30)} \frac{R_0}{s^2} = \frac{15R_0(s + 9)}{s^2(s + 16)(s + 30)}$$

$$= \frac{0.2813 R_0}{s^2} + \frac{0.0043 R_0}{s} + \text{(transient terms)}$$

Inverting,

$$c_{ss}(t) = 0.0043 R_0 + 0.2813 R_0 t$$

The response due to $q(t)$ is found through block diagram algebra:

$$C_d(s) = \frac{3(s+9)}{s+30} Q(s) = \frac{3(s+9)}{s+30} \frac{Q_0}{s} = \frac{0.9 Q_0}{s} + \text{(transient term)}$$

whence $c_{dss}(t) = 0.9 Q_0$. Then, reasoning as in Example 17.8, we have for the steady-state error:

$$e_{ss}(t) = r(t) - c_{ss}(t) + c_{dss}(t) = 0.7187 R_0 t - 0.0043 R_0 + 0.9 Q_0$$

17.20 For a system with block diagram given by Fig. 17-7(a), calculate the steady-state disturbance output (disturbance error), c_{dss}, for a disturbance input $q(t) = Q_0 u(t)$ if (a) $H(s) = 1$, $G_1(s) = K_1/s$, $G_2(s) = K_2/(s+1)$; (b) $H(s) = 1$, $G_1(s) = K_1$, $G_2(s) = K_2/s(s+1)$. ($K_1 K_2 > 1/4$.) (c) In both (a) and (b), the forward transfer functions are identical insofar as $R(s)$ is concerned; comment on the difference in these functions as seen by the disturbance input.

(a) By Problem 17.1 (with G replaced by $G_1 G_2$), the characteristic roots are the roots of the equation

$$1 + G_1(s) G_2(s) H(s) = 0$$

or $$s^2 + s + K_1 K_2 = 0$$

Under the assumption $K_1 K_2 > 1/4$, both roots have real part $-1/2$, and so the system is stable. By (1) of Example 17.8,

$$C_d(s) = \frac{\left(\dfrac{K_2}{s+1}\right)\left(\dfrac{Q_0}{s}\right)}{1 + \left(\dfrac{K_1}{s}\right)\left(\dfrac{K_2}{s+1}\right)} = \frac{K_2 Q_0}{s^2 + s + K_1 K_2}$$

The final-value theorem gives

$$c_{dss} = \lim_{s \to 0} s C_d(s) = 0$$

There is no steady-state error due to disturbance.

(b) The characteristic roots are the same as for the system of (a).

$$C_d(s) = \frac{\left[\dfrac{K_2}{s(s+1)}\right]\left(\dfrac{Q_0}{s}\right)}{1 + K_1 \left[\dfrac{K_2}{s(s+1)}\right]} = \frac{K_2 Q_0}{s(s^2 + s + K_1 K_2)}$$

$$c_{dss} = \lim_{s \to 0} s C_d(s) = \frac{Q_0}{K_1}$$

Or, there is a constant steady-state disturbance error.

(c) In (a), the disturbance sees an integrator in the feedback path; while in (b), the integrator appears in the forward path insofar as the disturbance input is concerned. However, the integrator acts in the forward path of the command signal in both cases, slowing down the system response. See Problem 17.21 for a better method of reducing disturbance error.

17.21 *Integral control* was introduced in Problem 17.20 to reduce steady-state disturbance error at the expense of increased transient decay time. *Proportional-plus-integral control* can be used to reduce steady-state error while allowing acceptable transient decay times. To illustrate, let $H(s) = 1$, $G_1(s) = K_3 + (K_1/s)$, and $G_2(s) = K_2/(s+1)$ in Fig. 17-7(a); assume that

$$K_1 K_2 > 0 \qquad \text{and} \qquad \sqrt{4 K_1 K_2} > K_2 K_3 + 1 > 0 \tag{1}$$

(a) Find the steady-state disturbance response (c_{dss}) given that $Q(s) = Q_0/2$, and compare with Problem 17.20(a). (b) Compare the time constants of decaying transient terms in this problem with those of Problem 17.20.

(a) As in Problem 17.20(a), one finds a quadratic equation satisfied by the characteristic roots:

$$s^2 + (K_2K_3 + 1)s + K_1K_2 = 0$$

Under the assumptions (1) on the coefficients, both roots are in the LHSP and the system is stable—indeed, it is underdamped, like the systems of Problems 17.20(a) and (b). By (1) of Example 17.8,

$$C_d(s) = \frac{\left(\dfrac{K_2}{s+1}\right)\left(\dfrac{Q_0}{s}\right)}{1 + \left(K_3 + \dfrac{K_1}{s}\right)\left(\dfrac{K_2}{s+1}\right)} = \frac{K_2Q_0}{s^2 + (K_2K_3 + 1)s + K_1K_2}$$

$$c_{dss} = \lim_{s \to 0} sC_d(s) = 0$$

The steady-state error due to a step disturbance is zero, which is identical to the result of Problem 17.20(a).

(b) For Problem 17.20, the roots of the characteristic equation are $s_{1,2} = -\frac{1}{2} \pm j\omega_d$, and the time constant is $\tau = 2$. For the present problem,

$$s_{1,2} = -\frac{K_2K_3 + 1}{2} \pm j\omega_d \qquad \text{and so} \qquad \tau = \frac{2}{K_2K_3 + 1}$$

Without violating (1), $K_2K_3 + 1$ can be chosen sufficiently large to reduce the time constant to any acceptable value; in Problem 17.20 no such freedom existed.

17.22 Given three s-domain quantities, $A(s)$, $B(s)$, and $C(s)$, such that $A(s)$ and $C(s)$ are related by a transfer function,

$$A(s) = F(s)\,C(s)$$

and such that $C(s)$ is insensitive to variations in $B(s)$. Show that

$$S_B^A = S_B^F$$

By (17.21),

$$S_B^A = S_B^{FC} = \frac{B}{FC}\frac{d}{dB}(FC) = \frac{B}{FC}C\frac{dF}{dB} = \frac{B}{F}\frac{dF}{dB} = S_B^F$$

17.23 The two positioning control systems of Fig. 17-13 were laboratory checked and found to track perfectly in the steady state for $\tau = 1$, $A = 3$, and $\theta_c(t) = 10\,u(t)$. When placed in field service at extremely low temperatures, the amplifier gain for each system increased so that $A = 3.333$; all else was unchanged. Use the static sensitivity function to predict the percentage change in output of each system in field service over that observed in the laboratory.

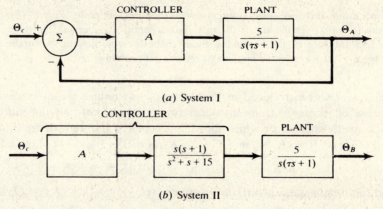

(a) System I

(b) System II

Fig. 17-13

For system I, with $\tau = 1$, the transfer function is

$$F_\text{I}(s) = \frac{5A}{s^2 + s + 5A}$$

Form the static sensitivity function, using (17.21) and Problem 17.22.

$$S_A^{\Theta_A(s)}\bigg|_{s=0} = S_A^{F_\text{I}(s)}\bigg|_{s=0} = \frac{A}{F_\text{I}(s)}\frac{dF_\text{I}(s)}{dA}\bigg|_{s=0} = \left\{\frac{s^2 + s + 5A}{5}\frac{d}{dA}\left[\frac{5A}{s^2 + s + 5A}\right]\right\}_{s=0} = 0$$

and

$$\frac{\Delta\Theta_A}{\Theta_A}\bigg|_{s=0} = S_A^{\Theta_A}\bigg|_{s=0}\left(\frac{\Delta A}{A}\right) = 0\%$$

The transfer function of system II is given by

$$F_\text{II}(s) = \frac{5A}{s^2 + s + 15}$$

from which

$$S_A^{\Theta_B(s)}\bigg|_{s=0} = S_A^{F_\text{II}(s)}\bigg|_{s=0} = \left\{\frac{s^2 + s + 15}{5}\frac{d}{dA}\left[\frac{5A}{s^2 + s + 15}\right]\right\}_{s=0} = 1$$

and

$$\frac{\Delta\Theta_B}{\Theta_B}\bigg|_{s=0} = S_A^{\Theta_B}\bigg|_{s=0}\left(\frac{\Delta A}{A}\right) = 1\left(\frac{3.333 - 3}{3}\right) = 11.1\%$$

17.24 For the canonical system of Fig. 17-6(a), (a) derive a formula for the sensitivity of the output to changes in the feedback path $H(s)$. (b) Apply this formula to the system of Fig. 17-13(a), with $A = 3$, to find the static sensitivity. Interpret the result.

(a) Making use of Problem 17.22,

$$S_{H(s)}^{C(s)} = S_{H(s)}^{F(s)} = \frac{H(s)}{F(s)}\frac{dF(s)}{dH(s)} = \frac{H(s)}{\left[\dfrac{G(s)}{1 + G(s)H(s)}\right]}\frac{d}{dH(s)}\left[\frac{G(s)}{1 + G(s)H(s)}\right]$$

After carrying out the indicated differentiation and simplifying, we find:

$$S_{H(s)}^{C(s)} = \frac{-G(s)H(s)}{1 + G(s)H(s)} \qquad\qquad (1)$$

(b) For the system of Fig. 17-13(a), with $A = 3$, $H(s) = 1$ and $G(s) = 15/s(s + 1)$. The static sensitivity function results from substituting these expressions for $G(s)$ and $H(s)$ into (1) and evaluating at $s = 0$.

$$S_{H(s)}^{C(s)}\bigg|_{s=0} = \left[\frac{-15}{s(s + 1) + 15}\right]_{s=0} = -1$$

The result means that the output decreases by the same percentage that H increases, or vice versa.

Supplementary Problems

17.25 The fifth-degree characteristic equation of Example 17.4 was found to have two RHSP roots. If the roots in the LHSP are known to be $s_1 = -1$, and $s_{2,3} = -1 \pm j1$, find the roots that render the system unstable.   *Ans.* $+1 \pm j2$

17.26 Below are given factored characteristic polynomials for four unstable (or marginally stable) systems. State whether or not each characteristic polynomial passes the necessity test for stability.

$$(a) \quad (s-1)(s+2)(s+3) \qquad (c) \quad (s+j2)(s-j2)(s+2)$$
$$(b) \quad (s-1)(s-2)(s+3) \qquad (d) \quad (s-1-j5)(s-1+j5)(s+6)$$

Ans. (a) fails test with one negative coefficient; (b) fails test with one negative and one zero coefficient; (c) passes test; (d) passes test

17.27 Apply the Routh-Hurwitz criterion to $D(s) = s^5 + s^4 + 10s^3 + 72s^2 + 152s + 240 = 0$.
Ans. Unstable, with two roots in the RHSP. (The roots are -3, $-1 \pm j\sqrt{3}$, $+2 \pm j4$.)

17.28 Apply the Routh-Hurwitz criterion to $D(s) = s^4 + 2s^3 + 11s^2 + 18s + 18 = 0$.
Ans. Unstable: no RHSP roots, but roots of the auxiliary equation at $\pm j3$.

17.29 If

$$G(s) = \frac{K}{s(s^2 + s + 1)} \qquad H(s) = 1$$

for the control system of Fig. 17-6(a), find the range of K for stable operation.   *Ans.* $0 < K < 1$

17.30 Rework Problem 17.29 if $H(s) = s + 1$ (rate feedback).   *Ans.* $K > 0$

17.31 For a control system in the form of Fig. 17-6(c),

$$G_u(s) = \frac{4(s+1)}{s^2(\tau s + 1)}$$

Can the system become unstable if the time constant τ is too large? If so, find the range on τ for absolute stability.   *Ans.* yes; $0 \le \tau < 1$

17.32 A tank, initially empty $[y(0) = 0]$, is shown in Fig. 17-14. The tank has a constant cross-sectional area $A = 2 \text{ m}^2$. Water $(\gamma = 9800 \text{ N/m}^3)$ can escape through an outlet in the bottom that has hydraulic resistance [see (15.7)] $R_o = 0.49 \text{ MPa} \cdot \text{s/m}^3$. A constant input flow of $q_i = 0.04 \text{ m}^3/\text{s}$ is introduced by opening the valve at $t = 0$. (a) Treating the liquid level as the controlled variable, find $y(t)$ for $t \ge 0$, assuming h large. (b) If $h = 1 \text{ m}$, will the tank overflow, and if so, at what time?
Ans. (a) $y(t) = 2(1 - e^{-0.01t})$ (m); (b) overflow for $t > 69.3 \text{ s}$

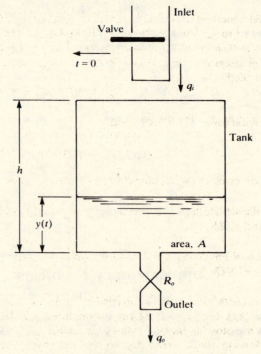

Fig. 17-14

17.33 Find the response for $t \geq 0$ of a simple second-order system to the impulse of Fig. 17-1(c), if the transfer function is given by (17.6) and (a) $\zeta = 1$, (b) $0 < \zeta < 1$, and (c) $\zeta > 1$. (d) From the foregoing results, what generalization can be made about the impulse response of an absolutely stable system?
Ans. (a) $c(t) = K\omega_n^2 t e^{-\omega_n t}$; (b) $c(t) = K\omega_d(1 - \zeta^2)e^{-\zeta\omega_n t}\sin \omega_d t$; (c) $c(t) = [K\omega_n^2/(\beta_2 - \beta_1)](e^{-\beta_1 t} - e^{-\beta_2 t})$, where $\beta_{1,2} = -\omega_n(1 \mp \sqrt{\zeta^2 - 1})$; (d) impulse response approaches zero as $t \to \infty$

17.34 A simple second-order system is formed by impressing an input voltage $v_1(t)$ across a series RLC circuit and taking as output the voltage $v_2(t)$ across the capacitor. (a) Show that the transfer function $F(s) \equiv V_2(s)/V_1(s)$ is of the form (17.6). (b) Determine ζ and ω_n for this circuit.
Ans. (a) $F(s) = \dfrac{(1/LC)}{s^2 + (R/L)s + (1/LC)}$; (b) $\omega_n = \sqrt{1/LC}$, $\zeta = \dfrac{R}{2}\sqrt{C/L}$

17.35 A simple second-order system has a normalized peak overshoot (NPO) of 0.6, achieved at $t = 0.02$ s. Using Fig. 17-10 and Problem 17.13(a), determine the characteristic equation.
Ans. $\zeta = 0.16$, $\omega_n = 159.13$ rad/s; $s^2 + 50.92 s + 25\,313 = 0$

17.36 The spring-mass-dashpot of Fig. 17-15 forms a simple second-order system which is to be position-commanded by force $f(t)$ so that $x_r(t) = Kt$ (m) for $t \geq 0$. If $M = 2$ kg, $\beta = 20$ N·s/m, and $k = 50$ N/m, find an expression for the controlled variable $x_c(t)$, assuming all initial conditions are zero.
Ans. $x_c(t) = K(t - 0.4) + K(t + 0.4)e^{-5t}$ (m) $(t \geq 0)$

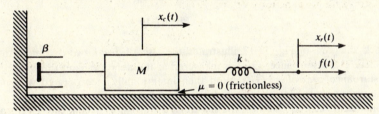

Fig. 17-15

17.37 (a) From the result of Example 17.6(a), derive an expression for the error of a simple first-order system with the input of Fig. 17-1(b). (b) Find the percentage difference between the error at $t = 5\tau$ and the error at $t = \infty$.   Ans. (a) $e(t) = K\tau(1 - e^{-t/\tau})$ $(t \geq 0)$; (b) 0.674%

17.38 Problem 17.20 offered a method to reduce steady-state disturbance error, which, however, increased the time for transient terms to die out. To illustrate this point, find the transient response of the system described by the block diagram of Fig. 17-7(a), if (a) $H(s) = 1$, $G_1(s) = 1$, and $G_2(s) = 2/(s + 1)$, there being no forward-path integrator; and (b) $H(s) = 1$, $G_1(s) = 1/s$, and $G_2(s) = 2/(s + 1)$, where a forward-path integrator is included.
Ans. (a) $c_{tr}(t) = Ae^{-3t}$ $(t \geq 0)$; (b) $c_{tr}(t) = Be^{-t/2}\cos(\sqrt{7}t/2)$ $(t \geq 0)$

17.39 For a system in canonical form, Fig. 17-6(a), with

$$G(s) = \frac{10}{s^2(s + 10)} \qquad H(s) = 3s^2 + 1$$

predict the steady-state error to the command $r(t) = 6t^2 u(t)$.   Ans. $e_{ass} = 12$

17.40 Under a step input, the settling time (t_s) of an underdamped, simple second-order system is defined as in Fig. 17-16. Show that $t_s < 3/\zeta\omega_n$.

17.41 For the control system of Fig. 17-12, if $r(t) = R_0 u(t)$ and $q(t) = Q_0 u(t)$, find the steady-state error.
Ans. $e_{ss} = 0.7188 R_0 + 0.9 Q_0$

17.42 A speed-control system with disturbance is modeled by the block diagram of Fig. 17-7(a). Let $G_1(s) = K$ (a proportional controller). $Q(s)$ is to represent a constant disturbance torque of $+20$ N·m. The plant, $G_2(s) = 1/(Js + \beta)$, is a motor for which $J = 15$ kg·m^2 and $\beta = 8$ N·m·s. Find the controller gain, K, such that the steady-state speed error due to the disturbance torque is no more than 2 rad/s.
Ans. $K \geq 2$

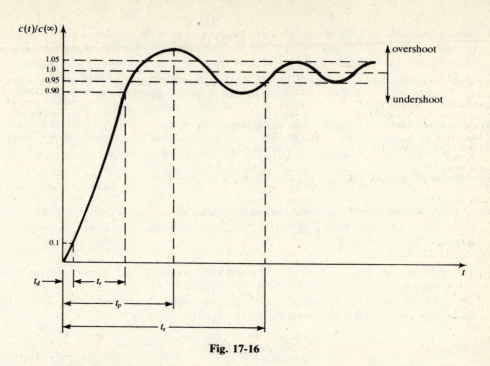

Fig. 17-16

17.43 The block diagram of Fig. 17-17 illustrates the use of *feedforward control* to reduce the effects of disturbance. The disturbance, $Q(s)$, is measured as it enters the plant (this can be difficult); operated on by a disturbance controller, $G_d(s)$; and then intentionally incorporated with negative polarity in the actuating error signal, $E(s)$, as indicated. By study of the plant error signal, $E_p(s)$, predict the form of the disturbance controller to minimize the effects of disturbance.   *Ans.* $G_d(s) = 1/G_c(s)$

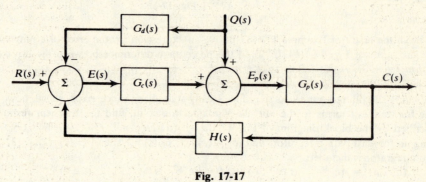

Fig. 17-17

17.44 For the two systems of Fig. 17-13, use the sensitivity function to predict the change in steady-state response when the plant (motor) time constant is changed from $\tau = 1$ to $\tau = 1.2$. Let $\theta_c(t) = 10\,u(t)$ and $A = 3$.   *Ans.*   0% for both systems

17.45 Solve Problem 17.23 without use of the sensitivity function, instead calculating the steady-state outputs at the original and the new value of the gain parameter.
Ans.   0% change for system I, 11.1% change for system II.

17.46 A control system with disturbance is modeled by Fig. 17-7(a). Find the sensitivity of the disturbance output $C_d(s)$ to changes in the plant $G_2(s)$.   *Ans.* $1/[1 + G_1(s)G_2(s)H(s)]$

17.47 Solve Problem 17.44 if $\theta_c(t) = A \sin \omega t$, where $\omega = 1$ rad/s. Use dynamic sensitivity at the forcing frequency.

Ans. $\dfrac{\Delta \Theta_A}{\Theta_A} = 0.014 \underline{/-4.08°}, \quad \dfrac{\Delta \Theta_B}{\Theta_B} = 0.1414 \underline{/-135°}$

Index